Frontiers in the History of Science

Series Editor

Vincenzo De Risi, Université Paris-Diderot – CNRS, PARIS CEDEX 13, Paris, France

Frontiers in the History of Science is designed for publications of up-to-date research results encompassing all areas of history of science, primarily with a focus on the history of mathematics, physics, and their applications. Graduates and post-graduates as well as scientists will benefit from the selected and thoroughly peer-reviewed publications at the research frontiers of history of sciences and at interdisciplinary "frontiers": history of science crossing into neighboring fields such as history of epistemology, history of art, or history of culture. The series is curated by the Series Editor with the support of an international group of Associate Editors.

Series Editor:
Vincenzo de Risi
Paris, France

Associate Editors:
Karine Chemla
Paris, France

Sven Dupré
Utrecht, The Netherlands

Moritz Epple
Frankfurt, Germany

Orna Harari
Tel Aviv, Israel

Dana Jalobeanu
Bucharest, Romania

Henrique Leitão
Lisboa, Portugal

David Marshal Miller
Ames, Iowa, USA

Aurélien Robert
Tours, France

Eric Schliesser
Amsterdam, The Netherlands

Angela Axworthy

Motion and Genetic Definitions in the Sixteenth-Century Euclidean Tradition

 Birkhäuser

Angela Axworthy
Max Planck Institute for the History of Science
Berlin, Germany

ISSN 2662-2564   ISSN 2662-2572   (electronic)
Frontiers in the History of Science
ISBN 978-3-030-95816-9   ISBN 978-3-030-95817-6   (eBook)
https://doi.org/10.1007/978-3-030-95817-6

Funding Information:- This publication was made possible thanks to the support of the Max Planck Institute for the History of Science independent research group "Modern Geometry and the Concept of Space" and of the Department I of the Max Planck Institute for the History of Science.

Mathematics Subject Classification: 01A40, 51-03, 03-03

This book is published under the imprint Birkhäuser, www.birkhauser-science.com by the registered company Springer Nature Switzerland AG
The registered company address is: Gewerbestrasse 11, 6330 Cham, Switzerland

To Eliott and Olivier

Acknowledgement

The research that gave rise to this book was mainly carried out during a postdoctoral fellowship within the MPIWG independent research group *Modern Geometry and the Concept of Space*. I would especially like to thank its director, Vincenzo De Risi, for having inspired me to undertake this research, and for his continuous support and insightful guidance throughout the whole process of conceptualisation and production of this book.

I wish furthermore to express my most sincere gratitude to Jürgen Renn, Matteo Valleriani and the Department I of the MPIWG for their support while I was preparing this book for publication.

I would also like to thank all the scholars who provided me with useful advice and theoretical or technical comments on earlier versions of this work, or with whom I led fruitful exchanges on related topics: Tawrin Baker, Delphine Bellis, Ariana Borelli, Karine Chemla, Stephen Clucas, Davide Crippa, Valérie Debuiche, Jackie Feke, Judith V. Field, Michael Friedman, Daniel Garber, Eduardo Giannini, Tal Glezer, Eberhard Knobloch, Antoni Malet, Hannes Ole Matthiessen, Thomas Morel, Richard Oosterhoff, Marco Panza, Stephen Pumfrey, David Rabouin, Sabine Rommevaux, Eleonora Sammarchi, Skúli Sigurdsson, Marius Stan, Tzuchien Tho, Henry Zepeda, as well as several anonymous reviewers. My thanks go in particular to Richard Arthur, not only for his helpful remarks on my work at various stages of its conception, but also for providing me with preliminary versions of his own work on the topic.

I am also thankful to the staff of the MPIWG library, for their availability and invaluable help in obtaining the sources I needed during the research phase of this project, to JoAnn Palmeri and Urte Brauckmann, for their help in obtaining the rights and the files of the images used in this book, as well as to Sarah Kempf for her positive reception of my publication proposal and for her assistance during the first phases of the publication process.

I would, additionally, like to thank my family, for their help, and their boundless support and patience during the conception, completion and publication of this book.

Contents

List of Figures

Introduction

1

1.1   Motion in Geometry: A Useful Yet Controversial Notion from Antiquity to the Early Modern Era

This study aims to explore the place and treatment, as well as the ontological and epistemological status, that were reserved to genetic definitions and, more generally, to motion in the framework of sixteenth-century commentaries on the geometrical books of Euclid's *Elements*. In classical geometry, a genetic definition (or a definition by genesis) corresponds to a definition that characterises a geometrical object (such as the line, the surface, the solid, the circle or the sphere) through its mode of generation rather than through its essential attributes. On the other hand, a definition which simply states the attributes of a geometrical object corresponds to a definition by property.[1] A genetic definition of the line, for example, defines the line as what results from the flow or motion of a point[2] rather than as a breadthless length.[3]

---

[1] The terminology "definition by genesis" and "definition by property" that is employed here is taken up from A.G. Molland (1976, p. 25) and J. Fauvel (1987, p. 3).

[2] Such a definition of the line may be found, for instance, in the commentary on the first book of the *Elements* by Proclus of Lycia (412–485), in Proclus (Friedlein 1873, p. 97) and Proclus (transl. Morrow 1992, p. 79): "Some define it [the line] as the 'flowing of a point'."

[3] This is the definition of the line provided by Euclid (Df. I.2) in the *Elements*: Euclid (transl. Heath 1956, p. 153): "A line is breadthless length". One may also compare the genetic definition of the circle, as provided in the *Definitions* attributed to Hero of Alexandria ((Heiberg 1974, pp. 32–33), Df. 27: "A circle is produced when a straight line, remaining in the same plane, while one of its extremities stays fixed, is carried around by the other extremity, until it is brought back to the place where it started to move.") with the non-genetic definition (or definition by property) of the circle provided by Euclid in the *Elements* (Df. I.15) (Euclid (Heiberg 1883, p. 4) and Euclid (Heath 1956, I,

© The Author(s), under exclusive license to Springer Nature Switzerland AG 2021
A. Axworthy, *Motion and Genetic Definitions in the Sixteenth-Century Euclidean Tradition*, Frontiers in the History of Science,
https://doi.org/10.1007/978-3-030-95817-6_1

The fact of attributing motion, understood in terms of transport (local motion) or of generation, to geometrical objects in the context of definitions was not an unusual practice in ancient Greek geometry.[4] In Euclid's *Elements*, which was regarded as the necessary introduction for the study of geometry in the middle ages and in the Renaissance, motion is introduced in Book XI, in the definitions of the sphere, the cone and the cylinder. In these definitions, the sphere is defined as the figure resulting from the rotation of a semicircle around its diameter.[5] The cone and the cylinder are defined according to a similar mode, the cone being defined as the figure resulting from the rotation of a right-angled triangle around one of its sides remained fixed[6] and the cylinder as the figure resulting from the rotation of a rectangle around one of its fixed sides.[7]

Beside definitions, motion was also assumed by Euclid (according to certain interpretations[8]) in Prop. I.4, I.8 and III.24, which demonstrate the congruence of triangles

---

p. 153): "A circle is a plane figure contained by one line such that all the straight lines falling upon it from one point among those lying within the figure are equal to one another.").

[4] For a more detailed description of the uses and characterisations of motion in ancient Greek geometry, see Molland (1976), Vitrac (2005a), and Rashed (2013) and *infra*, pp. 18–23.

[5] Euclid (Heiberg 1885, p. 4) and (Heath 1956, III, p. 261), Df. XI.14: "When, the diameter of a semicircle remaining fixed, the semicircle is carried round (περιενεχθέν, from περιφέρω) and restored again to the same position from which it began to be moved (φέρεσθαι, from φέρω), the figure so comprehended is a sphere". As I will not give the Greek version of the passages that are quoted in the footnotes, I will provide in parentheses (as in the above-quoted passage) the relevant terms in the forms in which they appear in the original texts, so as to account for their different uses in their proper context and also to make them easier to find in the original text. I will indicate, to avoid confusion, the form in the first person singular present indicative (as given in the dictionary) at the first occurrence of a verb.

[6] Euclid (Heiberg 1885, p. 6) and (Heath 1956, III, p. 261), Df. XI.18: "When, one side of those about the right angle in a right-angled triangle remaining fixed, the triangle is carried round (περιενεχθέν) and restored again to the same position from which it began to be moved (φέρεσθαι), the figure so comprehended is a cone. And, if the straight line which remains fixed be equal to the remaining side about the right angle which is carried round (περιφερομένη), the cone will be right-angled; if less, obtuse-angled; and if greater, acute-angled."

[7] Euclid (Heiberg 1885, p. 6) and (Heath 1956, III, p. 262), Df. XI.21: "When, one side of those about the right angle in a rectangular parallelogram remaining fixed, the parallelogram is carried round (περιενεχθέν) and restored again to the same position from which it began to be moved (φέρεσθαι), the figure so comprehended is a cylinder."

[8] Euclid did not explicitly speak of the motion of figures in this context, but only of their superposition. Moreover, as Christoph Clavius would argue in his commentary on Theodosius' *Spherics* (1586) and in the second edition of his commentary on Euclid's *Elements*, the fact that these propositions correspond to theorems, where the superposition of figure is only assumed hypothetically, could allow one to deny that any motion is then held to take place (Axworthy 2018). See also the considerations of B. Vitrac on this issue, in Vitrac (1990, p. 295). More generally, on the interpretation of superposition as motion, see Euclid (Heath 1956, I, pp. 225–227 and 249) and (Vitrac 1990, I, pp. 202–203, 213, 293–299), Killing (1898, II, pp. 2–3), Goldstein (1972), Mueller

and semicircles through superposition.[9] In these propositions, superposition may indeed be understood as implying the transfer and superposition of one figure toward and onto another. In a different manner, generative motion may be understood to take place within constructions, as when a line is required to be led from a specific point to another or when an entire figure is conceived as intersecting another in its motion in the context of a problem.

Diverse uses of motion in geometry are found in the works and mathematical practice of Archytas of Tarentum (c. 428–c. 350 BC),[10] Archimedes (c. 287–c. 212 BC),[11] Apollonius

---

(2006, p. 23), Vitrac (2005a, pp. 5 and 49–50), De Risi (2016b, pp. 593, 632 and 661), Axworthy (2018) and, on later developments in the early modern period, Arthur (2021, section 3.5).

[9] Euclid (Heiberg 1883, pp. 16–18) and (Heath 1956, I, p. 247), Prop. I.4: "If two triangles have the two sides equal to two sides respectively, and have the angles contained by the equal straight lines equal, they will also have the base equal to the base, the triangle will be equal to the triangle, and the remaining angles will be equal to the remaining angles respectively, namely those which the equal sides subtend. [. . .] If the triangle ABC be applied (ἐφαρμοζομένου, from ἐφαρμόζω) to the triangle DEF, and if the point A be placed (τιθεμένου, from τίθημι) on the point D and the straight line AB on DE, then the point B will also coincide with E, because AB is equal to DE. Again, AB coinciding (ἐφαρμοσάσης) with DE, the straight line AC will also coincide with DF, because the angle BAC is equal to the angle EDF; hence the point C will also coincide with the point F, because AC is again equal to DF. But B also coincided with E; hence the base BC will coincide with the base EF. [For if, when B coincides with E and C with F, the base BC does not coincide with the base EF, two straight lines will enclose a space: which is impossible. Therefore the base BC will coincide with EF] and will be equal to it. Thus the whole triangle ABC will coincide with the whole triangle DEF, and will be equal to it. And the remaining angles will also coincide with the remaining angles and will be equal to them, the angle ABC to the angle DEF, and the angle ACB to the angle DFE."; (Heiberg 1883, pp. 26–28) and (Heath 1956, I, p. 261), Prop. I.8: "If two triangles have the two sides equal to two sides respectively, and have also the base equal to the base, they will also have the angles equal which are contained by the equal straight lines. [. . .]"; (Heiberg 1883, pp. 224–226) and (Heath 1956, II, p. 53), Prop. III. 24: "Similar segments of circles on equal straight lines are equal to one another. [. . .]".

[10] According to Eutocius' commentary on the first proposition of the second book of Archimedes' *On the Sphere and the cylinder*, Archytas appealed to the generation of a semi-torus and of a right cone by the revolution of a semicircle and of a right-angled triangle, respectively, in order to solve the problem of finding two mean proportionals to two given line-segments (Huffman 2005, pp. 342–343). See *infra*, n. 83, p. 19. See also Heath (1981a, pp. 246–249), Vitrac (2005a, p. 8), Menn (2015) and Masià (2016).

[11] The most canonical example, for Archimedes, is his definition of the spiral, which also introduces a notion of time (Archimedes, *On Spirals*, Df. 1 (Heiberg 1881, pp. 50–52). See *infra*, n. 84, p. 19.). The notion of time is important as it is a defining factor of the physical concept of motion and required, in order to be applied to the definition of a mathematical object, to be deprived of any notion of velocity, being therefore only associated with uniform motion.

of Perga (c. 262–c. 190 BC),[12] Hero of Alexandria (c. 10–c. 75 AD),[13] and Pappus of Alexandria (c. 290–c. 350),[14] among others.[15]

Yet, the fact of presenting geometrical objects as endowed with motion, mobility and generability regularly raised doubts in Antiquity and in subsequent eras,[16] as it tended to overthrow the ontological distinction between the physical and the mathematical spheres of reality, given that, in the philosophical doctrines of Plato (c. 428–347 BC) and Aristotle (384–322 BC), only physical substances were considered as intrinsically capable of motion and generation. Moreover, the definition of the line, which is a continuous magnitude, as resulting from the generative motion of an indivisible point raised issues with regard to the composition of the continuum. The fact of defining geometrical objects by their mode of generation or even to attribute to them a non-generative local motion (as in the context of superposition) was called into question by Plato,[17] Aristotle,[18] Sextus Empiricus (c. 160–c. 210 AD)[19] or even later, in the Arabic tradition, by 'Umar al-Khayyām (1048–1131)[20] and Nasir al-Din al-Tusi (1201–1274).[21]

Nevertheless, motion and genetic definitions of geometrical objects remained in use throughout the premodern period, both in a mathematical and in a philosophical context, as shown by the mathematical work of Thābit ibn Qurra (c. 836–901) and al-Ḥasan ibn

---

[12] Apollonius, *Conics*, Df. I.1 (Heiberg 1891, p. 6) (see *infra*, n. 85, p. 19).

[13] In Hero's *Definitiones*, motion is introduced to define the line, the surface, the solid and the circle, in Df. 1–2, Df. 8, Df. 11 and Df. 27 (Heiberg 1974, pp. 14–17, 20–21, 22–23 and 32–33). See *infra*, n. 86, p. 20.

[14] Motion is, for instance, introduced by Pappus when presenting the generation of the conchoid of Nicomedes and of the quadratrix. Pappus, *Collection*, IV.22, § 39 and IV.24, § 45 (Hultsch 1876–1878, pp. 242–244 and 250–252). See also IV.30, § 57–59 (Hultsch 1876–1878, p. 270). See *infra*, n. 87, p. 20.

[15] Molland (1976) and Vitrac (2005a).

[16] On this issue, see Jaouiche (1986, pp. 50–52 and 68–72), Vitrac (2005a, pp. 9–18) and Rashed (2013). See also De Risi (2007, pp. 279–280), Dye and Vitrac (2009) and Vinel (2010). An analysis of these reservations concerning the use of motion in geometry is also provided *infra*, pp. 28–33.

[17] Plato, *Republic* VII, 527a–b (Emlyn-Jones and Preddy 2013, pp. 150–153). See *infra*, n. 127, p. 29.

[18] Aristotle, *Physics* II.2, 193b32–194a6 (transl. Barnes 1995, I, p. 331), *Metaphysics* I.8, 989 b32–33 (Barnes 1995, II, p. 1565), VI.1 1026a8–10 and 16–19 (Barnes 1995, II, p. 1619) and K.7, 1064a31–33 (Barnes 1995, II, p. 1681). See *infra*, n. 123, p. 28.

[19] Sextus Empiricus, *Adversus Mathematicos*, § 28. See *infra*, n. 25, p. 109.

[20] Al-Khayyām, *Epistle on the explanation of the problematic premises of the book of Euclid*. See *infra*, n. 135, p. 32. See also Jaouiche (1986, pp. 68–70), Vitrac (2005a, pp. 2–4) and Rashed (2013, pp. 58–60).

[21] Jaouiche (1986, pp. 68–70).

al-Haytham (c. 965–c. 1040), in the Arabic tradition,[22] or in the philosophical discourse of Albertus Magnus (c. 1200–1280), in the Latin middle ages.[23]

In the early modern period, the kinematic understanding of geometrical objects was increasingly employed by mathematicians and teachers of mathematics to investigate and express geometrical concepts,[24] including those set forth in Euclid's plane geometry. In the seventeenth century, motion was not only regarded as a tool to define and study geometrical figures, but came also to be considered by certain mathematicians and philosophers as essential to geometrical definitions and as foundational to geometry in general.[25] Among those who made an extensive use of genetic definitions within their mathematical practice and teaching are Bonaventura Cavalieri (1598–1647), René Descartes (1596–1650), Evangelista Torricelli (1608–1647), Gilles de Roberval (1602–1675) and Isaac Barrow (1630–1677).[26] Barrow, in particular, asserted the fundamental importance of motion in geometry.[27] He thus joined the philosophers, and above all Thomas Hobbes (1588–1679) and Baruch Spinoza (1632–1677), who claimed the epistemic superiority of genetic definitions in geometry given that these, and only these, would offer a properly scientific foundation (in the Aristotelian sense) to geometrical demonstrations, enabling them to exhibit "the causes and generations of things".[28] This attitude was taken up and developed

---

[22] Jaouiche (1986, pp. 49–68) and Rashed (2013). Other authors from this tradition are quoted in Rashed (2005).

[23] Chase (2008).

[24] Breger (1991, pp. 28–33), De Gandt (1995, pp. 202–221) and Mancosu (1996, pp. 94–100).

[25] Mancosu (1996, pp. 94–100). See also De Angelis (1964, pp. 82–98), Moretto (1984, pp. 94–102), De Risi (2007, p. 278) and De Risi (2016b, pp. 632–633).

[26] Molland (1976), Mancosu (1996, pp. 95–97) and Breger (1991, pp. 28–33). Other examples are given by F. de Gandt, in De Gandt (1995, pp. 202–221). See also Molland (1994).

[27] Barrow (1734, XII, 223): "No Geometrician refuses to define Figures by any Motion by which they can be generated with Ease. [. . .] Which Definitions are not only the most lawful, but the best: For they not only explain the Nature of the Magnitude defined, but, at the same time, shew its possible Existence, and evidently discover the Method of its Construction: They not only describe what it is, but prove by Experiment, that it is capable of being such; and do put it beyond doubt how it becomes such." On Barrow's discussion of generative motion in geometry, see Mahoney (1990, pp. 203–213).

[28] Hobbes (Molesworth 1839, I, p. 82): "To return, therefore, to definitions; the reason why I say that the cause and generation of such things, as have any cause or generation, ought to enter into their definitions, is this. The end of science is the demonstration of the causes and generations of things; which if they be not in the definitions, they cannot be found in the conclusion of the first syllogism, that is made from those definitions; and if they be not in the first conclusion, they will not be found in any other conclusion deduced from that; and, therefore, by proceeding in this manner, we shall never come to science; which is against the scope and intention of demonstration." On Hobbes's approach to geometry, see Sacksteder (1980), Sacksteder (1981), Jesseph (1999, chap. 3, in part. section 3.1) and Arthur (2021, pp. 298–299). On both Hobbes and Spinoza, see Gueroult (1974, pp. 482–486), Bernhardt (1978) and Medina (1985). The similarities between the positions of Hobbes and of Spinoza on this issue are summarised by Gueroult (1974, p. 484, n. 68) in this manner: "Tous les traits par lesquels Spinoza caractérise la géométrie (. . .) sont ceux que souligne Hobbes (. . .): 1) les

further by Barrow's student and successor as holder of the Lucasian Professorship of mathematics at Cambridge, Isaac Newton (c. 1642–c. 1627). Newton not only asserted the foundational role of motion in the constitution and definition of geometrical objects,[29] but also founded this kinematic understanding of geometrical objects, and of the whole of geometry thereby, on the science of mechanics.[30] For this reason, no distinction was to be made, for him, between the motion considered by mathematicians and the motion of physical bodies.

This attitude certainly did not remain undiscussed in this period, as shown by John Wallis' (1616–1703) remarks against Hobbes' use of motion in the reformulation of Euclidean definitions.[31] Doubts were raised, in particular, about the rigour of mathematical demonstrations based on kinematic notions, even among the mathematicians who appealed

---

causes des *universalia* sont connues de soi (. . .). 2) Elles se réduisent à une: le mouvement. 3) De la variété des mouvements naît la variété des figures (...). 4) Le mouvement permet la définition *per generationem* (. . .), c'est-à-dire par la cause. 5) Par là est possible la science (. . .). 6) Du mouvement du point naît la ligne, du mouvement de la ligne la surface, du mouvement de la surface le corps (. . .). 7) Ainsi, par le mouvement s'obtiennent toutes les définitions génétiques (. . .). 8) Ces définitions, contrairement aux définitions statiques, rendent compte de la possibilité de la chose (. . .). Elles en font connaître toutes les propriétés (. . .)." (This note is fully quoted in Mancosu 1996, p. 99, n. 23).

[29] Newton, *Quadrature of curves*, I, 2, § 2, in Newton (Whiteside 1964, p. 141): "I don't here consider Mathematical Quantities as composed of Parts *extremely small*, but as *generated by a continual motion*. Lines are described, and by describing are generated, not by any apposition of Parts, but by a continuous motion of Points. Surfaces are generated by the motion of Lines, Solids by the motion of Surfaces, Angles by the Rotation of their Legs, Time by a continual flux, and so in the rest. These *Geneses* are founded upon Nature, and are every Day seen in the motions of Bodies." See Arthur (2021, p. 297). More generally on the relation between geometry and mechanics in Newton's work, see, for instance, Dear (1995a, pp. 211–216), De Gandt (1995, pp. 209–221) and Guicciardini (2009, pp. 293–305).

[30] Newton (1687, p. i); Newton (transl. Cohen and Whitman 1999, pp. 381–382), *Preface to the reader*: "For the description of straight lines and circles, which is the foundation of *geometry*, appertains to *mechanics*. *Geometry* does not teach how to describe these straight lines and circles, but postulates such a description. For *geometry* postulates that a beginner has learned to describe lines and circles exactly before he approaches the threshold of *geometry*, and then it teaches how problems are solved by these operations. To describe straight lines and to describe circles are problems, but not problems in *geometry*. *Geometry* postulates the solution of these problems from *mechanics* and teaches the use of the problems thus solved. And *geometry* can boast that with so few principles obtained from other fields, it can do so much. Therefore *geometry* is founded on mechanical practice and is nothing other than that part of *universal mechanics* which reduces the art of measuring to exact propositions and demonstrations. But since the manual arts are applied especially to making bodies move, *geometry* is commonly used in reference to magnitude, and *mechanics* in reference to motion." (The emphasis is proper to the translation.) See Arthur (2021, p. 297) and Guicciardini (2009, pp. 293–299).

[31] Wallis (1655, pp. 6–7). *Cf.* Wallis (1656, pp. 52–54). On the debates that opposed Wallis to Hobbes over the status of genetic definitions, and on the nature of mathematics more generally, see Jesseph (1999, chaps. 3 and 4, in part. pp. 78–82 and 132–135).

to it within their mathematical practice, as Roberval.[32] One may also mention the attitude of Gottfried Wilhelm Leibniz (1646–1716), who, while attributing a foundational place to motion in the constitution of his *analysis situs*,[33] considered necessary to demonstrate (rather than simply assume) the possibility of the motions required for the generation of geometrical figures in order to allow them to be used as principles in mathematical demonstrations.[34] Still, it remains that, although the use of motion in geometry remained an object of debate, a much greater use was made of kinematic notions in geometry in the seventeenth century than before, also in the framework of the Euclidean tradition. This reflects what H. Breger expressed as the development of mechanistic styles in mathematical thought.[35]

1.2 Motion in Geometry and the Sixteenth-Century Euclidean Tradition

The contributions of certain sixteenth-century mathematicians to this development, and above all of commentators of Euclid's *Elements* such as Christoph Clavius (1538–1612), has been pointed out by historians of early modern mathematics.[36] However, no comprehensive assessment of the contributions of sixteenth-century commentators of Euclid to the changes in the uses and status of motion in geometry has been carried out up to now. The present study thus attempts to provide such an assessment, since, beyond Clavius, this tradition represents a privileged context for the investigation of the evolution of the status

---

[32] Roberval (1730, pp. 1–89). On Roberval's kinematic treatment of magnitudes, see De Gandt (1995, pp. 207–208) and on his hesitation concerning the admissibility of his mechanical approach to geometry, see Cléro and LeRest (1981, pp. 71–72 and 87–88).

[33] On the status and place of motion in Leibniz' *analysis situs*, see De Risi (2007, pp. 278–279).

[34] Leibniz, *Characteristica geometrica*, § 84, in Leibniz (Echeverría 1995, ix, p. 218) and Leibniz (transl. Arthur 2021, p. 306): "If, when two points A.B. of the trace A.C.B. remain at rest, the trace itself is moved, the line which its point C describes when it is moved is called circular. But *whether it is possible for some trace to be moved with two points remaining at rest must not be assumed, but should be defined by a demonstration*." (My emphasis.) *Cf. ibid.*, § 83 (Leibniz 1995, ix, pp. 216–218). On this issue, see De Risi (2007, p. 278) and Arthur (2021, p. 306). Interestingly, as shown by V. De Risi and R. Arthur, Leibniz demonstrated the possibility of the generation of geometrical figures through the concept of congruence, which was first defined independently from the notion of motion. On this, see *Characteristica geometrica*, § 108 (Leibniz 1995, ix, p. 228) and (Arthur 2021, p. 358): "When two things are perceived to exist simultaneously in space, by that very fact a path is perceived from one to the other. And since they are congruent, by that very fact is conceived the path of one into the place of the other. Now, two points are congruent to one another. So what is perceived when two points are simultaneously perceived is thus a Line, that is, the path of a point." *Cf. ibid.*, § 60–64 and 100, Leibniz (Echeverría 1995, ix, pp. 202–204 and 224).

[35] Breger (1991, pp. 15–33).

[36] See, for instance, De Angelis (1964, pp. 85–86); Dear (1995a, pp. 217–222); Dear (1995b); Mancosu (1996, pp. 94–100); Jesseph (1999, p. 80); Malet (2006).

of motion in geometry from the premodern to the early modern period. One reason to this is that a great number of editions and commentaries on the *Elements* were written and printed over this period in Europe.[37] Moreover, the sixteenth-century Euclidean tradition represents a context of transition between the pre- and early modern period in the Western history of mathematics with regard to the reception and reassessment of ancient Greek mathematical culture and practice. It therefore opened the path for seventeenth-century interpretations of ancient geometry. This tradition benefited indeed from the rediscovery of the commentary of Proclus on the first book of Euclid's *Elements*[38] and of the *Mathematical Collection* of Pappus,[39] which both contributed to the popularisation of the kinematic interpretation of geometry, as well as to the debates over the admissibility of motion in geometry. Pappus' work, in particular, was central to Descartes' transformation of geometry.[40] Furthermore, sixteenth-century Europe witnessed important developments in practical mathematics and in the theorisation of physico-mathematical sciences. This led to a reinterpretation of the nature of mathematical knowledge and practice and to a reassessment of the boundaries and relations between mathematics and the study of nature.

Generally speaking, the Euclidean tradition played an instrumental role in the discussions about the content, methods and epistemological status of geometry from Antiquity to the nineteenth century. Euclid's *Elements* represented also, for a great part of its pre- and early modern history, a standard work for the teaching of mathematics and a necessary introduction to all parts of mathematics (theoretical and practical), in addition to a canon of scientific knowledge. It was as such the most popular and extensively circulated mathematical treatise from Antiquity to the modern era, in Europe and beyond. In the sixteenth century, it was nearly considered a requirement for any mathematician or professor of mathematics worthy of name to provide a commentary on the *Elements*. This gave commentators not only the opportunity to prove their competence as mathematicians, but also to adapt the content of this work to their intended readership and epistemological or pedagogical agenda and to display their conception of mathematics. Hence, the commentaries on the *Elements* written in the sixteenth century offer us a privileged insight into the changes that took place in this context with respect to the status of mathematical objects and to the definition of mathematical knowledge, teaching and

---

[37] There were, from the first printed edition of Campanus' commentary by E. Ratdolt, in 1482 (Campanus 1482) to the early seventeenth-century Dutch version by Jan Pieterszoon Dou (1572/73–1635) from 1606, more than thirty different printed editions, translations or commentaries, many of which were reprinted several times. Though incomplete, one may consult, on these editions, the bibliographical inventories by P. Riccardi (1974) and M. Steck (1981).

[38] Proclus (1533). On the rediscovery of this work in the Renaissance, see De Pace (1993, pp. 121–185), Maierù (1999), Rommevaux (2004), Higashi (2007) and Kessler (1995).

[39] Commandino (1588).

[40] On Descartes and his use of Pappus, see, for instance, Bos (2001, pp. 271–283) and Mancosu (1996, pp. 68–71). Other sources on Descartes are mentioned *supra*, p. 251 sq., where a brief analysis of Descartes' views on geometrical motion are considered.

practice, as shown here through the particular issue of the status of motion and of genetic definitions within geometry.

1.3   The Significance of Genetic Definitions to the Assessment of the Status of Geometrical Motion

Genetic definitions, more than constructions and congruence theorems that appeal to superposition,[41] play an essential role in the assessment of the place and status of motion in geometry within the sixteenth-century Euclidean tradition. For one thing, given that Euclid rarely appealed to genetic definitions, and not at all within his plane geometry, the fact that some of his sixteenth-century commentators added such definitions in his first books is revealing of the importance that this type of definition was progressively gaining in the early modern interpretation of ancient Greek geometry. It is important to note, in this regard, that it is chiefly to genetic definitions, taken as principles and not as solutions to geometrical problems, that Barrow, Hobbes and Spinoza attributed a foundational role in geometry.[42]

Broadly speaking, genetic definitions played a key role in the sixteenth-century assessment of the status of motion because of the importance given to definitions themselves for the constitution and transmission of scientific knowledge in premodern Western epistemology, when science was chiefly defined according to the principles of Aristotle's doctrine of science.[43] According to this epistemological model, definitions were considered to state the essential relation of a substance to its properties and therefore give us access to the essence of the object under investigation.[44] In this framework, genetic definitions would be held as assertions that motion is an essential feature of geometrical objects, which could raise issues with regard to the ontological status that was attributed to mathematical objects by ancient philosophers, notably Plato and Aristotle. Since motion was then used to define geometrical objects, it could be more straightforwardly understood as inherent to geometrical objects themselves and as originating in them. On the other hand, the motion involved in constructions, and even that which is implied by a kinematic understanding of superposition, could be more easily conceived as externally-induced processes, that is, as expressing instrumental operations performed by the geometer (in an idealised form).

---

[41] Apart from a few notable examples, such as the debate concerning geometrical superposition between Peletier and Clavius, rarely did the motion involved in constructions or congruence demonstrations raise questions or doubts. When it did raise an issue, the problem did not lie in the fact that they appeal to motion, but in the possibility for this motion to be determined rationally or geometrically. On the debate over superposition between Peletier and Clavius, see Axworthy (2018). On illegitimate motions in geometrical problems, see *infra*, pp. 24 and 217–219.

[42] See *supra*, pp. 5–6.

[43] Serene (1982). See also Gilbert (1960, pp. 7–8).

[44] On this issue, see notably Demoss and Devereux (1988).

This would lead us to consider motion as an accidental rather than as an essential feature of geometrical objects. As such, genetic definitions would be more likely to raise doubts concerning the legitimacy of motion in geometry from a philosophical point of view, given that they would tend to attribute motion to geometrical objects on an essential level and thus overthrow the Aristotelian distinction between mathematical and physical entities. Indeed, Aristotle mainly distinguished mathematical objects from natural substances by denying the former the causal and essential connection with matter and motion, as well as the independent existence that characterised the latter.[45] Given the dominance of Aristotelian philosophy in the arts curriculum of the universities and colleges up to the seventeenth century, the fact of appending genetic definitions to Euclid's definitions by property, notably in a pedagogical context, therefore points to the erosion, in the context of sixteenth-century philosophy of mathematics, of the allegiance to Aristotelian ontology,[46] particularly among members of the Jesuit order such as Clavius.[47]

Now, concerning the status of motion in geometry, one of the main consequences of this erosion was a merging of the motion referred to in genetic definitions and in mechanical constructions in the seventeenth century, and which may be canonically illustrated by the approach to geometry adopted by Descartes[48] and Newton.[49] Descartes' connection between the generation of curves and instrumental processes, like Newton's foundation of geometry on mechanics,[50] will indeed contribute to render obsolete the difference between intrinsic and extrinsic motion in geometry, that is, between a motion that is attributed to geometrical objects by essence and one that would only be attributed to them accidentally.

In this framework, the different terminologies that were used to express motion in genetic definitions, in Antiquity as in the early modern period, could be quite diverse and equivocal, leading to different, and potentially conflicting, interpretations of the nature and ontological status of the motion that was then attributed to geometrical objects. As will be shown, these terminological differences within the sixteenth-century Euclidean tradition is itself revealing of variations and changes in the ontological status of motion in geometry.

---

[45] See *supra*, n. 18, p. 4. A similar situation occurred in Plato's philosophical doctrine, at least insofar as Plato placed mathematical entities among intelligible substances, which are not subjected to change. On Plato's position on this issue, see *infra*, pp. 29–30.

[46] On the place of Aristotle's ontology in the sixteenth-century philosophy of mathematics, see Higashi (2018).

[47] The teaching of Aristotle's treatises and its peripatetic commentaries represented indeed the core of the Jesuit educational program in philosophical matters, as shown in Lohr (1995) and Pereira (Casalini and Pavur 2016, pp. 211–213). On the Jesuit educational program, notably in relation to the teaching of mathematics, see Romano (1999, pp. 43–83) and Sasaki (2003, pp. 17–18).

[48] On Descartes' approach to geometry, see *infra* p. 251 *sq.*, along with the references provided there.

[49] On Newton's foundation of geometry on mechanics, see Dear (1995a, pp. 211–216) and Guicciardini (2009, pp. 293–305).

[50] Newton (1687, p. i) and (Cohen and Whitman 1999, pp. 381–382). See *supra*, p. 6.

Moreover, the fact of defining geometrical objects through their mode of generation, instead of through their spatial properties, raised the question of the function and status of definitions in the scientific discourse. Indeed, such definitions could be considered as assertions, or even demonstrations, of the existence of the *definiendum*, or at least of the possibility of its existence, which could he held as contradicting Aristotle's primary notion of definition as a scientific principle. In Aristotle's *Posterior analytics*, which (as said) was a central source for the determination of the scientific method from the middle ages to the early modern period, the definitions on which a science is founded would not *per se* say anything about the existence, and therefore about the mode of causation, of the *definiendum*.[51] A definition in the proper sense would merely state the proper attributes of the defined thing and give the meaning of its name, whereby it was later called a nominal definition.[52] In this context, a definition that would assert the existence of an object would, on the other hand, correspond to a hypothesis.[53] Hence, to a certain extent, a genetic definition, since it shows how a magnitude is produced, would be closer to a hypothesis than to a definition in the proper sense.

In this regard, it is important to note that Proclus' commentary on the first book of the *Elements*, which was instrumental to the reinterpretation of Euclidean geometry in the sixteenth century and which partially followed Aristotle's *Posterior analytics* concerning

---

[51] On the relation between essence, existence and definitions in Aristotle's *Posterior analytics*, see Demoss and Devereux (1988).

[52] Aristotle, *Posterior analytics* I.2, 72a22–26 (transl. Tredennick, 1989, pp. 33–35): "A definition is a kind of thesis <or laying-down>, because the arithmetician lays it down that to be a unit is to be quantitatively indivisible; but it is not a hypothesis, because to define the nature of a unit is not the same as to assert its existence." See also II.10, 93b29–32 and 93b38–94a3 (Tredennick 1989, pp. 206–207): "Since definition means 'an account of what a thing is,' obviously (1) one kind of definition will be an explanation of the meaning of the name, or of an equivalent denomination; e.g., it will explain the meaning of 'triangularity'. [. . .] The above is one definition of definition; but (2) in another sense definition is a form of words which explains why a thing exists."

[53] Aristotle, *Posterior analytics* I.10, 76b29–40 (Tredennick 1989, pp. 72–73): "Thus any provable proposition that a teacher assumes without proving it, if the student accepts it, is a hypothesis [. . .]. Definitions are not hypotheses, because they make no assertion of existence or non-existence. Hypotheses have their place among propositions, whereas definitions only need to be understood; and this does not constitute a hypothesis, unless it is claimed that listening is a kind of hypothesis. Hypotheses consist of assumptions from which the conclusion follows in virtue of their being what they are." See also I.2, 72a19–26 (Tredennick 1989, pp. 32–35): "A thesis which assumes one or the other part of a proposition, i.e., that something does, or does not exist, is a hypothesis; a thesis which does not do this is a definition. A definition is a kind of thesis <or laying-down>, because the arithmetician lays it down that to be a unit is to be quantitatively indivisible; but it is not a hypothesis, because to define the nature of a unit is not the same as to assert its existence." F. Romano (in Romano 2010) qualifies the distinction between definitions and hypotheses in Aristotle by pointing to the fact that Aristotle did not clearly distinguish definition and hypothesis as for their respective functions, since the definition may sometimes imply existence, while the hypothesis, in particular in mathematics, may set forth the meaning of the defined term at the same time as its existence. On this issue, see also Demos and Devereux (1988).

Euclid's principles, called all geometrical definitions hypotheses.[54] Now, given that Proclus counted the generations of geometrical objects among the objects of the geometer,[55] this could tend to attribute to genetic definitions and definitions by property the same status and function in the geometer's discourse.

Later discussions on the nature of scientific premisses would bring forth new conceptions and terminologies to describe the difference and relation between the various types of definitions, from those that are merely nominal or that simply posit the existence of the defined thing to those that also exhibit its proper cause and turn the nominal definition into an essential or real definition.[56] Such conceptions allowed genetic definitions to be regarded as legitimate, and even as foundational, in geometry, enabling geometrical concepts to be known in a scientific manner and to be established as real and even causal definitions.

Hence, the fact of expanding the place of genetic definitions in commentaries on Euclid's *Elements* brought to the fore the question of the nature and function of definitions in scientific knowledge, particularly in mathematics. This was all the more crucial in this context that mathematical demonstrations, and notably Euclidean geometrical demonstrations, held an important place in the discussions on the structure and parts of scientific demonstrations in the pre- and early modern era.[57]

Lastly, in certain medieval interpretations of Aristotle's theory of scientific demonstrations and of the status and role of axioms, postulates and definitions in this

---

[54] Proclus (Friedlein 1873, p. 76) and (Morrow 1992, pp. 62–63): "Axiom, postulate, and hypothesis are not the same thing, as the inspired Aristotle somewhere says. [...] When the student does not have a self-evident notion of the assertion proposed but nevertheless posits it and thus concedes the point to his teacher, such an assertion is a hypothesis. That a circle is a figure of such-and-such a sort we do not know by a common notion in advance of being taught, but upon hearing it we accept it without a demonstration." Given Proclus' allegiance to the Platonic ontological conception of mathematical beings, this has been interpreted either as the statement of the ontological necessity of geometrical objects (M. Caveing, introduction to Vitrac (1990, I, p. 122)), or as referring to the passage of the *Republic* according to which mathematicians would take for granted the existence of their objects without accounting for their principles (Heath 1956, I, p. 122).

[55] Proclus (Friedlein 1873, p. 57) and (Morrow 1992, p. 46): "Let us next speak of the science itself that investigates these forms. Magnitudes, figures and their boundaries, and the ratios that are found in them, as well as their properties, *their various positions and motions*—these are what geometry studies, proceeding from the partless point down to solid bodies, whose many species and differences it explores, then following the reverse path from the more complex objects to the simpler ones and their principles." (My emphasis.)

[56] Early modern developments of these discussions are displayed in De Angelis (1964, pp. 82–98).

[57] Although mathematics, which was chiefly represented by Euclid's *Elements* in this context, did not offer demonstrations that were conform to the model of the scientific syllogism such as defined by Aristotle in the *Posterior analytics* (Gilbert 1960, pp. 8–9), the assimilation between the form of mathematical demonstrations and that of scientific syllogisms was rather common at the time. For an overview of this question in the context of Renaissance mathematical writings, see Cifoletti (1995, pp. 1397–1398), Cifoletti (2009), Dear (1995b), Rommevaux (2005, p. 52) and Gatto (2006).

framework, one finds the idea that the truth of axioms, which Aristotle defined as necessary, self-evident, but also immediate and indemonstrable truths,[58] may be founded and therefore demonstrated from the definitions, at least insofar as the evidence of axioms would derive from the comprehension of the terms contained in their formulation.[59] In other words, definitions would allow in a certain way to demonstrate these principles that Aristotle had deemed undemonstrable. Hence, according to this medieval interpretation of ancient axiomatic, which was chiefly defined according to the model provided by Euclid's *Elements*, definitions were considered as the true principles and foundation of science. In this framework, genetic definitions could be taken as the foundation of the first three postulates of Euclid, which assert the possibility of leading a line between two given points and to extent a line further, as well as to draw a circle of any centre and of any radius.[60]

1.4   Sources and Research Questions

The body of sources on which this study focuses consists of six different commentaries on Euclid's *Elements* published between 1536 and 1574 (with reprints until 1612) in France, England and Italy, namely, those of Oronce Fine (1494–1555) (Paris, 1536, repr. 1544 and 1551);[61] Jacques Peletier (1517–1582) (Lyon, 1557; repr. 1610; transl. 1611 and 1628);[62] François de Foix, Count of Candale (1512–1594) (Paris, 1566; repr. 1578);[63] Henry

---

[58] Aristotle, *Posterior analytics* I.2 71b20–22 (Tredennick 1989, pp. 30–31): "Now if knowledge is such as we have assumed, knowledge must proceed from premises which are true, primary, immediate, better known than, prior to, and causative of the conclusion." and I.3 72b22–24 (Tredennick 1989, pp. 38–39): "We, however, hold that not all knowledge is demonstrative; the knowledge of immediate premises is not by demonstration. It is evident that this must be so; for if it is necessary to know the prior premises from which the demonstration proceeds, and if the regress ends with the immediate premises, the latter must be indemonstrable."

[59] Thomas Aquinas, Commentary on the first book of Aristote's *Posterior Analytics*, Lectio 5, in Aquinas (1989, pp. 23–26). I was made aware of this source by Vincenzo De Risi, when he presented his on-going research on the topic during the Oberwolfach Workshop on *Mathematics and its Ancient Classics worldwide* (May 31–June 4, 2021). I thank him for providing me with the precise reference to his main source on this question. On this topic, see also Netz (1999, p. 103) and De Risi (2016b).

[60] Euclid (Heiberg 1883, p. 8) and (Heath 1956, I, p. 154), Post. 1: "To draw a straight line from any point to any point"; Post. 2: "To produce a finite straight line continuously in a straight line." and Post. 3: "To describe a circle with any centre and distance."

[61] Oronce Fine, *In sex priores libros Geometricorum elementorum Euclidis Megarensis demonstrationes*. Paris: Simon de Colines, 1536.

[62] Jacques Peletier, *In Euclidis Elementa Geometrica Demonstrationum Libri sex*. Lyon: Jean de Tournes, 1557.

[63] François de Foix-Candale, *Euclidis Megarensis mathematici clarissimi Elementa, libri XV ad germanam geometriae intelligentiam è diversis lapsibus contractis restituta [...]. His accessit decimus sextus liber de solidorum regularium sibi invicem inscriptorum collationibus, tum etiam coeptum opusculum de Compositis regularibus solidis planè peragendum*. Paris: Jean Royer, 1566.

Billingsley (d. 1606) (London, 1570);[64] Federico Commandino (1509–1575) (Pesaro, 1572; transl. 1575);[65] and Christoph Clavius (Rome, 1574, 1589, 1591, 1603, 1607, 1611–1612, 1644 and 1654).[66] The *Mathematicall praeface* written by John Dee (1527–1608/1609) for Billingsley's English translation and commentary of Euclid will also be considered.

Beside those commentaries on the *Elements* and Dee's preface, I will also analyse certain complementary sources written by some of their authors, such as the *De geometria libri duo* (*Geometria*) of Oronce Fine,[67] the *Oeuvres intitulées louanges* of Jacques Peletier,[68] the commentary on the *Poimandres* (*Le Pimandre*) by Foix-Candale,[69] the *Monas hieroglyphica* by John Dee,[70] as well as Clavius' commentary on Sacrobosco's *De Sphaera*.[71] These will enable to clarify the ontological and epistemological conceptions underlying their approach to motion and genetic definitions in their respective commentaries on Euclid.

I have chosen these seven particular authors because their commentaries on the *Elements* were among those that included comments on Euclid's principles and because of their conspicuous and significant use of genetic definitions when commenting on Euclid's definitions.[72] Indeed, although there were other mathematicians or humanists who commented on Euclid's *Elements* in premodern Europe, not all commentators of Euclid added any discussion on the definitions, axioms and postulates, or at least not in a systematic manner,[73] focusing rather on the propositions. And among those that did

---

[64] Henry Billingsley and John Dee, *The elements of geometrie of the most auncient philosopher Euclide of Megara. Faithfully (now first) translated into the Englishe toung, by H. Billingsley, citizen of London. Whereunto are annexed certaine scholies, annotations, and inuentions, of the best mathematiciens, both of time past, and in this our age. With a very fruitfull praeface made by M. I. Dee, specifying the chiefe mathematicall scie[n]ces, what they are, and wherunto commodious: where, also, are disclosed certaine new secrets mathematicall and mechanicall, untill these our daies, greatly missed.* London: John Daye, 1570.

[65] Federico Commandino, *Euclidis elementorum: libri XV. Unà cum scholiis antiquis.* Pesaro: Camillo Francischino, 1572.

[66] Christoph Clavius, *Euclidis elementorum libri XV: Accessit XVI.* Rome: Vincenzo Accolto, 1574.

[67] Oronce Fine, *De geometria libri duo*, in *Protomathesis: Opus varium, ac scitu non minus utile quàm iucundum, nunc primum in lucem foeliciter emissum.* Paris: Gérard Morrhe, 1532.

[68] Jacques Peletier, *Euvres poétiques intituléz Louanges.* Paris: R. Coulombel, 1581.

[69] François de Foix-Candale, *Le Pimandre de Mercure Trismegiste de la philosophie Chretienne.* Bordeaux: Simon Millanges, 1579.

[70] John Dee, *Monas hieroglyphica.* Antwerp: G. Sylvius, 1564.

[71] Christoph Clavius, *In Sphaeram Ioannis de Sacrobosco Commentarius.* Rome: Helianus, 1570.

[72] In both aspects, these works more or less followed the model of Proclus' commentary on Euclid.

[73] For instance, the commentary of Campanus of Novara (c. 1220–1296), which was the first printed version of Euclid's *Elements* (Campanus 1482 and Busard 2005, I, p. 32) and which represented a model for several of the commentaries published in the first half of the sixteenth century, only contains comments on the definitions of Books II, V, VI and XI.

comment on the definitions, not all mentioned genetic definitions as an alternative to the definitions by property given by Euclid. However, it is important to note that such cases remain quite marginal, as practically all the sixteenth-century commentaries that did contain an exposition on the definitions introduced genetic definitions. One notable exception is the Italian translation and commentary of Niccolò Tartaglia's (1500–1557), which was first published in 1543 and reprinted in 1565.[74] Indeed, apart from the genetic definitions Euclid himself included in Book XI,[75] Tartaglia only mentioned the process of drawing a line or a figure in the imagination or concretely, by instrumental means,[76] without appealing to a more abstract form of genetic definition through which geometrical objects could themselves be attributed motion.

Within the considered sources, the nature, role and status of genetic definitions, and of the motion that is then attributed to geometrical objects, was discussed either when a genetic definition was added to one of Euclid's definitions by property, or at the occasion of a commentary on the few genetic definitions provided by Euclid in Book XI.[77] In the former case, this mostly concerned the first definitions of Book I, which present the most fundamental geometrical objects (point, line, surface, boundary, figure, circle, square. . .). In the latter case, such discussions only came forth in relation to Euclid's definition of the sphere (Df. XI.14), to which commentators generally added, in turn, a definition by property.

As will be shown throughout this study, the introduction of genetic definitions in the commentary on Euclid's plane geometry was not only an occasion for commentators to set forth their position on the status and role of motion in geometry, as well as on the nature and function of geometrical definitions, but also to touch upon other issues crucial to pre and early modern philosophy of mathematics, such as the composition of the continuum and the relation between arithmetic and geometry. These questions generally arose at the occasion of a commentary on the definitions of Book II,[78] in which a kinematic understanding of geometrical objects was often presented by the considered authors. This approach allowed commentators to relate the mode of generation of geometrical figures

---

[74]Tartaglia (1565). One may also mention the German commentary by Wilhelm Holtzman (Xylander, 1532–1576) (Holtzman 1562) and the French commentary by Pierre Forcadel (1500–1572) (Forcadel 1564).

[75]Tartaglia (1565, fol. 237r).

[76]See, for instance, Tartaglia (1565, fol. 8v), Df. I.2: "le linee fatte dell'arte, overo a caso sono fatte volontariamente, overo a caso dall'operante Geometrico con qualche stilletto pontito, overo con qualche materia colorata, in qualche spatio, come per esempio (in varij modi, si come etiam varij modi possono accadere) havemo designato di sopra."

[77]See *supra*, n. 5, 6 and 7, p. 2.

[78]Euclid, Df. I.2 (1956, I, p. 370): "Any rectangular parallelogram is said to be contained by the two straight lines containing the right angle." On the changing relation between numbers and magnitudes in the medieval and early modern Euclidean traditions, notably concerning Book II of the *Elements*, see Malet (2006), Malet (2012) and Corry (2013).

to the mode of generation of numbers, in particular by appealing to the Pythagorean notion of figurate or geometrical number.[79]

The groups of questions I will attempt to answer when considering these sources are the following:

1. Which place and degree of importance were given to genetic definitions in these commentaries? How were these genetic definitions formulated and according to which terminology was motion then referred to?
2. Was any discourse presented at this occasion concerning the nature and validity of motion as a means of definition in geometry? Was any doubt raised about the validity of genetic definitions, or about the introduction of motion more generally? And, if so, on which basis? Were there any conditions stated for motion to be admitted in geometry? Did these authors draw arguments from ancient or medieval authors to either restrict or justify the use of motion as a means of defining the properties and relations of geometrical objects? How were genetic definitions compared and related to definitions by property in this context?
3. How did the introduction of genetic definitions in Euclid's plane geometry contribute to change the concept of magnitude and its relation to discrete quantity?
4. Was there a consensus or, on the contrary, discernible differences in the reception, conception and treatment of motion as a means of definition from one commentary to the other and, in the latter case, can a clear evolution be outlined on this question from Fine to Clavius?

---

[79] Book II deals, for a great part, with relations of equivalence between the areas of rectangles that are delimited by the segments of a sectioned or prolonged line and which are therefore easily translatable into simple algebraic expressions. On this interpretation and the difficulties it raises, see B. Vitrac's discussion on the topic, in Vitrac (1990, I, pp. 323–324, 366–376). As mentioned then by B. Vitrac (1990, I, pp. 368–369), ulterior arithmetical treatments of Book II are found in later texts, such as in Hero's commentary on Euclid or in the *Arithmetica* of Diophantus (third c. AD), as well as in later scholia and in a certain number of Renaissance commentaries on the *Elements*, but it is not impossible that the Pythagorean doctrine of geometrical numbers may have incited mathematicians to propose an arithmetical treatment of comparable geometrical problems at an earlier date. Euclid used indeed a similar language in Book II and Book VII when dealing with geometrical areas and with geometrical numbers, respectively, both types of objects being said to result from and to be measured by the sides that contain them. Compare Df. II.1, Euclid (Heiberg 1883, p. 118) and (Heath 1956, I, p. 370): "Any rectangular parallelogram is said to be *contained* by the two straight lines containing the right angle." and Df. VII.18, Heiberg (1884, p. 186) and (Heath 1956, II, p. 278): "A square number is equal multiplied by equal, or a number which is *contained* by two equal numbers" and Df. VII.19 (Heiberg 1884, p. 188) and (Heath 1956, II, p. 278): "And a cube is equal multiplied by equal and again by equal, or a number which is *contained* by three equal numbers." The fact for two line-segments to contain a quadrilateral in Book II may therefore be interpreted as the fact for these line-segments to measure the area of this figure and, by a translation in arithmetical terms, may be held to represent the multiplication of two numbers.

What I will show through the analysis of the above-stated sources and questions is that, during the sixteenth century, which was a key period for the reception and interpretation of ancient mathematical and philosophical texts, the choice to introduce genetic definitions to teach geometrical notions did not go without a certain caution or reservation, and that this attitude was for a great part determined by the ontological status that was attributed to mathematical objects. What this study will also reveal is that the attitude of Euclid's commentators evolved in this respect throughout this period toward a greater acceptance of kinematic notions and genetic definitions in geometry. Their use was indeed to be extended as a means of explaining geometrical notions, and genetic definitions came to be more and more considered as essential to the study and practice of geometry. In this process, abstract and concrete interpretations of the generation of geometrical figures were brought closer together, marking the growing importance of practical mathematics in the early modern mathematical culture.

In order to properly understand the ways in which Renaissance mathematicians interpreted and made use of genetic definitions in their teaching of Euclidean geometry, and also to discern the doubts or issues which could be raised concerning the introduction of kinematic notions in geometry, it will be necessary in the following pages to briefly look at the way motion was introduced and characterised in ancient mathematical texts. I will consider here both definitions and propositions, so as to seize the continuities and variations between these two contexts and to better assess the specificities of the motion referred to within definitions. I will focus in particular on the terminology used to designate motion in different contexts and the ontological implications these terms could be endowed with by geometers and philosophers. These designations, as well as the kinematic notions and the practices to which they referred, were indeed those which philosophers and later mathematicians would have had in mind when evaluating the value of genetic definitions and of motion in geometry, even if their interpretation of ancient mathematical practice was necessarily marked by the philosophical, pedagogical and scientific traditions through which they gained access to these texts. This being done, I will go through the main reservations or doubts raised (explicitly and implicitly) since Antiquity concerning the introduction of motion in geometry, and in geometrical definitions in particular, as well as the conditions stated to legitimate its use in this context.

1.5   The Uses and Designations of Motion in Ancient Geometrical Texts

As was previously mentioned, Euclid appealed to motion in Book XI of the *Elements*, for his definitions of the sphere, the cone and the cylinder.[80] In this context, motion is referred to through the use of the verbs $\varphi \acute{\varepsilon} \rho \omega$ and $\pi \varepsilon \rho \iota \varphi \acute{\varepsilon} \rho \omega$ (in the middle passive form $\varphi \acute{\varepsilon} \rho \varepsilon \sigma \theta \alpha \iota$ and $\pi \varepsilon \rho \iota \varphi \acute{\varepsilon} \rho \varepsilon \sigma \theta \alpha \iota$), which respectively mean to carry, lead, convey, move or go, and to carry round, move round, or rotate. These verbs are related to the substantive $\varphi o \rho \acute{\alpha}$, which translates as motion, movement or carrying.

In the context of Euclid's Prop. I.4, I.8 and III.24, motion is not directly mentioned, but could be assumed as a condition for the superposition of figures.[81] The notion of superposition is then conveyed through the action of placing a figure on top of another figure, angle to angle and side to side, an action described through the use of the verbs $\tau \acute{\iota} \theta \eta \mu \iota$ (placing) and $\acute{\varepsilon} \varphi \alpha \rho \mu \acute{o} \zeta \omega$ (apply, fit on or fit to).

---

[80]Euclid, *Elements*, Df. XI.14, 18 and 21 (Heiberg 1885, pp. 4 and 6) and (Heath 1956, III, pp. 261–262). See *supra*, n. 5, 6 and 7, p. 2. See also Euclid (Heiberg 1885, p. 292) and (Heath 1956, III, p. 470), XIII.13: "If then, KL remaining fixed, the semicircle be carried round ($\pi \varepsilon \rho \iota \varepsilon \nu \varepsilon \chi \theta \acute{\varepsilon} \nu$) and restored to the same position from which it began to be moved ($\varphi \acute{\varepsilon} \rho \varepsilon \sigma \theta \alpha \iota$), it will also pass through the points F, G, since, if FL, LG be joined, the angles at F, G similarly become right angles; and the pyramid will be comprehended in the given sphere"; (Heiberg 1885, p. 298) and (Heath 1956, III, p. 475), XIII.14: "And for the same reason, if, LM remaining fixed, the semicircle be carried round ($\pi \varepsilon \rho \iota \varepsilon \nu \varepsilon \chi \theta \acute{\varepsilon} \nu$) and restored to the same position from which it began to be moved ($\varphi \acute{\varepsilon} \rho \varepsilon \sigma \theta \alpha \iota$), it will also pass through the points F, G, H, and the octahedron will have been comprehended in a sphere"; (Heiberg 1885, p. 302) and (Heath 1956, III, p. 479), XIII.15: "If then, KG remaining fixed, the semicircle be carried round ($\pi \varepsilon \rho \iota \varepsilon \nu \varepsilon \chi \theta \acute{\varepsilon} \nu$) and restored to the same position from which it began to be moved ($\varphi \acute{\varepsilon} \rho \varepsilon \sigma \theta \alpha \iota$), the cube will be comprehended in a sphere" and (Heiberg 1885, p. 312) and (Heath 1956, III, p. 485), XIII.16: "And if, XZ remaining fixed, the semicircle be carried round ($\pi \varepsilon \rho \iota \varepsilon \nu \varepsilon \chi \theta \acute{\varepsilon} \nu$) and restored to the same position from which it began to be moved ($\varphi \acute{\varepsilon} \rho \varepsilon \sigma \theta \alpha \iota$), it will also pass through Q and the remaining angular points of the icosahedron, and the icosahedron will have been comprehended in a sphere."

[81]Euclid (Heiberg 1883, pp. 16–18) and (Heath 1956, I, p. 247). See *supra*, n. 8, p. 2. Superposition is also used by Hero, in the *Metrica* (Schöne 1903, p. 4), by Archimedes, in *On Conoids and Spheroids*, Prop. 18 and in *On the Equilibrium of Planes*, 1, Post. 4 and Prop. 9–10 (Archimedes, Heath 1897, pp. 128, 189 and 194–195), by Apollonius in the *Conics* (Rashed 2009, IV, p. 90 and Toomer 1990, I, p. 264) and by Pappus in the *Collection*, III, § 83; IV, § 39 and § 45–46; VI, § 44 (Hultsch 1876–1878, pp. 138, 244, 252 and 524). For other cases, see Mugler (1958–1959, II, p. 208). On the use of superposition by Apollonius, see also Brigaglia (2012).

Beside the terms used by Euclid, other terms[82] were used to designate motion in geometry in the works of the previously mentioned authors, that is, Archytas of Tarentum[83] (αγωγή; φορά; κίνησις, to which may be added γραφή and ποίησις as the effect of these motions); Archimedes (φορά; αγωγή);[84] Apollonius of Perga (φορά; γραφή; γένεσις);[85]

---

[82] The list of notions proposed here for each of the quoted authors does not claim to be exhaustive, as it is mainly based on a selection of exemplary passages.

[83] Archytas (through Eutocius) (Huffman 2005, pp. 342–343): "When this semicircle is rotated (περιαγόμενον, related to περιάγω) from Δ to B, with the endpoint A of the diameter remains fixed, it will cut the cylindrical surface in its rotation (περιαγωγῇ) and will describe (γράψει, related to γράφω) a line on it. And again, if, while ΑΔ remains fixed, the triangle ΑΠΔ is rotated (περιενεχθῇ) in an opposite motion (κίνησιν, from κίνησις) to that of the semicircle, it will make (ποιήσει, related to ποιέω) the surface of a cone with the line ΑΠ, which as it is rotated (περιαγομένη) will meet the line on the cylinder in a point. At the same time the point B will also describe (περιγράψει, related to περιγράφω) a semicircle on the surface of the cone." On this proposition and Archytas's geometrical work, see also (Heath 1981a, pp. 246–249), Menn (2015) and Masià (2016). Masià (2016, pp. 196–197) makes an interesting point concerning the motion involved in Archytas's construction, as well as in Greek geometry in general, which is that Greek geometers were not interested in the motion itself, but in the result of the motion, which would be marked by the common use of kinematic terms (such as ἄγω, φέρω, γράφω and their derivatives) in perfective forms.

[84] Archimedes (Heiberg 1881, pp. 50–52) and (transl. Heath 1897, p. 165), *On Spirals*, Df. 1: "If a straight line drawn in a plane revolve (περιενεχθεῖσα) at a uniform rate about one extremity which remains fixed and return to the position from which it started, and if, at the same time as the line revolves (περιαγομένᾳ), a point move (φέρηταί) at a uniform rate along the straight line beginning from the extremity which remains fixed, the point will describe a spiral in the plane." Archimedes also used περιφορά in his treatise *On the Sphere and Cylinder*, Prop. 11, Df. 3 and Prop. 20, among other occurrences.

[85] Apollonius, *Conics*, Df. I.1 (Heiberg 1891, p. 6) and (transl. Densmore 1998, p. 3): "If from a point a straight line is joined to the circumference of a circle which is not in the same plane with the point, and the line is produced in both directions, and if, with the point remaining fixed, the straight line being rotated (περιενεχθεῖσα) about the circumference of the circle returns (φέρεσθαι) to the same place from which it began, then the generated (γραφεῖσαν) surface composed of the two surfaces lying vertically opposite one another, each of which increases indefinitely as the generating (γραφούσης) straight line is produced indefinitely, I call a conic surface, and I call the fixed point the vertex, and the straight line drawn (from the vertex to the centre of the circle I call the axis." He also used γένεσις in *Conics* I.51.

Hero of Alexandria ($\varphi o \rho \acute{\alpha}$; $\dot{\rho} \acute{v} \sigma \iota \varsigma$; $\kappa \acute{\iota} \nu \eta \sigma \iota \varsigma$);[86] and Pappus of Alexandria ($\varphi o \rho \acute{\alpha}$; $\kappa \acute{\iota} \nu \eta \sigma \iota \varsigma$; $\gamma \acute{\varepsilon} \nu \varepsilon \sigma \iota \varsigma$).[87]

---

[86] Hero of Alexandria, *Definitiones*, Df. 1 (Heiberg 1974, pp. 14–15, my translation from the Greek and from the German translation by Heiberg): "as the unit is the origin of number, the point is the origin of geometrical objects, and it is indeed the beginning of a progression ($\check{\varepsilon} \kappa \theta \varepsilon \sigma \iota \nu$), but not as if it were a part of the line as the unit is a part of number, but rather as intellectually presupposing it; then the line is conceived from the motion of a point ($\kappa \iota \nu \eta \theta \acute{\varepsilon} \nu \tau o \varsigma$, related to $\kappa \iota \nu \acute{\varepsilon} \omega$) or rather from the intellection of the point in the state of flowing ($\dot{\rho} \acute{v} \sigma \varepsilon \iota$, related to $\dot{\rho} \acute{v} \sigma \iota \varsigma$), and just as the point is the origin of the line, so the surface is the origin of the solid body"; (Heiberg 1974, pp. 14–17), Df. 2: "A line is a length without breadth and depth or that which first comes to existence within magnitude, or what has extension and is divisible according to one dimension, and it is produced when a point flows ($\dot{\rho} \upsilon \acute{\varepsilon} \nu \tau o \varsigma$, related to $\dot{\rho} \acute{\varepsilon} \omega$) downward from above according to the notion of continuity and is enclosed and limited by points, while itself is the limit of a surface."; (Heiberg 1974, pp. 20–21), Df. 8: "A surface [. . .] is produced by the flow ($\dot{\rho} \acute{v} \sigma \varepsilon \iota$) of a line, which flows ($\dot{\rho} \upsilon \varepsilon \acute{\iota} \sigma \eta \varsigma$) latitudinally from right to left."; (Heiberg 1974, pp. 22–23), Df. 11: "Any solid body is bounded by surfaces and is produced when a surface moves from front to back ($\dot{\varepsilon} \nu \varepsilon \chi \theta \varepsilon \acute{\iota} \sigma \eta \varsigma$)") and (Heiberg 1974, pp. 32–33), Df. 27: "A circle is produced when a straight line, remaining in the same plane, while one of its extremities stays fixed, is carried around ($\pi \varepsilon \rho \iota \varepsilon \nu \varepsilon \chi \theta \varepsilon \tilde{\iota} \sigma \alpha$) by the other extremity, until it is brought back to the place where it started to move ($\varphi \acute{\varepsilon} \rho \varepsilon \sigma \theta \alpha \iota$)." Hero also used $\varphi o \rho \acute{\alpha}$ in *Metrica* II, introduction.

[87] Pappus (Hultsch 1876–1878, pp. 242–244) and (transl. Sefrin-Weis 2009, p. 126), IV.22, § 39: "For the duplication of the cube a certain line is introduced by Nicomedes and it has a genesis ($\gamma \acute{\varepsilon} \nu \varepsilon \sigma \iota \nu$, from $\gamma \acute{\varepsilon} \nu \varepsilon \sigma \iota \varsigma$) of the following sort. Set out a straight line AB, and a straight line CDZ at right angles to it, and take a certain point E on CDZ as given. And assume that, while the point E remains in the place where it is, the straight line CDEZ travels ($\varphi \varepsilon \rho \acute{\varepsilon} \sigma \theta \omega$) along the straight line ADB, dragged via the point E in such a way that D travels ($\varphi \acute{\varepsilon} \rho \varepsilon \sigma \theta \alpha \iota$) on the straight line AB throughout and does not fall outside while CDEZ is dragged ($\dot{\varepsilon} \lambda \kappa o \mu \acute{\varepsilon} \nu \eta \varsigma$, from $\check{\varepsilon} \lambda \kappa \omega$) via E. Now, when such a motion takes place ($\kappa \iota \nu \acute{\eta} \sigma \varepsilon \omega \varsigma$ $\gamma \varepsilon \nu o \mu \acute{\varepsilon} \nu \eta \varsigma$) on both sides, it is obvious that the point C will describe a line such as LCM is, and its symptoma is of such a sort that, whenever some straight line starting from the point E toward the line meets it, the straight line cut off between the straight line AB and the line LCM is equal to the straight line CD." Pappus (Hultsch 1876–1878, pp. 250–252) and (Sefrin-Weis 2009, pp. 131–132), IV.24, § 45: "For the squaring of the circle a certain line has been taken up by Dinostratus and Nicomedes and some other more recent (mathematicians). It takes its name from the symptoma concerning it. For it is called "quadratrix" by them, and it has a genesis ($\gamma \acute{\varepsilon} \nu \varepsilon \sigma \iota \nu$) of the following sort. Set out a square ABCD and describe the arc BED of a circle with centre A, and assume that AB moves ($\kappa \iota \nu \varepsilon \acute{\iota} \sigma \theta \omega$) in such a way that while the point A remains in place, the point B travels ($\varphi \acute{\varepsilon} \rho \varepsilon \sigma \theta \alpha \iota$) along the arc BED, whereas BC follows along with the travelling ($\varphi \varepsilon \rho o \mu \acute{\varepsilon} \nu \omega$) point B down the straight line BA, remaining parallel to AD throughout, and that in the same time both AB, moving uniformly ($\kappa \iota \nu o \upsilon \mu \acute{\varepsilon} \nu \eta$), completes the angle BAD, i.e.: the point B completes the arc BED, and BC passes through the straight line BA, i.e. the point B travels ($\varphi \varepsilon \rho \acute{\varepsilon} \sigma \theta \omega$) down BA. Clearly it will come to pass that both AB and BC reach the straight line AD at the same time. Now, while a motion ($\kappa \iota \nu \acute{\eta} \sigma \varepsilon \omega \varsigma$) of this kind is taking place ($\gamma \iota \nu o \mu \acute{\varepsilon} \nu \eta \varsigma$), the straight lines BC and BA will intersect each other during their travelling ($\varphi o \rho \tilde{\alpha}$) in some point that is always changing its position together with them. By this point a certain line such as BZH is described in the space between the straight lines BA and AD and the arc BED, concave in the same direction as BED, which appears to be useful, among other things, for finding a square equal to a given circle." See also Pappus' classification of problems according to the mode of generation of lines, within plane or solid surfaces, or in a more complex manner, through the composition of two motions, in Pappus (Hultsch

In the works of these authors, motion is either introduced in genetic definitions or in propositions. In the latter context, motion is mostly referred to as a means of construction. In both cases, the represented motion is generative, having the function of producing or extending magnitude in one, two or three dimensions. The magnitude or figure that is thereby generated is conceived as either undetermined or determined in quantity and position (that is, in the latter case, relatively to other given or constructed magnitudes). In the context of definitions, the generation of magnitudes is mostly undetermined, as it represents the mode of generation of a class of objects, rather than that of a particular object. Non-generative local motion of figures (i.e. the simple displacement of a geometrical object from one place to the other without implying that another magnitude or figure is traced out in this process) could, on the other hand, be assumed to intervene within congruence theorems.

The verbs that were most commonly used in ancient Greek to refer to a kinematic process in geometry, φέρω and περιφέρω (related to the term φορά),[88] were typically used, both in definitions and in propositions, to designate the rotation or transversal motion of a line-segment or of a plane figure from one point to another in order to generate a magnitude or a figure of higher dimension. The term ρύσις (flow or flux) or the related verb ρέω were specifically used in the context of definitions to designate the generation of a line or a surface by the flow of a point or a line, as in the *Definitions* attributed to Hero of Alexandria.[89] Terms related to κίνησις (motion, change) were used to describe the motion

---

1876–1878, p. 270) and (Sefrin-Weis 2009, pp. 144–145), IV 30, § 57–59: "We say that there are three kinds of problems in geometry, and that some <of the problems> are called "plane," others "solid," and yet others "linear." Now, those that can be solved by means of straight line and circle, one might fittingly call "plane." For the lines by means of which problems of this sort are found have their genesis (γένεσιν) in the plane as well. All those problems, however, that are solved when one employs for their invention either a single one or even several of the conic sections, have been called "solid." For it is necessary to use the surfaces of solid figures—I mean, however, (surfaces) of cones—in their construction. Finally, as a certain third kind of problems the so-called "linear" kind is left over. For different lines, besides the ones mentioned, are taken for their construction, which have a more varied and forced genesis (γένεσιν), because they are generated (γεννώμεναι) out of less structured surfaces, and out of twisted motions (κινήσεως). Of such a sort, however, are both the lines found on the so-called loci on surfaces and also others, more varied than those and many in number, which were found by Demetrius of Alexandria in the "linear constitutions," and by Philo of Tyana, from the twisting of both plectoids and all sorts of other surfaces on solids and which have many astonishing symptomata about them. And some of them were deemed, by the more recent geometers, worthy of rather extensive discussion, and a certain one of them is the line that was also called "the paradox" by Menelaus and of this same kind <i.e., the linear kind> are also the other spiral lines, the quadratrices and the conchoids and the cissoids." On this classification and its foundation on the mode of generation of lines, see Vitrac (2005a, pp. 26–32). As noted in Mugler (1958–1959), Pappus also used περιφορά in IV, pp. 31 and 34; κίνησις in IV, pp. 32 and 45; and γένεσις in IV, pp. 30 and 32, among other occurrences.

[88] See *supra* n. 80 and 83–87, pp. 18–20.

[89] Hero (Heiberg 1974, p. 14). See *supra*, n. 86, p. 20.

of a semicircle generating a semi-torus in Eutocius' account of Archytas' solution to the problem of finding two mean proportionals to two given straight lines.[90] Pappus also used the term γένεσις (genesis, generation) when presenting the mode of construction of a curve.[91]

Other verbs that were commonly used in a mathematical context in ancient Greece and that denoted motion are ἄγω (lead, carry, bring) and γράφω (draw, delineate, describe, paint, write), which are often used in the context of propositions to designate the process by which a line is drawn or led from one point to another (as in Euclid's first postulate)[92] or by which a circle is produced or described.[93] These terms, as marked by the actions and processes they first designate, tend to imply the intervention of an exterior agent, who would draw or lead a line from one point to another or delineate a circle, by instrumental means or in the imagination. These terms therefore have a more operative connotation than those mentioned above.

Such as used in Euclid's genetic definitions (Df. XI.14, 18 and 21[94]), the notion of φορά presents geometrical figures as generated by the local displacement of a point or of a geometrical object of lower dimension[95] throughout a delimited spatial extension or *locus* that is determined by the configuration and dimensions of the moving figure. The resulting figure thus corresponds to the trace left by the moving figure.

The term ῥύσις, on the other hand, would suggest that the generating object (e.g. a point) produces a magnitude (e.g. a line) not by moving (or by being moved) locally, but by letting it flow or emanate from itself continuously, as would a stream of water or a ray of light flow from its source. This notion had, as we will see, also a metaphysical or theological meaning.

---

[90] Archytas (Huffman 2005, pp. 342–343). See *supra* n. 83, p. 19.

[91] Pappus (Hultsch 1876–1878, pp. 242–244), IV.22, § 39. See *supra* n. 87, p. 20.

[92] Euclid (Heiberg 1883, p. 8) and (Heath 1956, I, p. 154), Post. 1: "To draw (ἀγαγεῖν) a straight line from any point to any point". See also, for example, a proposition such as Prop. XI.11 (Heiberg 1885, p. 32) and (Heath 1956, III, p. 292): "From a given elevated point to draw (ἀγαγεῖν) a straight line perpendicular to a given plane."

[93] See, for instance, Euclid (Heiberg 1883, pp. 226–228) and (Heath 1956, II, pp. 54–55), Prop. III.25: "Given a segment of a circle, to describe (προσαναγράψαι, related to προσαναγράφω) the complete circle of which it is a segment. [. . .] the circle drawn (γραφόμενος, related to γράφω) with centre E [. . .]". As indicated by Mugler (1958–1959, I, pp. 107–109), γράφω was generally used for non-straight lines. The term used for the drawing of straight lines was ἄγω (Mugler, 1958–1959, I, pp. 40–41). Indeed, in Euclid's *Elements*, the verb γράφω is specifically used for the generation of circles and curved lines (as in the examples given here) and the drawing of straight lines (as in the examples provided in n. 92) is rather designated by the verb ἄγω. Hence, both terms seem to have an equivalent and mutually complementary function in this context, each being related to one of the two instrumental procedures relevant to Euclidean constructions, that is, the use of the straightedge for the drawing of a straight line and the use of the compass for the drawing of a circumference.

[94] See *supra* n. 5, 6 and 7, p. 2.

[95] That is, a line in relation to a surface or a surface in relation to a solid.

Terms such as κίνησις and γένεσις tend to evoke, more directly than φορά, physical or mechanical processes such as the local motion of an animate or inanimate material body or the diminution or augmentation of the quantity of a natural substance. As suggested above, the term ῥύσις, according to one of its primary meanings, may also be related to physical processes, such as the flowing of water from a spring.

Generally speaking, the appearance of the terms γένεσις, κίνησις and ῥύσις in mathematical texts (as in Hero and Pappus' works),[96] which (to my knowledge) are not used by Euclid in the *Elements* and which primarily evoke physical motions,[97] points to a greater freedom in the designation of kinematic processes in a mathematical context. In this framework, a confusion or a shift in meaning between the mathematical and physical realms would have been more likely to occur.

1.6   The Ontological Implications of Ancient Kinematic Notions

The motion attributed to geometrical objects may be interpreted as self-induced, as when the point, the line or the figure are said to move in order to generate a line, a surface or a solid independently from the intervention of an exterior agent, or it may be conceived as resulting from the action of the geometer. This is mostly the case for the processes by which lines or figures are understood to be constructed in the framework of problems and theorems.

Hence, whilst ῥύσις, φορά, κίνησις and γένεσις, especially in the context of a genetic definition, may be primarily taken to represent a motion or a generation of magnitudes independent from an external agent, γραφή, and ἀγωγή to a certain extent,[98] may more straightforwardly evoke a constructive process, which would find its cause in the will and action of the geometer. These could be understood as produced either in the imagination or concretely, by the means of an instrument, such as a straightedge or compass.[99]

But in a general manner, if an exterior agent is presupposed in any of these cases, this is not made explicit, and it would seem that there was no need in a mathematical context, at least for the above-mentioned ancient authors, to establish an ontological discrimination

---

[96] See *supra* n. 86–87, p. 20.

[97] See for example, Aristotle, *Physics*, III.1, 200b12 and *Metaphysics* II.2, 994a5 for κίνησις and *ibid.* 994b1 for γένεσις.

[98] The terms ἀγωγή and φορά also seem interchangeable in certain cases, as in Archimedes's definition of the spiral (see *supra*, n. 84, p. 19), where περιάγω is used as a synonym of περιφέρω.

[99] Γράφω is however also used in certain texts in a way that would suggest that the geometrical object has the intrinsic property to move and generate a line or figure in its motion. See, for example, Archimedes's definition of the spiral, where the point is said to describe (γράψει) the curve in its motion. Archimedes (Heiberg 1881, p. 52); Archimedes (Heath 1897, p. 165), *On Spirals*, Df. 1: "the point will *describe* a spiral in the plane." (My emphasis.) On the formulation of this definition and the way it expresses the mode of production of the spiral, see Netz (1999, pp. 154–155).

between self-induced or externally-induced motions. The possibility to make a distinction between these two levels of interpretation of geometrical motion would then mainly depend on the context of the discourse, according to whether a construction is required to be performed on a given object (concretely or imaginarily) in a specific aim (e.g. to solve a particular geometrical problem), or whether a universal property or relation of geometrical objects is stated.

Thus, on a purely mathematical level, the origin and ontological status of motion, whether it is considered as realised in the imagination, instrumentally, or even independently from the action and will of the geometer, would not have been relevant to Greek geometers, as what would have mattered is the construction of the required magnitude or figure, as well as the knowledge of the properties and relations made possible through the considered motion. Pappus' *Mathematical Collection*, for instance, shows us that there was not always a great difference in intention between a motion that is imaginarily generated (or even intellectually assumed) and a motion that is carried out instrumentally (by means of the compass and straightedge, but also by more complex instruments, as was the conchoid[100]), since instruments and the curves they allow us to produce would then be held as idealised or considered *in abstracto*.[101] The instruments then simply represent a type of motion or combination of motions of one or several points, line-segments or plane figures, by which the considered curve or figure is defined as generated. Hence, in this context, the introduction of motion did not call for any justification.[102] At most was the generation of the quadratrix[103] called into question and considered more mechanical than geometrical by Sporus of Nicaea (c. 240–c. 300), and by Pappus after him, because of the ignorance of the ratio between the two combined motions that are necessary for its construction, that is, of a circular and a rectilinear motion of line-segments.[104]

---

[100] Heath (1981a, pp. 238–240). For Pappus' treatment of the conchoid, see *Collection*, IV.23, § 40–43 (Hultsch 1876–1878, pp. 244–248).

[101] As noted by B. Vitrac (Vitrac 2005a, pp. 8–9), this approach of geometry, which involves the attribution of motion to geometrical objects and the use of instruments to generate curves, even if only considered *in abstracto*, is what the Greek notion of "mathematical mechanics" (found in Diogenes Laërtius's account of Archytas' life) may have pointed to, but it may also have been used to designate the study of machines or mechanical devices (μηχαναί): Diogenes Laërtius (transl. Hicks 1925, pp. 394–396), VIII, § 83: "He [Archytas] was the first to bring mechanics (τὰ μηχανικά) to a system by applying mathematical principles; he also first employed mechanical motion (κίνησιν ὀργανικήν) in a geometrical construction, namely, when he tried, by means of a section of a half-cylinder, to find two mean proportionals in order to duplicate the cube." See also Vitrac (2009).

[102] Molland (1976), Vitrac (2005a, p. 32) and Rashed (2013, p. 61).

[103] A representation of the construction of the quadratrix (by Christoph Clavius) is provided *infra* p. 218.

[104] Pappus (Hultsch 1876–1878, pp. 252–254) and (Sefrin-Weis 2009, pp. 132–133), IV.25, § 46: "Sporus, however, is with good reason displeased with it, on account of the following <observations.> For, first of all, he takes into the assumption the very thing for which it <i.e., the quadratrix> seems to be useful. For how is it possible when two points start from B, that they move

Yet, on a philosophical level, that is, when considering the ontological status of mathematical objects and the epistemic value of mathematics for the description of the world, the difference between an internally and an externally-induced motion would matter, since, in one case, motion is properly attributed to geometrical objects as an inherent and essential property, and, in the other, it is not.

In this regard, it is important to note that, in Antiquity, the notion of ῥύσις σημείου or of flow of the point to represent the mode of generation of the line mostly appeared in a philosophical context. It appears, for instance, in the texts of Aristotle,[105] of Iamblichus (c. 242–c. 325)[106] and of Proclus,[107] but also in those of Theon of Smyrna (70–135)[108] and of Sextus Empiricus.[109] To these occurrences, may be added that found in Hero's

---

(κινεῖσθαι), the one along the straight line to A, the other along the arc to D, and come to a halt <at their respective end points> at the same time, unless the ratio of the straight line AB to the arc BED is known beforehand? For the velocities of the motions (κινήσεων) must be in this ratio, also. Also, how do they think that they come to a halt simultaneously, when they use indeterminate velocities, except that it might happen sometime by chance; and how is that not absurd? Furthermore, however, its endpoint, which they use for the squaring of the circle, i.e.: the point in which it intersects the straight line AD, is not found <by the above generation of the line>. Consider what is being said, however, with reference to the diagram set forth. For when the <straight lines> CB and BA, traveling, come to a halt simultaneously, they will <both> reach AD, and they will no longer produce an intersection in each other. For the intersecting stops when AD is reached, and this <last> intersection would have taken place as the endpoint of the line, the <point> where it meets the straight line AD. Except if someone were to say that he considers the line to be produced, as we assume straight lines <to be produced>, up to AD. This, however, does not follow from the underlying principles, but <one proceeds> just as if the point H were taken after the ratio of the arc to the straight line had been taken beforehand. Without this ratio being given, however, one must not, trusting in the opinion of the men who invented the line, accept it, since it is rather mechanical (μηχανικοῖς). Much rather, however, one should accept the problem that is shown by means of it." On this issue, see also *infra*, pp. 217–219.

[105] Aristotle, *De anima* I. 4, 409a4–6 (Barnes 1995, I, p. 652): "they say a moving line generates a surface and a moving point a line." See *infra*, n. 122, p. 28.

[106] Iamblichus, *In Nicomachi arithmeticam* IV, § 4 (Pistelli 1894, pp. 57, 6–8): "the point, whose flow (ῥύσιν) geometers consider the line to be, is the principle and element of length" (My translation from the Greek and from the French translation by N. Vinel, in Vinel (2014, p. 127).

[107] Proclus (Friedlein 1873, pp. 96–97) and (Morrow 1992, p. 79): "*A line is length without breadth.* [. . .] The line has also been defined in other ways. Some define it as the 'flowing of a point', others as 'magnitude extended in one direction'." See also *infra*, p. 48.

[108] Theon (Dupuis 1892, p. 136) and (transl. Lawlor 1979, p. 55), § 31: "As for the point, it is neither by multiplication nor by addition that it forms the line, but by a continuous motion, in the same way that the line forms a surface and the surface a solid."

[109] Sextus Empiricus (transl. Bury 1971, p. 259), *Adversus Mathematicos* (Book III = *Adversus Geometras*), § 28: "Eratosthenes is accustomed to say that the sign [the point] neither occupies any space nor measures out the interval of the line, but by flowing (ῥυὲν) makes the line. But this is inconceivable. For flowing (ῥεῖν) is conceived as extension from a place to a place, as water extends. And if we shall imagine the sign to be something of that sort, it will follow that it is not like a thing without parts, but of the opposite sort, abounding in parts".

Definitions,[110] since this work consists for a great part of a list of commented definitions and could therefore be attributed a philosophical scope. Indeed, as Theon of Smyrna[111] and Sextus Empiricus,[112] the author of the *Definitions* sought to convey the essence of the straight line and the intellectual nature of its mode of generation.[113] He also distinguished the mode of generation of the line from the mode of generation of numbers from the unit,[114] addressing the philosophical issue of the composition of the continuum by explaining how the line can be held to originate from a point without being conceived as composed of indivisibles.[115]

---

[110] See *supra*, n. 86, p. 20.

[111] Theon (Dupuis 1892, p. 136). See *supra*, n. 108, p. 25.

[112] Sextus Empiricus, *Adversus Mathematicos*, § 28. See *supra*, n. 109, p. 25.

[113] Hero of Alexandria, *Definitiones*, Df. 1 (Heiberg 1974, pp. 14–15): "the line is conceived from the motion of a point or rather from the intellection ($\nu o \eta \theta \acute{\epsilon} \nu \tau o \varsigma$) of the point in the state of flowing." On this point, see Vitrac (2005a, p. 21).

[114] Hero (Heiberg 1974, pp. 14–15): "as the unit is the origin of number, the point is the origin of geometrical objects, and it is indeed the beginning of a progression ($\check{\epsilon} \kappa \theta \epsilon \sigma \iota \nu$), but not as if it were a part of the line as the unit is a part of number."

[115] See, for instance, some of the objections raised by Aristotle, in *Physics* VI, 231a20–29 (Barnes 1995, I, pp. 390–391): "Now if the terms 'continuous', 'in contact', and 'in succession' are understood as defined above—things being continuous if their extremities are one, in contact if their extremities are together, and in succession if there is nothing of their own kind intermediate between them—nothing that is continuous can be composed of indivisibles: e.g. a line cannot be composed of points, the line being continuous and the point indivisible. For the extremities of two points can neither be *one* (since of an indivisible there can be no extremity as distinct from some other part) nor *together* (since that which has no parts can have no extremity, the extremity and the thing of which it is the extremity being distinct)." and Sextus (Bury 1971, pp. 260–263): "Consequently, the line is not one single sign.—Nor yet is it a number of signs placed in a row. For these signs are conceived either as touching one another or as not touching. If as not touching one another, being intercepted they will be separated by certain spaces, and being separated by spaces they will no longer form one line. And if they are conceived as touching one another, they will either touch wholes as wholes or parts with parts. But if they shall touch parts with parts, they will no longer be without dimensions and without parts for the sign which is conceived—shall we say?—as midway between two signs will touch the sign in front with one part, and that behind with another, and the plane with a different part, and the other place with yet another, so that in very truth it is no longer without parts but with many parts. And if the signs as wholes should touch wholes, it is plain that signs will be contained in signs and will occupy the same place; and thus they will not be placed in a row, so as to form a line, but if they occupy the same place they will form one point." See also Iamblichus, *In Nicomachi arithmeticam*, IV.5–6 (Pistelli 1894, pp. 57, 16–20): "as one times one does not produce more than one, thus points brought together with one another do not produce more than a point. For the line is not an aggregation of several points; if they are in contact, there is no dimension and, if they assume a dimension, there is no contact, so much that the point is no longer a part of the line." (My translation from the Greek and from the French translation by N. Vinel, in Vinel (2014, p. 127)). More generally on this issue, see Iamblichus (Vinel 2014, 35–41), Dye and Vitrac (2009, 183–188) and *infra*, pp. 65–66.

In this context, the notion of ῥύσις is certainly presented as a properly spatial and local process, from which are derived the various dimensions, and according to which the line is generated by the flow of a point from bottom to top ("γίνεται δὲ σημείου ῥυέντος ἄνωθεν κάτω ἐννοίᾳ τῇ κατὰ τὴν συνέχειαν") and the surface by the flow of a line from right to left ("γίνεται δὲ ῥύσει ὑπὸ γραμμῆς κατὰ πλάτος ἀπὸ δεξιῶν ἐπ᾽ ἀριστερὰ ῥυείσης").[116] Yet, as shown by N. Vinel, this notion of ῥύσις σημείου or of flow of the point as a means of generating the line finds its origin in Ancient Pythagoreanism, where it was not initially intended as a spatial and kinematic process, but rather as the metaphysical or intelligible procession or emanation of the multiple from the divine One, in conformity with the Pythagorean theology of numbers.[117] In this context, this notion was used to set forth the correspondence between the generation of numbers from the unit, the generation of magnitude from the point and the production of the cosmos from the divine One. The earliest known appearance of this notion to establish a correspondence between numbers and magnitudes is in the work of Philo of Alexandria (first century AD), who explained the Pythagorean correspondence between the unit and the point, as well as between the number two and the line, by appealing to the analogy between the ῥύσις of the unit in arithmetic and the ῥύσις of the point in geometry.[118]

As said, the notion of ῥύσις was not generally used in ancient Greek mathematical texts and is *a fortiori* absent from Euclid's *Elements*, where the line, but also the surface and the point itself for that matter, are rather defined by a successive process of suppression of the dimensions from the body.[119] The order of Euclid's definitions however follows the order of generation of the different types of magnitude (point, line, surface, solid).[120] Yet, the

---

[116] Hero (Heiberg 1974, pp. 16 and 20), Df. 2 and 8 respectively. See *supra*, n. 86, p. 20. The notion of ῥύσις disappears when presenting the solid as produced by the surface, which is not said to flow, but to move (ἐνεχθείσης from φέρω, related to φορά) from back to front: Hero (Heiberg 1974, p. 22), Df. 11: "περατοῦται δὲ πᾶν στερεὸν ὑπὸ ἐπιφανειῶν καὶ γίνεται ἐπιφανείας ἀπὸ τῶν πρόσω [ἔμπροσθεν] ἐπὶ τὰ ὀπίσω ἐνεχθείσης" (my emphasis) (see *supra*, n. 86, p. 20). As noted by G. Dye and B. Vitrac (Dye and Vitrac 2009, pp. 176–177), Hero's definitions of the line, surface and solid establish the relation between extension, spatial disposition and generation by flow, to which we may also add generation by local motion (φορά).

[117] Vinel (2010).

[118] Philo of Alexandria, *De Opificio mundi*, § 49 (transl. Colson and Whitaker 1994, pp. 36–37): "For under the head of 1 what is called in geometry a point falls, under that of 2 a line. For if 1 extend itself (ῥύσει), 2 is formed, and if a point extend (ῥύσει) itself, a line is formed: and a line is length without breadth". See also Vinel (2010, pp. 111–112).

[119] The suppression of depth from the body produces the surface (Df. I.5), the suppression of breadth from the surface produces the line (Df. I.2), and the suppression of length from the line produces the point (Df. I.1).

[120] Euclid, Df. I.1 (1956, I, p. 153): "A point is that which has no part"; I.2 (*ibid.*): "A line is breadthless length"; I.5 (*ibid.*): "A surface is that which has length and breadth only" and Df. XI.1 (1956, III, p. 260): "A solid is that which has length, breadth, and depth." On the difference between the two modes of definition of magnitudes (by suppression and by acquisition of dimensions), see the introduction of M. Caveing to Vitrac (1990, I, pp. 129–130).

notions of flow of the point, of the line and of the surface, as respectively generating the line, the surface and the solid, regularly appeared from late Antiquity in philosophical or mathematical texts to explain the properties of geometrical objects, notably in relation to the definitions of Euclid. And, as will be shown here, it appeared in several sixteenth-century commentary on Euclid, where it was at times an occasion to discuss the ontological status of geometrical motion.

1.7   Reservations and Objections Against Geometrical Motion from Antiquity to the Middle Ages

As was previously mentioned, the attribution of motion to geometrical objects, even if relatively common in the practice of ancient Greek mathematicians, was regularly challenged by philosophers since Antiquity.[121] In Aristotle and Sextus Empiricus, the attacks on motion in geometry were specifically aimed at generative motions, as instantiated in particular by the notion of line as generated by the flow of a point. Although Aristotle acknowledged the Pythagorean origin of this notion, he interpreted it explicitly as a spatial process and as a motion in the proper sense ($\kappa i \nu \eta \sigma \iota \varsigma$) and was followed in this respect by Sextus.[122] The interpretation of $\rho \dot{\upsilon} \sigma \iota \varsigma$ as $\kappa i \nu \eta \sigma \iota \varsigma$ by Aristotle and Sextus questioned, first of all, the conception of the point and of geometrical objects as immobile substances,[123] but also the mereological relation between the point and the line. Indeed, for

---

[121] See *supra*, p. 4.

[122] Aristotle, *De anima* I. 4, 409a4–7 (Barnes 1995, I, p. 652): "Further, since they say a moving line ($\kappa \iota \nu \eta \theta \varepsilon \widetilde{\iota} \sigma \alpha \nu \ \gamma \rho \alpha \mu \mu \dot{\eta} \nu$) generates a surface and a moving point a line, the movements of the unit ($\mu o \nu \dot{\alpha} \delta \omega \nu \ \kappa \iota \nu \dot{\eta} \sigma \varepsilon \iota \varsigma$) must be lines (for a point is a unit having position, and the number of the soul is, of course, somewhere and has position)" and Sextus (Bury 1971, pp. 258–259): "Eratosthenes is accustomed to say that the sign [the point] neither occupies any space nor measures out the interval of the line, but by flowing ($\rho \upsilon \dot{\varepsilon} \nu$) makes the line. But this is inconceivable. For flowing ($\rho \varepsilon \widetilde{\iota} \nu$) is conceived as extension from a place to a place, as water extends. And if we shall imagine the sign to be something of that sort, it will follow that it is not like a thing without parts, but of the opposite sort, abounding in parts". See also Vinel (2010) and Dye and Vitrac (2009).

[123] Aristotle, *Physics* II.2, 193b32–194a6 (Barnes 1995, I, p. 331): "Now the mathematician, though he too treats of these things, nevertheless does not treat of them as the limits of a physical body; nor does he consider the attributes indicated as the attributes of such bodies. That is why he separates them; for in thought they are separable from motion ($\chi \omega \rho \iota \sigma \tau \grave{\alpha} \ \tau \widetilde{\eta} \ \nu o \acute{\eta} \sigma \varepsilon \iota \ \kappa \iota \nu \dot{\eta} \sigma \varepsilon \acute{\omega} \varsigma$), and it makes no difference, nor does any falsity result, if they are separated. The holders of the theory of the Forms do the same, though they are not aware of it; for they separate the objects of natural science, which are less separable than those of mathematics. This becomes plain if one tries to state in each of the two cases the definitions of the things and of their attributes. Odd and even, straight and curved, and likewise number, line, and figure, do not involve motion; not so flesh and bone and man—*these* are defined like snub nose, not like curved"; *Metaphysics* I.8, 989 b32–33 (Barnes 1995, II, p. 1565): "The 'Pythagoreans' use stranger principles and elements than the natural philosophers (the reason is that they got the principles from non-sensible things, for the objects of mathematics, except those of

Aristotle, what is deprived of parts cannot be conceived as the principle of continuous quantity and cannot therefore be conceived as being in motion,[124] given that motion is itself a continuous quantity.[125] The kinematic conception of the point also challenged the ontological primacy of the things that are at rest over those that are in motion.[126]

Prior to Aristotle, Plato also challenged the kinematic treatment of geometrical objects. He did not however directly challenge the fact of attributing motion to geometrical objects, but rather criticised the introduction of kinematic notions within the practice and language of geometers. Plato considered that geometers should not manipulate or talk of manipulating geometrical objects as material realities, by constructing or moving them (with or without instruments), in order to study their properties and relations, given their essential immutability, intelligibility and ontological primacy over the physical world.[127]

---

astronomy, are of the class of things without movement (ἄνευ κινήσεως)" and VI.1 1026a8–10 and 16–19 (Barnes 1995, II, p. 1619): "it is clear, however, that it considers some mathematical objects *qua* immovable (ἀκίνητα) and *qua* separable from matter. [...] For natural science deals with things which are inseparable from matter but not immovable, and some parts of mathematics deal with things which are immovable (ἀκίνητα), but probably not separable, but embodied in matter; while the first science deals with things which are both separable and immovable" and K.7, 1064a31–33 (Barnes 1995, II, p. 1681): "Natural science deals with the things that have a principle of movement in themselves; mathematics is theoretical, and is a science that deals with things that are at rest, but its subjects cannot exist apart."

[124] Aristotle, *De anima* I. 4, 408b32–409a3 (Barnes 1995, I, p. 651): "Of all the opinions we have enumerated, by far the most unreasonable is that which declares the soul to be a self-moving number (ἀριθμὸν κινοῦνθ᾽ ἑαυτόν); it involves in the first place all the impossibilities which follow from regarding the soul as moved (κινεῖσθαι), and in the second special absurdities which follow from calling it a number. How are we to imagine a unit being moved (μονάδα κινουμένην)? By what agency? What sort of movement can be attributed to what is without parts or internal differences? If the unit is both originative of movement (κινητική) and itself capable of being moved (κινητή), it must contain difference." See also Aristotle, *Met.*, A, 9, 992 a10–21 (Barnes 1995, II, p. 1568): "When we wish to refer substances to their principles, we state that lines come from the short and long (i.e. from a kind of small and great), and the plane from the broad and narrow, and the solid from the deep and shallow. Yet how then can the plane contain a line, or the solid a line or a plane? For the broad and narrow is a different class of things from the deep and shallow. Therefore, just as number is not present in these, because the many and few are different from these, evidently no other of the higher classes will be present in the lower. But again the broad is not a genus which includes the deep, for then the solid would have been a species of plane. Further, from what principle will the presence of the points in the line be derived? Plato even used to object to this class of things as being a geometrical fiction. He called the indivisible lines the principle of lines—and he used to lay this down often. Yet these must have a limit".

[125] Aristotle, *Physics* VI.2.

[126] Aristotle, *Topics* VI.4, 142 a19–21 (Barnes 1995, I, p. 240): "Another form [of the failure] occurs if we find that the account has been rendered of what is at rest and definite through what is indefinite and in motion (κινήσει); for what is still and definite is prior to what is indefinite and in motion (κινήσει)".

[127] Plato, *Republic* VII, 527a–b (transl. Emlyn-Jones and Preddy 2013, pp. 150–153): "Therefore, I said, those who are experienced in the finer points of geometry will not dispute with us this at least: that this knowledge contains everything that's the opposite to the arguments put forward in its by

The fact that Plato's criticism focused on the way geometers deal with their objects and speak about them could also have aimed to condemn the attribution of motion to geometrical objects in the context of definitions.

Yet, it is unlikely that Plato, as the author of the *Timaeus*, would have been unaware of the Pythagorean understanding of the generative motion of the point, the line and the surface, which was non-spatial by nature and which was attributed a primordial role in the creation of the universe. This type of motion would have differed indeed, for Plato, from that which geometers commonly attribute to geometrical objects, as the latter corresponds to a properly spatial kind of motion, which is therefore comparable to the motion of physical bodies and which would contradict the suprasensible and eternal nature of mathematical entities.

It is perhaps in order to prevent the admission of a spatial type of motion in geometry that the Pythagoreans, in their influential classification of mathematics into arithmetic, music, geometry and astronomy[128] (transmitted to the Latin world by Boethius under the name *quadrivium*),[129] distinguished geometry from astronomy according to the relation of

---

those who engage in it.—How do you mean? he asked.—I think the way they argue is quite absurd and is forced on them: I mean, they talk as if they were doing something and making all their terms to fit their activity: they talk about making the square, applying and adding, and similarly with everything else; but in my view the subject as a whole is studied for the sake of knowledge.—I agree entirely, he said.—So do we still need to agree fully on this?—In what respect?—That it is the knowledge of the eternally real and not what comes into being and then passes away.—That's easy to do, he said. Geometry after all is the knowledge of the eternally real." The argument presented by Speusippus (c. 408–c. 339/8 BC) and Amphinomus (c. fourth century BC), according to Proclus' commentary on Euclid, in favour of reducing all geometrical propositions to theorems (therefore excluding problems), evokes this text of Plato insofar as they considered that geometrical problems, which chiefly deal with constructions of figures, do not account for the eternal and immutable essence of mathematical beings. Proclus (Friedlein 1873, pp. 77–78) and (Morrow 1992, pp. 63–64): "Some of the ancients, however, such as the followers of Speusippus and Amphinomus, insisted on calling all propositions 'theorems', considering 'theorems' to be a more appropriate designation than 'problems' for the objects of the theoretical sciences, especially since these sciences deal with eternal things. There is no coming to be among eternals, and hence a problem has no place here, proposing as it does to bring into being or to make something not previously existing—such as to construct an equilateral triangle, or to describe a square when a straight line is given, or to place a straight line through a given point. Thus it is better, according to them, to say that all these objects exist and that we look on our construction of them not as making, but as understanding them, taking eternal things as if they were in the process of coming to be." Generally speaking, on Plato's criticisms of the introduction of kinematic processes in geometry, notably instrumental, see Molland (1976) and Vitrac (2005a, pp. 9–11).

[128] On this ancient classification of mathematics, see Vitrac (2005b).

[129] Boethius (Friedlein 1867, p. 9) and (Masi 1983, pp. 72–73): "of these types, arithmetic considers that multitude which exists of itself as an integral whole; the measures of musical modulation understand that multitude which exists in relation to some other; geometry offers the notion of stable magnitude; the skill of astronomical discipline explains the science of moveable magnitude. [. . .] This, therefore, is the *quadrivium* by which we bring a superior mind from knowledge offered by the senses to the more certain things of the intellect."

their objects to motion.[130] In this framework, geometry was specifically defined as the science that deals with *immobile magnitude*, and astronomy, as the science dealing with *magnitude in motion*. This designation of geometry as the science of immobile magnitude thus represented a clear assertion of the essential separation of geometrical objects from motion, such as conceived in the context of astronomy, that is, as a local motion of celestial bodies.

Plato took up this classification of mathematical sciences in the *Republic* and adapted it by distinguishing plane and solid geometry.[131] In this Platonic version of the Pythagorean classification, the study of solid geometry, which studies three-dimensional immobile magnitudes, would prepare for the study of astronomy, serving as a stepping stone to go from the more abstract objects of plane geometry to the mobile magnitudes of astronomy. But, in doing so, it would also establish the boundary between the magnitudes that are at rest and those that are in motion.

Proclus, as well as Iamblichus, did not reject the use of motion in geometry, but they were aware of the metaphysical origin and dimension of the notion of flow of the point and of the problems that arise from the notion of a moving point as the principle of continuous quantity.[132] This is very likely why they distinguished the flow (ῥύσις) from the motion (κίνησις) of the point, the latter being what geometers would produce in their imagination while attempting to reproduce the intelligible flow of the point in a spatial manner.[133]

---

[130] Vitrac (2005a, pp. 11–13).

[131] Plato, *Republic* 528a–b. On the Platonic adaptation of the Pythagorean classification of mathematics, see Vitrac (2005b).

[132] Vinel (2010, p. 119). See also Vitrac (2005a, p. 46).

[133] Proclus (Friedlein 1873, p. 97) and (Morrow 1992, pp. 79–80): "The [. . .] definition [. . .] which calls it [the line] the flowing of a point appears to explain it in terms of its generative cause and sets before us not <the> line in general, but the material line. This line owes its being to the point, which, though without parts, is the cause of the existence of all divisible things; and the 'flowing' (ῥύσις) indicates the forthgoing (πρόοδον) of the point and its generative power that extends to every dimension without diminution and, remaining itself the same, provides existence to all divisible things." and Proclus (Friedlein 1873, p. 185) and (Morrow 1992, p. 145): "The drawing of a line from any point to any point follows from the conception of the line as the flowing (ῥύσιν) of a point and of the straight line as its uniform and undeviating flowing (ῥύσιν). For if we think of the point as moving (σημεῖον κινούμενον) uniformly over the shortest path (κίνησιν), we shall come to the other point and so shall have got the first postulate without any complicated process of thought." N. Vinel (Vinel 2010, p. 119) compares, according to this interpretation, Iamblichus, *In Nicomachi arithmeticam* IV, § 4 (Pistelli 1894, pp. 57, 6–8): "the point, whose flow (ῥύσιν) geometers consider the line to be, is the principle and element of length" and *ibid.* IV, § 25 (Pistelli 1894, pp. 61, 12–18): "It will be made clear afterwards that the unit governs the generation of solid figures and that it admits their whole properties in a potential manner, and, besides, that it returns to itself, as if it were moving (κινηθεῖσα) from itself and around itself, and that accordingly the circle returns to itself by going from a point, around a point and by maintaining an equal distance." (my translation from the Greek and the French translation by Nicolas Vinel, in Vinel (2014, pp. 127 and 131). As noted by N. Vinel at this occasion,

The Persian mathematician and philosopher 'Umar al-Khayyām is an exemplary case among medieval Arabic mathematicians for his critical attitude toward the use of motion in geometry.[134] To him, genetic definitions notably raised the problem of the causal order between what is first and what is second in the order of existence.[135] Indeed, since only three-dimensional objects can be found in nature, the notions of point, line and surface are ontologically posterior to that of body and it would therefore be absurd to present the point as the origin of the body, through the successive generation of the line and of the surface.

As noted by B. Vitrac,[136] one may also read, behind the distinction made by Abū al-Sijzī (c. 945–c. 1020)[137] and Abū Ja'far al-Khāzin (c. 900–c. 971)[138] between a mobile and a fixed geometry, a certain prudence in considering the latter as geometry in the strict sense and to consider motion as an intrinsic feature of geometrical objects.

The fact of attributing motion to geometrical figures was also challenged in the Latin middle ages, for instance within commentaries on the *Tractatus de sphaera* of Johannes de Sacrobosco (c. 1195–c. 1256). This thirteenth-century treatise of spherical astronomy starts with the quotation of the two definitions of the sphere by Euclid and by Theodosius of Bithynia (c. 160–c. 100 BC),[139] the former defining the sphere by its mode of

---

while the notion of ῥύσις was used for the generation of the first dimension from the point, κίνησις was used for the generation of a specific figure.

[134] See Vitrac (2005a, pp. 3 and 6) and Rashed (2013).

[135] Al-Khayyām, *Epistle on the explanation of the problematic premisses of the book of Euclid*, in Djebbar (2001, pp. 86–87): "This is a discourse that has nothing to do with geometry and this is so for several reasons, among which are the following: how can the line move on two lines while maintaining its verticality and which proof [is there] that this is possible? Or: what relation is there between geometry and motion, and what is the meaning of motion? Or else: it is obvious that, for the experts, the line is a breadth which cannot be anywhere else than in a surface and this surface in a body, or that it is itself in a body without there being a surface beforehand. How would then motion be attributable to it if it is abstracted from its place? Or else: how can the line result from the motion of the point since it precedes the point in essence and in existence?" (my translation from the French version of A. Djebbar).

[136] Vitrac (2005a, p. 36).

[137] Al-Sijzī, *Treatise on the trisection of the rectilinear angle* (Woepcke 1851, p. 120): "Proposition solved by one of the ancient [authors] by the means of the straightedge and of mobile geometry, but which we must solve by the means of fixed geometry" (my translation from the French version of F. Woepcke).

[138] Abū Ja'far al-Khāzin, in Knorr (1989, p. 311): "On the construction of two lines between two lines so that <all of them> follow in proportion, according to the method of fixed geometry, by the Shaik Abū Ja'far Muḥammad ibn al-Ḥusain, may Allah have mercy on him".

[139] Sacrobosco (Thorndike 1949, p. 118): "A sphere is thus described by Euclid: A sphere is the transit of the circumference of a half-circle upon a fixed diameter until it revolves back to its original position. That is, a sphere is such a round and solid body as is described by the revolution of a semicircular arc. By Theodosius a sphere is described as such: A sphere is a solid body contained within a single surface, in the middle of which there is a point from which all straight lines drawn to the circumference are equal, and that point is called the 'center of the sphere'. Moreover, a straight line passing through the center of the sphere, with its ends touching the circumference in opposite

generation[140] and the latter by its spatial properties.[141] This text therefore incited its readers to determine which definition is more adequate to define the geometrical sphere. One of the arguments provided by Pierre d'Ailly (c. 1350–1420) on this issue was that "a mathematical notion should not be defined by terms that introduce motion", since "mathematics abstracts [its objects] from motion and matter".[142] In this tradition, Euclid's kinematic definition of the sphere was thus regularly presented as an improper definition in view of its appeal to motion, questioning the suitability of motion as a means of defining geometrical objects on account of their essential immutability. This comparison between Euclid's definition of the sphere and that of Theodosius, as well as some of the arguments then presented in favour of Theodosius' definition, reappeared in certain sixteenth-century commentaries on Euclid's *Elements* (as will be shown later), confirming that the introduction of kinematic notions in geometrical definitions still raised concern at the time. Indeed, in the sixteenth century, when the rediscovery of Proclus' commentary on the first book of Euclid's *Elements* contributed to popularise the notion of flow of the point (*puncti fluxus*) as a means of explaining the Euclidean definition of the line, the status of geometrical motion still raised questions, even in a mathematical context.[143]

The flow of the point was then regularly assimilated in its function and status to the motion involved in the generation of plane and solid figures, such as the circle, the sphere, the cone and the cylinder, for which Euclid and other Ancient geometers mostly used the term φορά and which was straightforwardly used to designate a local process. Now, this contributed to blur the initial (and in some cases ignored) distinction between the flow of the point and other types of genetic processes appealed to in geometry on a metaphysical or ontological level. Doubts raised concerning one of these processes (e.g. the rotation of the semicircle in the definition of the sphere) were therefore transferred onto the other (e.g. the flow of the point in the genetic definition of the line) and vice versa.

---

directions, is called the 'axis of the sphere'. And the two ends of the axis are called the 'poles of the world'."

[140] See n. 5, p. 2.

[141] Theodosius (Nizze 1852, p. 1) and Heath (1981b, p. 247): "The sphere is a solid figure contained by one surface such that all the straight lines falling upon it from one point among those lying within the figure are equal to one another".

[142] D'Ailly (Sacrobosco 1531, fol. 146v): "Quaeritur primo utrum diffinitio sphaerae sit bona quam dat autor in textu scilicet sphaera est transitus circunferentiae dimidii circuli quotiens fixa diametro quousque ad locum suum redeat circunducitur. [...] Secundo sic quia *terminus mathematicalis non debet diffiniri per terminos importantes motus modo sphaera est terminus mathematicalis & transitus est terminus importans motum igitur consequentia est nota maior patet quia mathematica abstrahit a motu & a materia minor nota est*". (My emphasis.)

[143] The philosophical treatment of Euclid's first book of the *Elements* by Proclus, along with the medieval tradition of commentaries of scientific works such as Sacrobosco's *Sphaera*, encouraged early modern commentators of Euclid to lead philosophical discussions on the nature and methods of mathematics in their exposition of the *Elements*.

1.8   Conditions and Justifications for the Admission of Motion in Geometry

In response to the objections or doubts raised by philosophers concerning the legitimacy of kinematic notions in geometry, several arguments were brought forth by both mathematicians and philosophers from Antiquity to the early modern period. These arguments aimed either to justify the use of motion in geometry or to restrict the conditions under which it could be used.

Hero of Alexandria, for instance, took care to present the motion or flow of the point generating the line as a noetical process ("κινηθέντος ἤ μᾶλλον νοηθέντος ἐν ῥύσει νοεῖται γραμμή"),[144] that is, as a process taking place in thought, as opposed to concretely.[145] Hence, although he conceived the flow of the point as a properly spatial process, since he held it as the proper cause of dimension (διάστασις) in the context of geometry, the spatiality of this process would not be that of physical motion, but rather that of the imagination.

A similar condition was also established by medieval Arabic mathematicians, such as Thābit ibn Qurra and Ibn al-Haytham, who insisted on the imaginary (as opposed to physical) nature of the motion admitted by geometers in the frame of their definitions and demonstrations.[146] These authors also stated that motion is admissible in geometry on the condition that it is uniform and simple.[147] Actually, for Thābit ibn Qurra, the motion involved in the generation of the circle (conceived as produced by the rotation of a line-segment around one of its fixed extremities), as well as the motion potentially implied by the superposition of figures, was not only admissible, but played a fundamental role in geometry, as it represented a necessary means of establishing the equality or inequality of figures.[148] Ibn al-Haytham also showed that the motion involved in the generation and

---

[144] Hero (Heiberg 1974, p. 14), Df. 1: "the line is conceived from the motion of a point or rather from the intellection of the point in the state of flowing" (see *supra* n. 86, p. 20).

[145] Vitrac (2005a, pp. 20–21). On the ontological relation between intellection (νόησις) and generative motion in geometry, notably in the context of Euclid's study of the properties of the solids of revolution in Book XII of the *Elements*, see Lachtermann (1989, pp. 87–91).

[146] Jaouiche (1986, pp. 58–60, 152–153 and 70) and Vitrac (2005a, pp. 47–48 and 52). See also Ighbariah and Wagner (2018).

[147] This argument also evokes the criteria established by Pappus in his *Collection* (Pappus 2009, pp. 132–133, Prop. 31) to distinguish the properly mathematical means of construction of geometrical objects from those which he called mechanical and which involved processes resulting from a combination of motions of indefinite ratio, such as the quadratrix. On Pappus' treatment of the quadratrix, see *supra*, n. 87, p. 20 and *infra*, pp. 217–218.

[148] Thābit ibn Qurra, *Small treatise on the fact that if two straight lines are led according to two angles less than two rights, they meet*, in Jaouiche (1986, p. 152): "However, if we also consider the first of the propositions which precede it and which precede all the other propositions of this book, and by examining the question, we will learn that the principle of its demonstration corresponds to the equality of the straight lines led from the centre of the constructed circle and that the validity of this,

transformation of figures guarantees the continuity of the resulting magnitudes and conditions the discovery of the invariant elements of figures, playing thereby a primordial role in the definition, as well as in the demonstration of the properties of geometrical objects.[149]

Although Proclus, as Iamblichus, distinguished the intelligible *flow* of the point ($\acute{\rho}\acute{\upsilon}\sigma\iota\varsigma$) and the actual *motion* of the point ($\kappa\acute{\iota}\nu\eta\sigma\iota\varsigma$), he nevertheless counted motion among the objects considered by the geometer.[150] He precisely legitimated the use of motion in the study of the essential properties of figures by stating that the extended and movable figures of the imagination, on which the geometer actually carries out his demonstrations, would lead to the knowledge of the purely intelligible, extensionless and motionless substances which the geometer aims to contemplate in a universal manner.[151] The imagination would provide geometrical objects with an "intelligible matter" through which they could be

---

regarding these [straight lines (?)], [depends] for us on nothing else than on what we have understood and what was established within us [concerning] the construction of the circle and of its production. And we understand this by imagining only one straight line of a [determinate] magnitude or some other thing which preserves magnitude and distance as it moves circularly from a given place until it returns to its starting point by remaining fixed by one of its extremities to a single immobile point. In this way, we perfectly understand and realise what we have said concerning the equality of the straight lines which proceed from the centre of the circle. We have indeed moved a straight line by the imagination in order to make it coincide with all these distances, [and this] by a continuous transfer." (my translation from the French version by Jaouiche). More generally, on Thabit's characterisations of the various uses of motion in geometry, see Jaouiche (1986, pp. 151–153).

[149] Rashed (2013, pp. 64–66).

[150] Proclus (Friedlein 1873, p. 57) and (Morrow 1992, p. 46): "Let us next speak of the science itself that investigates these forms. Magnitudes, figures and their boundaries, and the ratios that are found in them, as well as their properties, their various positions and motions—these are what geometry studies."

[151] Proclus (Friedlein 1873, pp. 185–186) and (Morrow 1992, pp. 145–146): "If someone should inquire how we can introduce motions into immovable geometrical objects and move things that are without parts—operations that are altogether impossible—we shall ask that he be not annoyed if we remind him of what was demonstrated in the Prologue about things in the imagination, namely, that our ideas inscribe there the images of all things of which the understanding has ideas and that this unwritten tablet was the lowest form of 'nous', the 'passive'. This statement, however, does not remove our difficulty, for the 'nous' that receives these forms from elsewhere receives them through motion. But let us think of this motion not as bodily, but as imaginary, and admit not that things without parts move with bodily motions, but rather that they are subject to the ways of the imagination. [...] Consequently the forms peculiar to geometrical objects are quite other than the things whose existence comes from them. The motion of bodies is one thing, the motion of objects conceived in imagination is something else; and the space of extended objects is other than the space of partless beings. We must keep them separate and not confuse them, lest we disarrange the natures of things." On Proclus' conception of motion in geometry, see Nikulin (2008).

divided and moved in a non-physical manner, allowing them to be clearly distinguished from physical beings.[152]

Thus, in Proclus' commentary on Euclid, the generations of geometrical objects were conceived as proper definitions, which would be complementary to their definitions by property.[153] As was noted by Vitrac, such a situation occurred in Apollonius' *Conics*, where both types of characterisations of geometrical objects (by properties and by generation) are presented, though not necessarily as means of definition.[154]

The fact that certain ancient Greek geometers, when they proposed a genetic definition of their objects, also provided a definition by property has certainly been interpreted as resulting from the will to avoid the introduction of motion in geometrical definitions.[155] Yet, it seems that, for these geometers, definitions by property and definitions by genesis were in fact mutually complementary.

---

[152]Proclus (Friedlein 1873, pp. 49–50) and (Morrow 1992, p. 40): "But if the objects of geometry are outside matter, its ideas pure and separate from sense objects, then none of them will have any parts or body or magnitude. For ideas can have magnitude, bulk, and extension in general only through the matter which is their receptacle, a receptacle that accommodates indivisibles as divisible, unextended things as extended, and motionless things as moving."

[153]Proclus (Friedlein 1873, p. 272) and (Morrow 1992, p. 212): "Nicomedes made use of conchoids—a form of line whose *construction*, kinds, *and properties* he has taught us, being himself the discoverer of their peculiarities—and thus succeeded in trisecting the rectilinear angle generally" (my emphasis) and Proclus (Friedlein 1873, p. 356) and (Morrow 1992, 277): "After a species has been constructed, the apprehension of its inherent and intrinsic property differentiates the thing constructed from all others." See Vitrac (2005a, pp. 29–30).

[154]Apollonius, book 4. Vitrac (2005a, p. 30).

[155]This is pointed out by Vitrac (2005a, p. 29) in reference to Jaouiche (1986, 69) and Fauvel (1987, p. 3). As Vitrac noted in particular (Vitrac 2005a, b, pp. 27–30), although Apollonius combined, in his treatment of the three fundamental types of conic sections, the generation, the symptomata (i.e. the quantitative properties of the conics) and the naming of the conic sections (Prop. I.11–13, Heiberg 1891, pp. 36–52), he clearly distinguished the generation and the demonstration of the symptoma within the preface to his first book: (Heiberg 1891, p. 4) and (transl. Densmore 1998, 1): "The first book contains the generation of the three sections and of the opposite branches, and the principal properties (τὰ ἀρχικὰ συμπτώματα) [...]". The symptomata, through which Apollonius differentiated the three main type of conic sections (parabola, hyperbola and ellipse), correspond to the quantitative relation (equality, excess and lack) between the base of the rectangle on the "abscissa" (the interval between the tangent to the diameter of the curve) and the "ordinate" (i.e. the segment that perpendicularly joins any point of the curve to its diameter), equal to the square on the ordinate, and a line-segment of determinate length, called the *latus rectum* or the parameter, situated on the tangent to the diameter at right angles to the abscissa. The parabola is the curve in which the rectangle on the abscissa and equal to the square on the ordinate has its base equal to the *latus rectum*; the hyperbola, the curve in which the base of the rectangle on the abscissa, equal to the square on the ordinate, exceeds the *latus rectum*; and the ellipse, the curve for which the base of the rectangle on the abscissa, equal to the square on the ordinate, is inferior to the *latus rectum*. On this topic and on the distinction between generation and symptomata of conic curves, see Fried and Unguru (2001, pp. 74–90) and Netz (1999, pp. 100–101).

In this regard, one of the arguments brought forth by al-Khayyām, who clearly rejected the use of motion as a means of definition in geometry,[156] is that the geometer's kinematic understanding of geometrical objects may be useful to determine the effective (and non-essential) construction of figures, though not their essence and true cause. The proper essence of geometrical objects could actually not be determined by the mathematician, but only by the philosopher. In this perspective, motion could be admitted in certain cases for didactic purposes, which is how al-Khayyām justified, or rather excused, Euclid's definition of the sphere in terms of motion.[157]

As we will see, similar arguments were also set forth in sixteenth-century Europe through the distinction between *definitio* and *descriptio*, that is, between the definition properly speaking, which states the essential properties of a geometrical figure, and the process through which a figure is generated in the framework of the geometer's investigation.[158] Thus, the notion of *descriptio*, which is implied, for instance, by the use of the verb *describere* in the Latin translation of Euclid's third postulate by Bartolomeo Zamberti (1473–1543),[159] was used to generically designate a definition that states the mode of generation or of construction of a given geometrical object,[160] as opposed to its essential properties.

* * *

In order to set forth the conceptions of the above-mentioned sixteenth-century commentators of Euclid on motion and genetic definitions in geometry, each of them will be dealt with separately, their conceptual divergences and convergences being presented along the way. A chronological order will also be observed in order to better set forth the evolution of the considered issues over time.

---

[156] See *supra*, p. 32.

[157] Al-Khayyām, *Epistle on the explanation of the problematic premises of the book of Euclid* (Djebbar 2001, p. 87): "To he who would say that Euclid defined the sphere at the beginning of the eleventh book [...] by saying that 'the sphere results from the rotation of a semicircle until it returns to the starting-point', we will answer by saying [this]: the true and clear definition of the sphere, which is that 'it is a solid figure surrounded by a single surface within which is a point such that all the straight lines going from it toward the surface that surrounds it are equal', is known. Euclid diverged from this description as for what he said by a general and simplifying approach given that, in the books where he deals with the bodies, he allowed himself many simplifications by counting on the training of the learner as he tackled them." (my translation from the French translation of Djebbar). See also Rashed (2013) and Vitrac (2005a, p. 55).

[158] The verb *describere* was commonly found in Latin geometrical works to designate the production of a line or figure in the context of a construction.

[159] Heiberg (1883, p. 8): "καὶ παντὶ κέντρῳ καὶ διαστήματι κύκλον γράφεσθαι" (Euclid 1956, I, p. 199: "To *describe* a circle with any centre and distance"), translated by Zamberti as: "Omni centro & intervallo: circulum describere" (Lefèvre 1516, fol. 5r). It may be noted that the translation circulated by Campanus is different, γράφεσθαι having been translated by *designare*: "Super centrum quodlibet, quantumlibet occupado spacium, circulum designare" (Lefèvre 1516, fol. 3v).

[160] W. Sacksteder (in Sacksteder, 1981, p. 577) defines what a *descriptio* was for Thomas Hobbes, that is, the result of the artificial and selected motion supposed by the mathematician in his imagination to obtain a figure possessing the required properties.

Oronce Fine

2.1 The Life and Work of Oronce Fine and the Significance of His Commentary on the *Elements*

Oronce Fine or Finé[1] (Briançon, 1494–Paris, 1555) was a French mathematician and professor of mathematics, but also a cartographer, an editor and engraver, a designer and maker of mathematical instruments.[2] He is mostly known for his role as first Royal lecturer in mathematics within the humanist college founded by François Ier in 1530 and which was later known as the Collège Royal (and much later as the Collège de France).[3] Fine kept this position until his death in 1555.[4] Before 1530, he taught mathematics at the Collège de Navarre, where he had studied, and at the Collège de Maître Gervais.[5]

---

[1] The question whether Oronce Fine's last name should be accentuated remains unsettled. See, for instance, Rochas (1856–1860, I, p. 384), Ross (1971, pp. 8–9), Poulle (1978) and Dupèbe (1999, II, p. 519). I have chosen to follow the non-accentuated form as it is privileged by historians of the Dauphiné (where Fine's family finds its origins) and also as it allows more flexibility.

[2] On Fine's cartographical work, see Gallois (1918), Hillard and Poulle (1971), de Dainville (1970), Karrow (1993, pp. 68–90) and Brioist (2009). On his design of mathematical instruments, see Destombes (1951), Eagleton (2009) and Turner (2009). On his work as an editor and engraver, see Brun (1934), Brun (1966), Johnson (1928), Ross (1971, pp. 32–58), Pantin (2009b), Pantin (2010), Pantin (2012), Pantin (2013), Oosterhoff (2014, 2016, 2017).

[3] On the history of the Collège de France, see Tuilier (2006).

[4] On Fine's role as a Royal lecturer in mathematics, see Tuilier (2006a, b), Pantin (2006), Pantin (2009a), Dhombres (2006) and Axworthy (2016, pp. 16–19). More generally, on Fine's life, education, and career, see Ross (1971, pp. 8–30), Poulle (1978) and Axworthy (2016, pp. 12–22).

[5] Dupèbe (1999, II, pp. 533 and 540–41), Boudet (2007), Pantin (2009a) and Pantin (2013).

Fine published a great number of works on various mathematical disciplines, theoretical and practical.[6] His best known work is the *Protomathesis*,[7] published in Paris in 1532, at the beginning of his career as a Royal lecturer. This work corresponds to a compendium of four mathematical treatises (dealing with practical arithmetic, practical and theoretical geometry, cosmography and gnomonics[8]) and which aimed to present and promote his teaching program for mathematics as a Royal lecturer. Each of the four treatises that compose the *Protomathesis* were in part or entirely reprinted several times throughout Fine's career, and the totality of the compendium was translated and published in Italian by Cosimo Bartoli (1503–1572) in 1587,[9] along with Fine's treatise on burning mirrors (*De speculo ustorio*), first published in 1551.[10] The geometrical part of this compendium, the *De geometria libri duo*, offers us, in the first book, an introduction to the basic notions and principles of Euclidean geometry, rudiments of spherical geometry and trigonometry, as well as an overview of different measurement units. The second book corresponds to a treatise of practical geometry.

Fine's commentary on the first six books of Euclid's *Elements* (*In sex priores libros geometricorum elementorum Euclidis*) was first published in 1536 in Paris and was reprinted in 1544 and 1551.[11] This work is the earliest of a long tradition of French commentaries on the Euclid's *Elements*,[12] which also includes the commentaries of Peletier and Foix-Candale. It was written with an explicitly pedagogical aim, as would have required his position of Royal lecturer. Fine claimed indeed in the preface to have produced this commentary in order to compensate for the lack of mathematical teaching at the university of Paris.[13] This is coherent with the primary aim of François Ier's college of Royal lecturers, which was to provide a teaching in domains useful to the study of ancient

---

[6] Bibliographical lists of Fine's works are proposed in Ross (1971, pp. 398–449), Hillard and Poulle (1971), Ross (1974), Pantin (2013) and Axworthy (2016, pp. 407–19).

[7] Fine (1532).

[8] *De arithmetica practica. Libri IIII; De geometria libri II; De Cosmographia, sive sphaera mundi libri V; De solaribus horologiis et quadrantibus. Libri IIII.*

[9] Fine (1587).

[10] Fine (1551b).

[11] Fine (1536, 1544, 1551c).

[12] Loget (2004).

[13] Fine (1536, sig. *2r–v): "Dum celebres illas et fidissimas artes, Francisce Rex invictissime, quae solae Mathematicae, hoc est, disciplinae merverunt adpellari, sub tuo felici profiterer nomine: raros admodùm offendi (etiam in numerosa auditorum multitudine) qui satis fido ac liberali animo, tam utile ac iucundum philosophandi genus, à limine (ut aiunt) salutare, ne dicam ad illius penetralia, penitioràque secreta, pervenire dignarentur [. . .] qui enim ad lauream adspirant philosophicam, iureiurando profitentur arctissimo, sese praenominatos Euclidis libros audivisse. An verò illius elementa, multis ab hinc annis, usque ad nostra viderint (ne dicam intellexerint) tempora (paucis forsitam exceptis, quos aequus amavit Iuppiter) non ausim honestè confiteri."

texts, but which were neglected by the university.[14] Hence, Fine's commentary on Euclid offers detailed and didactical explanations designed to facilitate the understanding of Euclid's principles and propositions, as well as the Greek text of the enunciations.[15]

Fine is an important author to consider here not only because of his appeal to genetic definitions in his commentary on Euclid, but also because he was one of the earliest sixteenth-century commentators of Euclid. He was as such instrumental in shaping the humanist approach to the *Elements*, at least in France, where several translations and commentaries of the first six books were published after his.[16] His commentary on the *Elements* was moreover known internationally and was referred to, within the investigated tradition, by Billingsley,[17] Commandino[18] and Clavius.[19]

His commentary on the definitions of Euclid's first book of the *Elements* shares content with the first book of the *Geometria*, including references to the generation of geometrical objects. These works, which complete each other in this regard, will therefore both be considered here.

This chapter will survey the uses and formulation of genetic definitions in Fine's teaching of Euclidean geometry, as well as the ontological and epistemological considerations he set forth on geometrical motion in general. It will also consider the way Fine used motion to determine the composition of continuous quantity, as well as the relation between arithmetic and geometry. This chapter will also be the occasion to lay out certain ancient conceptions that are relevant to the investigated tradition, notably those found in Proclus' commentary on Euclid, as well as the medieval distinction between *definitio* and *descriptio*.

As we will see, Fine made an extensive use of genetic definitions in his geometrical work. He also presented different positions on the epistemic status and function of genetic definitions in geometry, these being at times described as a pedagogical tool and, at other times, as foundational to the concepts and methods of geometry. We will also see that, in Fine's philosophy of mathematics, different interpretations of the ontological status of geometrical objects converged (Aristotelian, Neoplatonic, Pythagorean), explaining the

---

[14] The first royal lectureships were dedicated to Greek and Hebrew. On the creation of these lectureships and on their first appointees, see Irigoin (2006); Kessler-Mesguich (2006) and Tuilier (2006a). The Collège de Guyenne was also founded in Bordeaux in 1533 to cater for the pedagogical needs of humanists.

[15] The fact of adding the Greek text of the enunciations would be mostly motivated by the humanist milieu of the Royal lecturers in which Fine was evolving.

[16] Peletier (1557), Forcadel (1564) and Errard (1598).

[17] Billingsley (1570, fol. 95v), III.18: "An other demonstration after Orontius".

[18] Commandino (1572a, sig. *2v), preface: "Nam ut pauca de hac re loquar, Orontius quidem Phinaeus haud obscuri nominis auctor priores tantum sex libros nulla graeci codicis ratione habita edidit."

[19] See, for instance, Clavius (1611–1612, p. 116): "Ex Orontio" or (*ibid.* p. 218): "Orontius, & nonnulli alij interpretes, longe aliter definitionem hanc exponunt".

apparent tensions that arise in his definition of the status and function of motion in geometry.

2.2 Formulation and Distribution of Genetic Definitions in the *Geometria* and in Fine's Commentary on Euclid

In his *Geometria* and in his commentary on the *Elements*, Fine regularly appealed to genetic definitions of geometrical objects. He did so in order to show how the different dimensions (length, breadth, depth), as well as the various types of geometrical figures (lines, plane and solid figures), are derived from objects of lower dimension according to mathematicians. In the *Geometria*, the genetic definition of the line is provided at the end of the commentary on the definition of the point and is thus given before Euclid's definition of the line by property. The same pattern is applied to the definitions of the surface and of the solid: the genetic definition then precedes the definition by property and displays the causal relation between the object of lower dimension and that of higher dimension.

> We call a point that which cannot be divided, or which has no parts, having been imaginarily separated from the continuum (whose principle it is said to be). Mathematicians describe the line as caused by its *intelligible flow*, as if it left a trace, acquiring length, the first of the dimensions.[20]
>
> The line is therefore a breadthless length, deprived of breadth and thickness, whose limits are points. Some also call the latter signs. [...] Thereafter, the surface is correspondingly described by the *abstract flow of the line*, thus acquiring the second dimension, breadth, as length was previously obtained.[21]

---

[20] Fine (1532, fol. 51r), § 2: "Punctum id vocamus, quod partiri non potest, seu cuius pars nulla est, à continuo (cuius principium esse dicitur) imaginariè separatum. Ex cuius intelligibili fluxu, non secus ac si vestigium relinqueret, linea secundum mathematicos causari describitur: longitudinem dimensionum primariam acquirendo". *Cf.* Fine (1536, p. 1), Df. I.1: "Ex cuius quidem puncti abstracto defluxu, per infinitam suiipsius multiplicationem, longitudo dimensionum primaria conficitur: quae linea vocitatur" and (*ibid.*) Df. I.2: "*Linea verò, est longitudo latitudinis expers.* Hoc est, latitudine privata. Cum enim punctum omni careat dimensione: suo fluxu, seu transsumptivo motu, causat tantummodo longitudinem."

[21] Fine (1532, fol. 51r), § 3: "Linea igitur, est illatabilis longitudo, latitudine crassitieique privata: cuius limites sunt puncta, quae etiam à nonnullis signa vocitantur. [...] Ex lineae postmodum abstracto defluxu, superficies respondenter describitur: latitudinem dimensionum secundam, cum prius obtenta longitudine, consequenter adipiscendo." Cf. Fine (1536, p. 1), Df. I.3: "*Lineae autem limites, sunt puncta.* Incipit enim à puncto, & *ex infinitis conficitur punctis*, in punctúmque terminatur. Omnis porrò linea, vel recta, vel obliqua venit imaginanda." (My emphasis.) and (*ibid.*, p. 2), Df. I.5: "*Superficies est, quae longitudinem, latitudinemque tantùm habet.* Quae cùm exordiatur à linea, & ipsius lineae terminativa puncta, ad motum eiusdem, rectam vel obliquam lineam describant, in eadémque linea mota quiescat ipsa superficies" and (*ibid.*), Df. I.6: "*Superficiei extrema sunt lineae.* Porrò cùm linea, ad descriptionem mota superficiei, recta fuerit, atque in longum lineae rectae uniformiter, brevissimeque traducta: fit superficies, quae plana dicitur." See also Fine (1536, p. 4),

The surface is defined as that which has length and breadth only, bounding all solids and having lines as its extremities. [. . .] Finally, the solid is imaginarily represented by the same mathematicians as caused by *the flow of the surface*, obtaining thickness or depth, as length and breadth were previously acquired, in such way that thickness or depth will be the last dimension. That is why the solid is called the body contained by three dimensions, that is, as resulting from length, breadth and thickness or depth and as directly delimited by a single or several surfaces.[22]

The formulation of the genetic definition of the line in Fine's commentary on the *Elements* is also noteworthy. There, the flow of the point is described as a *transumptivus motus*, which may be here translated as a "translational motion", that is, as a motion which translates or transports the point from one place to another.[23]

The line is a length exempt from breadth, that is, deprived of breadth. Indeed, since the point lacks all dimensions, it only causes length by its *flow, or translational motion*.[24]

Df. I.13: "*Terminus est, quod cuius finis est*. Utpote, punctum ipsius lineae, linea superficiei, superficies denique solidi: quemadmodùm ex eorundem abstractiva descriptione facilè colligitur."

[22] Fine (1532, fol. 51r), § 4–5: "Superficies enim dicitur, quae longitudinem latitudinemque tantum habet, omnium solidorum terminativa: cuius extrema sunt lineae. [. . .] Ex superficiei denique fluxu, ipsum solidum ab eisdem mathematicis imaginatur phantasticè causari, crassitudinem, seu profunditatem, cum prius acquisitis longitudine atque latitudine finaliter obtinendo: eo quippe modo, ut ipsa crassitudo sive profunditas sit dimensionum ultima. Solidum itaque, dicitur corpus trina dimensione contentum, longitudine videlicet, latitudine, atque crassitie seu profunditate resultans, unica superficie, pluribusve superficiebus immediatè terminatum". *Cf.* Fine (1536, p. 2), Df. I.7: "*Plana superficies est, quae ex aequali suas interiacet lineas*. [. . .] Ex superficiei denique fluxu, solidum sive corpus trina dimensione, utpote, longitudine, latitudine, atque profunditate contentum, abstractivè describitur."

[23] The adjective *transumptivus*, which literally means "taken beyond" or "taken across" (being composed of *trans-* and *sumo*), may either connote a motion from a place to another, and in particular a motion in a straight line as opposed to a circular motion, in which sense it may be translated as "translational", or may designate a motion that is only metaphorically or improperly called motion, in which sense it may be translated as "metaphorical". As will appear more clearly in the following pages, both meanings could be applied to Fine's use of the term, since while he defined the flow of a point as a spatial process and as a translation of the point from one place to another, he also admitted that mathematicians do not speak of motion in the proper sense, that is in the way that natural philosophers do, since motion is only a quality that belongs to physical substances. It is, as such, possible that Fine played on the double meaning of this term. Yet, in this specific case, it seemed more conform to his intention to translate this term as a "translational", that is, as designating a motion that results in a change of position of the point. I would like to thank Vincenzo De Risi for his valuable comments on this issue.

[24] Fine (1536, p. 1), Df. I.2: "*Linea verò, est longitudo latitudinis expers*. Hoc est, latitudine privata. Cum enim punctum omni careat dimensione: suo fluxu, seu transsumptivo motu, causat tantummodo longitudinem."

In his commentary on Df. I.7 of the *Elements*, where Fine defined the flow of the surface as the cause of the solid, he also mentioned the motion of solid figures, stating that this motion does not add any supplementary dimension to continuous quantity, but rather modifies the size and the configuration of figures.

> Finally, from the flow of the surface, is abstractly described the solid or the body, which is contained by three dimensions, that is, length, breadth and depth. [...] It does not seem thereafter that the motion of the solid adds any dimension, but it increases the extent of the dimensions and modifies the figure.[25]

2.3   The Function of Genetic Definitions in Fine's Exposition of Euclid

In both the *Geometria* and the commentary on Euclid, genetic definitions are presented as useful imaginations that enable us to reach the intuition of a breadthless length and of a depthless surface, which are otherwise impossible to encounter in the physical world. They are also viewed as useful to understand the spatial relation between geometrical objects and their extremities, since the extremities of a magnitude are thereby identified as the beginning and end of its generation.

But more fundamentally, genetic definitions are described by Fine as properly relevant to the mathematician's conception and treatment of his objects (*secundum mathematicos*), for which he regarded them as fully appropriate to the teaching of the geometrical notions of line, surface and solid. Genetic definitions are indeed, to Fine, a condition of the synthetic gradation from the indivisible principle of continuous quantity, the point, to its most complex elements, solid figures, insofar as these definitions display the causal relation between magnitudes of immediately inferior and superior dimensions. The flow of the point (which is deprived of parts and dimensions) is accordingly defined as the *cause* of the line (which has only one dimension, that is, length); the flow of the line, as the *cause* of the surface and of plane figures (which possess only two dimensions, length and breadth); and the flow of the surface, as the *cause* of the body and of all solid figures (which have all three dimensions, length, breadth and depth). When dealing with the generation of the solid in particular, Fine shows that the fact that the solid contains three dimensions presupposes that it results from a three-fold generative process which starts from the motion of the point.

Through his systematic addition of genetic definitions to Euclid's definitions by property, and in particular by placing these before Euclid's definitions of the line, surface and

---

[25] Fine (1544, p. 2), Df. I.7: "Ex superficiei denique fluxu, solidum sive corpus trina dimensione, utpote, longitudine, latitudine, atque profunditate contentum, abstractivè describitur. [...] Solidum porrò motum, nullam videtur acquirere dimensionem: sed ipsas dimensiones augmentat, immutatque figuram." The 1536 edition does not contain the last sentence.

solid, Fine thus shows how, from the point, a quasi-infinite number of different plane and solid figures can be created. This is confirmed by his commentary on Euclid's Df. I.7, where he mentions the motion of solid bodies and presents it as the *cause* of solids of greater sizes and of more complex configurations.

> It does not seem thereafter that the motion of the solid adds any dimension, but it increases the extent of the dimensions and modifies the figure. Hence, in virtue of the variety of the lines and of the surfaces, and of their diverse motion or abstract flow, is drawn the various and practically infinite multitude of planes and solids, that is, of surfaces and bodies, designated by different names according to the diversity of their boundaries and angles.[26]

In this context, this gradation from the flow of the point to the constitution of lines, plane and solid figures mirrors the synthetic movement of geometrical knowledge, as described by Proclus in a passage of his *Commentary on the first book of Euclid's Elements* and which Fine takes up in the preface of the *Geometria*[27] (without acknowledging his source).[28] This movement, which goes from the point to the constitution of solid figures and which, at each step, derives knowledge of these objects from the principles of geometry (definitions, axioms and postulates) or from prior propositions, is then described as complementary to the movement of analysis, which allows the geometer to demonstrate the necessity of his conclusions. This analytical part of geometrical knowledge aims to display the dependence of the geometer's constructions and conclusions on prior

---

[26] Fine (1544, p. 2), Df. I.7: "Solidum porrò motum, nullam videtur acquirere dimensionem: sed ipsas dimensiones augmentat, immutátque figuram. Igitur pro linearum atque superficierum varietate, diversóque eorundem motu, seu abstracto defluxu: varia, & penè infinita tum planorum, tum etiam solidorum, hoc est superficierum & corporum abstrahitur multitudo, pro limitum & angulorum varietate, diversis expressa nominibus".

[27] Proclus (Friedlein 1873, p. 57) and (Morrow 1992, p. 46): "Let us next speak of the science itself that investigates these forms. Magnitudes, figures and their boundaries, and the ratios that are found in them, as well as their properties, their various positions and motions—these are what geometry studies, proceeding from the partless point down to solid bodies, whose many species and differences it explores, then following the reverse path from the more complex objects to the simpler ones and their principles. It makes use of synthesis and analysis, always starting from hypotheses and first principles that it obtains from the science above it and employing all the procedures of dialectic— definition and division for establishing first principles and articulating species and genera, and demonstrations and analyses in dealing with the consequences that follow from first principles, in order to show the more complex matters both as proceeding from the simpler and also conversely as leading back to them". On Fine's relation to Proclus, see Axworthy (2016, pp. 58–71).

[28] Fine may have quoted Proclus' text from the *De expetendis et fugiendis rebus opus* of Giorgio Valla (1447–1500), published in 1501 (Valla 1501), and which contained Latin translations of passages of Proclus' commentary on Euclid's first book of the *Elements* without referring them to their author. The first printed edition of Proclus' commentary on Euclid dates from 1533 (Proclus 1533) and is therefore posterior to the publication of Fine's *Protomathesis*. On this issue, see Axworthy (2016, pp. 59–65).

propositions and ultimately on the principles of geometry, setting forth thereby the well-foundedness of the relations and properties that are the objects of his demonstrations.

> Geometry provides knowledge of the ratios of magnitudes, of figures and of the limits which are in them, as well as of the properties, the various positions and the motions that are proper to them. It is derived from the indivisible sign or point until it reaches solid figures and carefully studies their multiform differences by a subtle examination, comparing the most complex things to the simplest and returning to their principles. Enfolded in the precepts of dialectic, since it appeals to a greater variety of principles, obtained from a prior teaching, it appears to be more certain and more rigorous than other sciences (apart from arithmetic, which has simpler principles).[29]

Hence, genetic definitions are not merely, for Fine, a useful tool to help provide an appropriate intuition of the spatial properties of geometrical objects in the context of teaching. They are, more importantly, both a condition of the knowledge of the essential properties of geometrical objects and, according to Proclus' constructive interpretation of Euclidean geometry which Fine takes up here, a condition of their constructibility (to avoid here the problematic notion of existence[30]).

This is confirmed by the fact that Fine also introduced the notion of line as resulting from the flow of a point in his commentary on the first two postulates of the *Elements*. He did so in order to explain how the construction of a line from one point to another, or the

---

[29] Fine (1532, fol. 50r): "Est itaque Geometria (. . .) quae magnitudinum, figurarum, & terminorum in his existentium, rationes indicat: affectiones insuper, variasque positiones, & motus haec concernentes. Quae rursum à signo, sive puncto divisionis experte deducta, ad solidas usque transgreditur figuras: & multiformia ipsarum discrimina, compositiora simplicioribus comparans, ad eorumque recurrens principia, subtili examine perpensat. Haec inquàm dialecticis obvoluta praeceptis, cum magis variantibus, à praevia sibi disciplina sumptis utatur principijs: caeteris scientiis (dempta Arithmetica, cuius principia sua excellunt simplicitate) certior, ac examinatior esse videtur". Cf. Fine (1536, sig. 2r), *Epistle to François Ier*: "Perscrutatur enim Geometria continuae, & prout immobilis est, quantitatis accidentia: nempe magnitudinum, & figurarum rationes, affectiones item, positionesque diversas: multiformia ipsarum discrimina subtili admodum examine discutiendo. Exordium praeterea sumit, à per sese, & vulgo notis principiis, & potissimis dialectices innixa praeceptis, ac collecta syllogismis, ad prima demonstrationum insurgit elementa. à quibus per mediorum ordinem discurrendo, atque simplicia compositis, & composita simplicibus comparando, progreditur ad ultima: ad propria tandem singula resolvendo principia."

[30] Proclus, in the commentary on Prop. I.4, considered that Euclid preceded the first theorems of the *Elements* with three problems because he felt the need to prove the existence of the objects on which he carried out his demonstrations: Proclus (Friedlein 1873, pp. 233–234) and (Morrow 1992, p. 183): "For unless he had previously shown the existence of triangles and their mode of construction, how could he discourse about their essential properties and the equality of their angles and sides?" This interpretation of Euclidean geometry, which was followed and applied to the interpretation of ancient geometrical practices by H. G. Zeuthen (1896), was refuted by Knorr (1983). On this topic, see also Harari (2003).

indefinite extension of a straight line, should be conceived and understood to be carried out in geometry.

> *To draw a straight line from any point to any point.* Indeed, any given point can describe a straight line by abstractly *flowing* according to the shortest path to any other point, which is also imagined in any place.[31]
> *To produce a finite straight line continuously and in a straight line.* For each terminating point of a given straight line can produce, by the same rectilinear *flow* of the point abstractly continued as much as desired, a given straight line that is longer, as is deduced from the given description of straight lines.[32]

And, similarly, according to the formulation of the second postulate in the *Geometria*:

> *To extend at will any finite straight line continuously and in a straight line.* Indeed, the extreme points of the line can *flow* in a straight line as much as desired.[33]

In this framework, the notion of flow of the point, conceived as the causal principle of the line, plays a foundational role within the construction practice of the geometer. It guarantees that the construction of a geometrical figure, such as required by the enunciation of a Euclidean problem, is carried out in conformity with the geometrical definition of the line.

Thus, Fine goes here from presenting the genetic definition of the line as a means of exhibiting the essential features of the geometrical line, notably in the context of teaching, to presenting it as the underpinning of the two first postulates and, thereby, of the construction procedures that depend on them. Genetic definitions are therefore endowed here with a foundational role in the argumentative structure of the *Elements* and in the constitution of geometry in general.

---

[31] Fine (1536, p. 8), Post. 1: "*Ab omni puncto in omne punctum, rectam lineam ducere.* Potest enim datum quodcunque punctum, in aliud quodlibet punctum, etiam ubilibet imaginatum, per viam abstractivè fluendo brevissimam, rectam describere lineam".

[32] Fine (1536, p. 8), Post. 2: "*Rectam lineam terminatam, in continuum rectumque producere.* Nam utrunque punctum ipsius datae rectae lineae terminativum, per rectum eiusdem puncti defluxum, quantumlibet abstractivè continuatum, potest ipsam datam lineam rectam efficere longiorem. quemadmodum ex data linearum rectarum colligitur descriptione".

[33] Fine (1532, fol. 54v), Post. 2: "*Omnem rectam lineam terminatam, in continuum, rectumque liberè prolongare.* Puncta enim ipsius lineae terminativa, possunt quantumcunque rectissimè defluere."

2.4   The Ontological Status of Geometrical Motion

The notion of flow of the point as related to the explanation of Euclid's first postulates was previously introduced by Proclus, in his commentary on the *Elements*. Now, this is the very passage[34] in which Proclus distinguished ῥύσις and κίνησις, stating that the local motion which the geometer confers to the point in his imagination in order to produce the line aims to imitate its intelligible flow.

> The drawing of a line from any point to any point follows from the conception of the line as the flowing of a point (ῥύσις σημείου) and of the straight line as its uniform and undeviating flowing. For if we think of the point as moving (κίνησις σημείου) uniformly over the shortest path, we shall come to the other point and so shall have got the first postulate without any complicated process of thought.[35]

Yet, even if Fine took up Proclus' commentary at several occasions (again, without acknowledging his source), notably the passage in which the Neoplatonic philosopher distinguished the intelligible and indivisible circle of the νοῦς and the imaginary and divisible circle of the imagination,[36] he did not make any distinction between ῥύσις and κίνησις (or rather between *fluxus* and *motus*) as for their ontological status and function in

---

[34] This passage was quoted earlier in n. 133, p. 31.

[35] Proclus (Friedlein 1873, p. 185) and (Morrow 1992, p. 145).

[36] Fine (1532, fol. 50r): "Cognoscit enim propter quid, & quia est, circum intellectilia versans, sensilia tamen attingendo: sententia namque animi, cum suas aspectu debiliter amplectatur rationes, à sensibus cognitionem ipsarum tentat abducere: aliam ab ea quae inspicitur concipiendo figuram, & circum aliam demonstrationes ostentans". *Cf.* Proclus (Friedlein 1873, pp. 54–55) and (Morrow 1992, pp. 43–44): "When, therefore, geometry says something about the circle or its diameter, or about its accidental characteristics, such as tangents to it or segments of it and the like, let us not say that it is instructing us either about the circles in the sense world, for it attempts to abstract from them, or about the form in the understanding. For the circle [in the understanding] is one, yet geometry speaks of many circles, setting them forth individually and studying the identical features in all of them; and that circle [in the understanding] is indivisible, yet the circle in geometry is divisible. Nevertheless we must grant the geometer that he is investigating the universal, only this universal is obviously the universal present in the imagined circles. Thus while he sees one circle [the circle in imagination], he is studying another, the circle in the understanding, yet he makes his demonstrations about the former. For the understanding contains the ideas but, being unable to see them when they are wrapped up, unfolds and exposes them and presents them to the imagination sitting in the vestibule; and in imagination, or with its aid, it explicates its knowledge of them, happy in their separation from sensible things and finding in the matter of imagination a medium apt for receiving its forms". See the rendering of this passage in Valla's *De expetendis et fugiendis rebus opus* (Valla 1501, sig. n1r): "Nam cum habeat animi sententia rationes debiliter aspectu complectendo ipsas tenues reddit subactasque versando ad phantasiam producens eas pro foribus collocat in illaque, aut etiam cum illa ipsarum revolutat cognitionem a sensibus amans abducere, verum phantastam convertit materiam idoneam ipsam comperiens ad formas suas excipiendas". On the elements Fine took up from this passage and the way he interpreted them, see Axworthy (2016, pp. 60–70).

geometry. Indeed, in Fine's commentary on Euclid's definitions of the point, line, surface and solid, he indifferently and synonymously used the terms *fluxus*, *defluxus* or *motus* to describe the mode of generation of these objects and attributed the same function to the processes designated by these different terms.[37] Moreover, these generative processes were indifferently described by Fine as abstract, intelligible or imaginary (*intelligibilis, abstractus, abstractivè continuatus, imaginarius, phantasticè causatus*). As these different characterisations were sometimes used to describe the same process (e.g. the generation of the line by the flow or motion of the point[38]), and in the same context, the use of these different terms would not convey any ontological distinction. His commentary on Euclid's Df. I.7, where he presented all geometrical objects as ultimately derived through the same synthetic progression, also confirms that Fine placed the generations of all geometrical magnitudes and figures on the same level, ontologically and epistemologically.[39]

Hence, it appears that, contrary to Proclus in the above-quoted passage, Fine interpreted the intelligible flow of the point as a spatial process, as it was the case in Hero's *Definitions*,[40] and not as a divine emanation of the line from the point. According to Fine, the point produces magnitude by abstractly flowing from one place to any other place or to any imaginable point situated within the spatiality of the imagination ("in aliud quodlibet punctum, etiam ubilibet imaginatum"[41]) and thereby by leaving a trace (*vestigium*) on the intelligible matter of the imagination, as would the pointed edge of a stylus on a wax tablet.[42] The same would apply to the generation of the surface and of the solid by the flow of a line and of a surface, respectively.

Indeed, in both the *Geometria* and the commentary on the *Elements*, the flow of the point, the line and the surface are explicitly described as the origin of διάστασις or dimension, in its three states, length, breadth and depth. More generally, these generative processes are presented as the cause of the variety of plane and solid figures, including

---

[37] See, for instance, Fine (1536, p. 1), Df. I.2: "*Linea verò, est longitudo latitudinis expers*. Hoc est, latitudine privata. Cum enim punctum omni careat dimensione: suo *fluxu, seu transsumptivo motu*, causat tantummodo longitudinem." See also Fine (1544, p. 2), Df. I.7: "Igitur pro linearum atque superficierum varietate, diversóque eorundem *motu, seu abstracto defluxu*: varia, & penè infinita tum planorum, tum etiam solidorum, hoc est superficierum & corporum abstrahitur multitudo, pro limitum & angulorum varietate, diversis expressa nominibus". See *supra* n. 20, p. 42 and n. 26, p. 45.

[38] Fine (1532, fol. 51r), § 2: "Ex cuius [puncti] *intelligibili* fluxu [. . .] linea secundum mathematicos causari describitur". *Cf.* Fine (1536, p. 1), Df. I.1: "Ex cuius quidem puncti *abstracto* defluxu [. . .] longitudo dimensionum primaria conficitur." (My emphasis).

[39] Fine (1544, p. 2), Df. I.7. See *supra*, n. 37.

[40] See *supra*, n. 86, p. 20.

[41] Fine (1536, p. 8), Post. 1. See *supra*, n. 31, p. 47.

[42] Fine (1532, fol. 51r), § 2: "Ex cuius [puncti] intelligibili fluxu, *non secus ac si vestigium relinqueret*, linea secundum mathematicos causari describitur." (My emphasis). The fact that Fine wrote, in his *Geometria*, that the point produces the line by its flow *as if* (*ac si*) it left a trace does not mean that he did not have a spatial conception of the flow of the point, but rather that he acknowledged the abstract or imaginary character of this process.

figures with specific dimensions, as is suggested when Fine says that the motion of the solid "increases the extent of the dimensions and modifies the figure".[43]

Furthermore, it does not seem that there was any difference, for Fine, between a type of motion that is self-induced, or intrinsic to the geometrical object, and one that is understood as initiated and performed by the geometer, either in the imagination or concretely. In fact, when Fine taught the mode of generation of figures, the flow of the point was also designated as what describes or draws out (*describo, describere*) the line.[44] This is also confirmed by the relation he established between the flow of the point and the process sanctioned by the first two postulates.[45]

2.5   *Definitio* and *Descriptio*

Now, even if Fine said that the line is drawn out (*describitur*) by the flow of the point, when dealing with the definition of the point and with the first postulate, he nevertheless clearly distinguished the "description" of the figure (*descriptio*), which here corresponds to its genetic definition, from its definition properly speaking (*definitio*), that is, its definition by property, which states the essential properties of the figure. This is shown in particular by his commentary on Euclid's definition of the circle, which is a definition by property[46] and to which he added, in both the *Geometria* and in the commentary on the *Elements*, the genetic definition of the circle. Indeed, in the *Geometria*, the genetic definition of the circle is explicitly designated in the margin as the "mathematical *description* of the circle" (*circuli descriptio mathematica*).

> *Mathematical description of the circle.* The circle is made when a straight line is carried around in the plane, one of its extremities remaining fixed, until it stops at the place where it started to be moved.[47]

---

[43] Fine (1544, p. 2), Df. I.7: "Solidum porrò motum, nullam videtur acquirere dimensionem: sed ipsas dimensiones augmentat, immutatque figuram". See *supra*, n. 25, p. 44.

[44] Fine (1532, fol. 51r), I.2, § 2: "Ex [puncti] intelligibili fluxu, non secus ac si vestigium relinqueret, linea secundum mathematicos causari *describitur*." *Cf.* Fine (1536, p. 8), Post. 1: "Potest enim datum quodcunque punctum, in aliud quodlibet punctum, [...] per viam abstractivè fluendo brevissimam, rectam *describere* lineam." (My emphasis.)

[45] See *supra*, pp. 46–47.

[46] Euclid (Heiberg 1883, p. 4) and (Heath 1956, I, p. 153), Df. I.15: "A circle is a plane figure contained by one line such that all the straight lines falling upon it from one point among those lying within the figure are equal to one another."

[47] Fine (1532, fol. 51v), III.3: "[In marg. *Circuli descriptio mathematica*]. Fit autem circulus, cum in plano recta quaedam linea, extremorum altero intra manente fixo circumducitur, quousque unde ferri ceperat ibidem quiescat".

In Fine's text, the definition of the circle by property, which is clearly designated in the margin as the *definition* of the circle (*circuli definitio*),[48] antecedes its description, or genetic definition. This indicates that, in the geometer's discourse, *descriptio* and *definitio* have different functions, even if both are presented together in his commentary on the definitions of the point, line and surface, in both the *Geometria* and in the commentary on the *Elements*.

Thus, while Fine placed the genetic definitions of the line, surface and solid before their definitions by property in both texts, this was not the case when he dealt with the circle. He then placed the genetic definition in second place, as if it corresponded to a reconstitution *a posteriori* of the mode of production of the circle based on the knowledge of its essential properties. In the commentary on Euclid, he wrote that the definition of the circle will be made clearer through its description ("Haec *diffinitio* ex data nuper (…) abstractiva circuli *descriptione* fit manifesta"[49]), implying that the genetic definition has an illustrative or didactic function, rather than a foundational role, with regard to the definition by property. As such, the function of the *descriptio* would be to explain why the circle possesses the properties enunciated in the *definitio*, namely, that it is bounded by a single line and that all the lines drawn from its centre to its circumference are mutually equal.

Conversely, when presenting Euclid's third postulate in the *Geometria*, Fine added in the commentary that the operation of describing a circle of any centre and of any radius is made clear by the mathematical definition of the circle.[50] Hence, if, when dealing with the definition of the circle, the *descriptio* is presented as subordinate to the *definitio* in the order of teaching, this situation is reversed in the exposition of the postulate. This would be due to the different functions of a definition and of a postulate, the former teaching the essential properties of the figure and the latter establishing the possibility of producing a circle of any size and any centre. In other words, if, when dealing with the definition of the circle, the *descriptio* would help us understand the meaning of the *definitio*, when dealing with the

---

[48] Fine (1532, fol. 51r–v), III.3: "[In marg. *Circuli diffinitio*.] Circulus est figura superficialis, unica linea (quae circunferentia dicitur) terminata: in cuius medio punctum adsignatur, centrum eiusdem circuli nominatum, à quo, ad ipsam circunferentiam omnes quae ducuntur rectae lineae sunt invicem aequales." *Cf.* Fine (1536, p. 5), Df. I.15: "*Haec diffinitio, ex data nuper (cùm de planis loqueremur angulis) abstractiva circuli descriptione fit manifesta. Cùm enim ab recta linea data, circum a punctum completè revolvitur: punctum b suo motu circumferentiam causat, & immotum punctum a in circuli centrum permutatur*". See also Fine (1536, p. 5), Df. I.16: "*Centrum verò ipsius circuli, punctum adpellatur. De puncto medio velim intelligas: ut punctum a, in obiecta circuli figura bcde. Lineae nanque limites sunt puncta: quorum immotum (circa quod videlicet alterum in circuli descriptione* circunducitur) in medio permanet, & centrum efficitur circuli." (My emphasis.)

[49] Fine (1536, p. 5), Df. I.15. See *supra*, n. 48.

[50] Fine (1532, fol. 54v): "Signato quocunque puncto, circa ipsum, occupato quantolibet intervallo, hoc est, sumpta libera semidiametri quantitate, circulum describere. *Id fit manifestum, ex mathematica circuli diffinitione*." (My emphasis.) *Cf.* Fine (1536, p. 8), Post. 3: "Omni centro & intervallo circulum describere."

postulate, it is the *definitio* that comes forth as a useful tool to understand the mode of production, or *descriptio*, of the circle.

Now, this latter situation is precisely that which occurred in Fine's commentary on Euclid's definitions of the line, surface and solid, where the *descriptio* preceded the *definitio*.[51] In this context, Fine did not merely intend to explain the definition, but also intended to display the ontological and logical dependence of the line, surface and solid on the motion of the point, line and surface, respectively. Thus, while, in his commentary on the definition of the circle, Fine mainly followed the order of knowledge or of teaching, in the other cases mentioned above, which focused on more general and primitive notions, he meant to follow the order of causation of geometrical objects.

Yet, since genetic definitions play a fundamental role in the constitution of geometry according to Fine, notably as they provide an ontological foundation to definitions by property, it may seem surprising that only the latter are properly called definitions in this context.

The distinction between *definitio* and *descriptio* may be found in medieval works, such as Pierre d'Ailly's *Questiones* on Sacrobosco's *Sphaera*,[52] in order to compare Euclid's definition of the sphere, which states the mode of generation of the sphere, and that of Theodosius, which states the spatial properties of the sphere. For Pierre d'Ailly, Euclid's definition of the sphere, insofar as it involves motion, did not correspond to a definition in an absolute sense, but rather to a *definitio descriptiva*,[53] which evokes the designation of genetic definitions as *descriptiones*, a term used, for instance, in Fine's *Geometria* to designate the genetic definition of the circle.[54] In D'Ailly's interpretation, the genetic definition would not correspond to a proper definition insofar as it attributes motion to geometrical objects, which are by essence abstracted from matter, and therefore from motion.[55]

---

[51] See *supra*, pp. 42–44.

[52] This work was popular in the Renaissance, having been first edited and published in 1498 by Pedro Sanchez Ciruelo (D'Ailly and Ciruelo 1498) and printed thereafter several times with other commentaries on Sacrobosco's *Sphaera* within large astronomical compendia such as Sacrobosco et al. (1531).

[53] D'Ailly (Sacrobosco et al. 1531, p. 147r): "Sciendum est [. . .] quod dupliciter capitur transitus. Primo modo capitur pro motu de aliquo loco in aliquem locum sed sic non capitur hic. Secundo modo capitur transitus pro corpore contento a circunferentia imaginata describi per transitum alicuius alterius circunferentiae & sic capitur hic. Iuxta quod est diffinitio sic exponenda. *Sphaera est transitus &c. idem est corpus contentum a superficie imaginata describi* ex transitu circunferentiae dimidii circuli quousque illa circunferentia est ista *diffinitio descriptiva*." (My emphasis.)

[54] See *supra*, n. 47, p. 50.

[55] See *supra*, n. 142, p. 33. On D'Ailly's interpretation of genetic definitions in mathematics, see also *supra*, p. 33.

It is very likely that Fine took up this terminology from this medieval comparison of Euclid's and Theodosius' definitions of the sphere.[56] Indeed, although he used the term *descriptio* to designate the generation of the line, surface and solid in general,[57] he only used it to explicitly mark out the distinction between the definition by property and the definition by genesis when he dealt with the definition of the circle, whose two definitions are respectively analogous to the definitions of the sphere in Theodosius' *Spherics* and in Euclid's *Elements*.[58] It may be noted, for that matter, that Campanus, when commenting on Euclid's definition of the sphere, wrote that the sphere is described (*describitur*) by the revolution of the semicircle.[59] As will be shown later, the term *descriptio*, when used to restrict the value of genetic definitions as definitions of geometrical objects,[60] will be later taken up and explained in more detail by Foix-Candale and Billingsley in their own commentaries on Euclid's *Elements* when dealing with Euclid's definition of the sphere,[61] also in reference to Sacrobosco's *Sphaera*.[62]

---

[56] He may have drawn it from Pierre d'Ailly's *Quaestiones*, but also from another commentary, as this distinction was present in other commentaries on Sacrobosco's *Sphaera*. On Fine's knowledge of Sacrobosco's *Sphaera* and his contributions to the *Sphaera* tradition, see Axworthy (2020). See also *infra*, n. 72, p. 57.

[57] See Fine (1532, fol. 51r), I.2, § 3: "Ex lineae postmodum abstracto *defluxu*, superficies respondenter *describitur*"; Fine (1536, p. 1), Df. I.1 (*in marg.*): "Ut linea ex puncto *describatur*"; Fine (1536, p. 2), Df. I.6: "Porrò cùm linea, ad *descriptionem* mota superficiei [. . .]" and Fine (1536, p. 2), Df. I.7: "Ex superficiei denique *fluxu*, solidum sive corpus trina dimensione [. . .] abstractivè *describitur*."

[58] While the definition of the sphere (XI.14) is formulated as a genetic definition in Euclid's *Elements*, his definition of the circle (I.15) corresponds to a definition by property. A genetic definition of the circle was presented however in Hero's *Definitiones*. Compare Theodosius, *Spherics*, I (transl. in Heath 1981b, II, p. 247): "The sphere is a solid figure contained by one surface such that all the straight lines falling upon it from one point among those lying within the figure are equal to one another" and Euclid (Heiberg 1883, p. 4) and (Heath 1956, I, p. 153), Df. I.15: "A circle is a plane figure contained by one line such that all the straight lines falling upon it from one point among those lying within the figure are equal to one another." Compare also Hero (Heiberg 1974, pp. 32–33), Df. 27: "A circle is produced when a straight line, remaining in the same plane, while one of its extremities stays fixed, is carried around by the other extremity, until it is brought back to the place where it started to move" (see *supra*, n. 86, p. 20) and Euclid (Heiberg 1885, p. 4) and (Heath 1956, III, p. 261), Df. XI.14: "When, the diameter of a semicircle remaining fixed, the semicircle is carried round (περιενεχθέν) and restored again to the same position from which it began to be moved (φέρεσθαι), the figure so comprehended is a sphere."

[59] Campanus (Lefèvre 1516, fol. 189v): "Super quamlibet lineam semicirculo descripto, si linea illa fixa semicirculus tota revolutione circunducatur: corpus quod *describitur*, sphaera nominatur. Cuius centrum: constat esse centrum semicirculi circunducti." (My emphasis.)

[60] As opposed to when it is used to simply designate the production of a line or figure in the context of a geometrical construction.

[61] See *infra*, chapters 4 and 5.

[62] That was only the case in Billingsley's commentary. See *infra*, n. 40, p. 157.

As we have seen, Fine fully considered genetic definitions as mathematically relevant and as complementary to definitions by property, in the sense that they would help provide an adequate representation of geometrical objects in the student's imagination and would enable us to acquire the knowledge of their essential properties. Genetic definitions would also be necessary, according to Fine, for the very constitution of the science of geometry, as it proceeds from the point to the most complex solid figures.

However, as will be shown in the following section, he did not (as Pierre d'Ailly) consider motion as a true property of geometrical objects, given that these are by essence deprived of matter and motion. Therefore, the difference he established, in his commentary on the definition of the circle in the *Geometria*, between *definitio* and *descriptio* could very well be due to the fact that the latter attributes motion to geometrical objects, which is not one of their essential characters.

Fine certainly distinguished the type of description which takes place *in concreto*, such as the procedure of drawing out a figure by instrumental means on a sheet of paper or on a wax tablet or even on a field with ropes and rods, and that which takes place in geometry, which is abstract and takes place in the imagination. Indeed, in the *Geometria*, Fine specifically called the genetic definition of the circle a "descriptio *mathematica*" and, in the commentary on Euclid, an "*abstractiva* descriptio",[63] as if to distinguish this type of description from one that is more concrete by nature. Nevertheless, Fine's use of the term *descriptio* points to the fact that genetic definitions, which provide the mode of production of magnitudes, cannot stand as definitions in the strict sense.

2.6   Mathematical Versus Physical Motion

If Fine did not, as Proclus, compare the generative motion of the geometrical point to the non-spatial flow of the suprasensible point, he did compare geometrical motion to physical motion, that is, to motion in the proper sense and which is carried out in the temporality and spatiality of natural substances. When doing so (in the *Geometria*), his aim was to distinguish these two types of motion from an ontological point of view, stating that the motion to which geometers appeal is only defined according to the category of position and therefore only consists in a translational motion.

> Mathematicians, as they abstractly generate all these [figures], seem to use motion in the same way as natural philosophers, but they use it however in a different way. For, truly, natural philosophers deal with motion insofar as it is subordinated to place and to ulterior perfection,

---

[63] See *supra*, n. 48, p. 51. This expression was also used to describe the generation of the figure in the commentary on definition I.13. Fine (1536, p. 4), Df. I.13: "*Terminus est, quod cuius finis est.* Utpote, punctum ipsius lineae, linea superficiei, superficies denique solidi: quemadmodùm ex eorundem *abstractiva descriptione* facilè colligitur." (My emphasis.)

while mathematicians only appeal to a translational motion (*transumptivus motus*), since they separate quantity from substance and from the other categories (except for position).[64]

In other words, since mathematicians consider objects that are deprived of physical matter, these objects cannot be attributed a type of motion that is determined by their material composition and substantial form (entelechy), as is the motion of natural substances according to Aristotle's physical doctrine.[65] They can only be attributed a motion defined in terms of change of position within a geometrically defined space, which is that of the imagination (*ubilibet imaginatum*). Thus, when Fine wrote that mathematicians only appeal to a motion that is translational (*transumptivus motus*), he intended to say that the type of motion to which mathematicians appeal in the course of their investigation on geometrical objects cannot be regarded as motion in the strict sense, since motion is by essence necessarily related to physical and changeable matter.[66] This is why he considered important to consistently designate this motion as abstract, intelligible or imaginary, being careful to add "as if" (*ac si*) when stating that the point produces the line by leaving a trace in its flow ("*ac si* vestigium relinqueret").[67]

Hence, although Fine clearly admitted the legitimacy of motion and of genetic definitions in geometry, the passage quoted here, in addition to his constant assertion of the abstract character of this motion and to his adoption of the more ancient distinction between definition and description, confirms his awareness of the difficulties that were raised by previous authors concerning the kinematic understanding of geometrical objects.

For that matter, it may be interesting to note that the above-quoted passage is substantially close to a passage found in the commentary on Sacrobosco's *Sphaera* by Bartolomeo Vespucci (1479–c. 1517). This commentary was published in 1508, side by side with the commentary on the *Sphaera* by Francesco (later Giovanni-Battista) Capuano de

---

[64] Fine (1532, fol. 54r), VII.6: "In quarum omnium abstractiva deductione, videntur ipsi mathematici motu, quemadmodum & naturales uti: differenter tamen. Eo nanque naturales, prout ad locum & ulteriorem perfectionem ordinatur: porrò mathematici trans[s]umptivo tantum utuntur motu. utpote, qui à substantia, caeterisque praedicamentis (dempto situ) ipsam videntur abstrahere quantitatem."

[65] See notably the concepts of motion presented in the *Physics* III.1, 201a25–26 (Barnes 1995, p. 343): "To begin then, as we said, with motion. Some things are in fulfilment only, others in potentiality and in fulfilment" and *De caelo*, I.2, 268b27–269a2 (Barnes 1995, I, p. 448): "Bodies are either simple or compounded of such; and by simple bodies I mean those which possess a principle of movement in their own nature, such as fire and earth with their kinds, and whatever is akin to them. Necessarily, then, movements also will be either simple or in some sort compound—simple in the case of the simple bodies, compound in that of the composite—and the motion is according to the prevailing element."

[66] In this sense, the term *transumptivus* may be taken as meaning both "translational" and "metaphorical". But whether Fine consciously intended this term to convey both meanings is not made clear at any point. On this issue, see *supra* n. 23, p. 43 and *infra* n. 95, p. 63.

[67] Fine (1532, fol. 51r), I.2, § 2: "Ex cuius intelligibili fluxu, non secus *ac si* vestigium relinqueret." (My emphasis.)

Manfredonia (late fifteenth c.), in a compendium containing various other astronomical treatises, including Pierre d'Ailly's *Questiones*.[68] This passage came forth as Vespucci commented on Euclid's definition of the sphere, which (as said) was quoted by Sacrobosco at the beginning of the *Sphaera* beside the definition of the sphere by property proposed by Theodosius in the *Spherics*.[69]

> Concerning the first definition of the sphere, note that some blame Euclid for the fact that he seems to define mathematical figures through motion, since motion should only be considered by the natural philosopher. To whom we answer that motion is taken by the mathematician and by the natural philosopher in different manners, for the latter considers a motion that is subordinated to ulterior perfection and to completion, of which Aristotle speaks in the *Physics*, while the mathematician does not consider any motion besides that which is translational and which is only according to the imagination. He considers indeed quantity by abstracting it from the substance and from the other categories, except for position [...].[70]

Vespucci continues:

> wherefrom if void could be given in nature, the mathematician would demonstrate the properties of his figures within it. Therefore, the mathematician does not consider motion as does the natural philosopher, but just as some imaginary flow to this or that place, given that the different sorts of figures would be produced in various ways by the converging of lines toward each other. And this is because, among mathematicians, the line is usually said to be caused by the flow of a point, the surface, by the flow of line, the body, by the flow of a surface, although this only takes place within the intellect and only bears truth for the intellect.[71]

---

[68] Francesco (Giovanni-Battista) Capuano de Manfredonia, *Expositio in sphaera*, and Bartolomeo Vespucci, *Annotationes nonnullae eiusdem Bartholomei Vespucii hinc idem intersertae*, in Sacrobosco et al. (1508, fols. 1r–54r). Vespucci's annotations on Sacrobosco's *Sphaera* were later appended by Luca Gaurico (1475–1558) to the commentary on Sacrobosco's *Sphaera* by Prosdocimo de' Beldomandi (1370/80–1428), in Sacrobosco et al. (1531, fols. 1r–56r). On Capuano's commentary on Sacrobosco, see Nenci (2020).

[69] See *supra* n. 139, p. 32.

[70] Vespucci (Sacrobosco et al. 1531, fol. 2v): "Circa primam sphæræ diffinitionem adverte quod aliqui in hoc culpant Euclidem cum ipse figuras mathematicas per motum diffinire videatur, cum motus non ab alio quam a naturali consideretur, quibus respondemus, alium esse motum a mathematico consideratum, alium a naturali, nam naturalis considerat motum qui est ad ulteriorem perfectionem et ad terminum, de quo Aristoteles loquitur in physicis, sed mathematicus non considerat nisi quendam motum transumptivum, secundum imaginationem tantum, considerat enim quantitatem abstrahendo a substantia ac cæteris prædicamentis, præter quam a situ [...]".

[71] Vespucci (Sacrobosco et al. 1531, fol. 2v): "unde si vacuum posset dari in rerum natura mathematicus in illo suas figuras demonstraret, considerat igitur motum non ut facit naturalis, licet pro quodam fluxu imaginario ad hunc vel ad illum situm, cum ex lineis aliter et aliter invicem concurrentibus figurarum species diversæ producantur. Et hoc est quid apud mathematicos dici solet ex fluxu puncti lineam, ex lineæ fluxu superficiem, ex superficiei fluxu corpus causari, cum hoc tantum apud intellectum locum ac veritatem habeat".

Referring here to the debates raised, notably in the tradition of the *Sphaera*, over the suitability of genetic definitions in geometry, and in particular of Euclid's definition of the sphere, Vespucci explains, in terms quite similar to those later used by Fine in his *Geometria*, that the motion mathematicians attribute to their objects is only a motion that may be defined according to the category of position, as it merely consists in a translational motion (*transumptivus motus*). On the other hand, the motion considered by natural philosophers (again represented by Aristotle) is regarded as an essential attribute of natural substances, as it is subordinated to the form and matter of the individuated being and takes place in the temporality and spatiality of the physical world (notably toward or against the substance's natural place in the cosmos). Vespucci adds that, since this motion is abstracted from the substance and from all categories except for position, the only place within which it may be considered to occur is that of the imagination. As a clear assertion of its ontological distinction from physical motion, Vespucci states that, if void existed in nature, mathematicians would demonstrate the properties of their objects within it. Fine did not go as far as to formulate this in these terms, but, as Vespucci, he intended to defend the use of genetic definitions in geometry against the philosophers who considered motion as the sole object of natural philosophy. In order to do this, both presented the motion introduced in geometrical definitions as distinct in its ontological status and mode of consideration from motion in the proper sense, that is, physical motion, therefore restricting geometrical motion to imaginary transport.

It may be possible that Fine and Vespucci were then quoting a common source, which I have not yet identified. It is nevertheless very likely (as said above) that the considerations provided by Fine on geometrical motion and its difference with physical motion came from the medieval tradition of commentaries on Sacrobosco's *Sphaera*, a text which was part of the mathematical curriculum from the late middle ages and to which Fine actively contributed from the beginning of his career.[72] In this context, this discussion would also contribute to setting the limit between what pertains to geometry and what pertains to natural philosophy within astronomy. It would indeed point to the discussion of the question "whether astronomy is a part of natural science" that was raised by Aristotle, in the second chapter of the second book of the *Physics*,[73] and which contributed to the

---

[72] Fine published an edition of Sacrobosco's *Sphaera* in 1516 (Fine 1516). On Fine's contributions to the *Sphaera* tradition, see Axworthy (2020).

[73] Aristotle, *Physics* II.2, 193b25–33 (Barnes 1995, I, p. 331): "Further, is astronomy different from natural science or a department of it? It seems absurd that the student of nature should be supposed to know the nature of sun or moon, but not to know any of their essential attributes, particularly as the writers on nature obviously do discuss their shape and whether the earth and the world are spherical or not. Now the mathematician, though he too treats of these things, nevertheless does not treat of them as the limits of a natural body; nor does he consider the attributes indicated ast he attributes of such bodies. That is why he separates them; for in thought they are separable from motion."

medieval discussions on the notion of *scientia media* or subalternate science,[74] among which was placed astronomy in view of its physical object and of its mathematical methods of investigation.[75] Such a discourse on the nature of mathematical motion was all the more important in Fine's *Geometria* that this treatise was published in the *Protomathesis*, where it was immediately followed by his *Cosmographia, sive sphaera mundi.*[76] This treatise, which corresponded to the central part of Fine's quadripartite mathematical compendium, may be considered as an updated and augmented adaptation of Sacrobosco's teaching of spherical astronomy.[77]

On a more fundamental level, this distinction between physical and mathematical motion in the context of a teaching on spherical astronomy points to the link that would have existed in early Greek science, as noted by B. Vitrac,[78] between the modelisation of celestial motions and the use of kinematic concepts in geometry. In this context, the notion of a point moving at equal distance from another point which remains fixed could be understood as an abstract representation of a star carried circularly by an ethereal sphere

---

[74] Gagné (1969), Ribeiro do Rinascimento (1974, pp. 33–95), Laird (1987, pp. 147–169) and Livesey (1989, pp. 20–53). See also, on the status of the *scientiae mediae* in the fifteenth- and sixteenth-centuries, Mandosio (1994). Interestingly, Capuano discussed this issue in the prologue of his commentary on Sacrobosco's *Sphaera*, quoting Aristotle's *Physics* II.2 and stating that astronomy should be considered as a science intermediary to mathematics and natural philosophy: Capuano, in Sacrobosco et al. (1531, fol. 59r): "quod intelligit Arist. 2 physi. tex. 19 quando astronomiam dicit esse mediam inter naturalem & mathematicam, ut infra patebit." On the notion of *scientia media* or subalternate science in Fine's mathematical work, see Axworthy (2016, pp. 301–353).

[75] On the status of astronomy in Fine's epistemological thought, see Axworthy (2016, pp. 193–248).

[76] It is notable that, in the *Geometria*, Fine added definitions of geometrical objects that would be more relevant to astronomy than to geometry, such as the orb: Fine (1532, fol. 54r): "Orbis autem, est figura solida, duabus rotundis sphaericisve superficiebus terminata: utpote, interiori quae concava dicitur, & extrinseca quae convexa nominatur [. . .]."

[77] On the place of the *Cosmographia* in Fine's *Protomathesis*, see Axworthy (2020, pp. 226–227).

[78] Vitrac (2005a, p. 7–8). Vitrac attributes this interpretation to Aristophanes, *The Clouds* 200–217. This interpretation may also be corroborated by passages of Proclus' commentary on Euclid, where the geometrical circle is related to the circular motion of heavenly bodies and is said to have a pole, a term used in celestial mechanics when dealing with celestial circles and spheres (although the circle's pole, as Proclus explains, was disregarded in Euclid's teaching of plane geometry): Proclus (Friedlein 1873, p. 147) and (Morrow 1992, p. 117), Df. I.15–16: "The first and simplest and most perfect of the figures is the circle. [. . .] whether you analyze the cosmic or the supercosmic world, you will always find the circle in the class nearer the divine. If you divide the universe into the heavens and the world of generation, you will assign the circular form to the heavens and the straight line to the world of generation [. . .]. It is because of the circular revolution of the heavens that generation returns in a circle upon itself and brings its unstable mutability into a definite cycle." and Proclus (Friedlein 1873, p. 152) and (Morrow 1992, p. 121): "And since this one point from which all the lines to the circumference are equal lies either within or outside the circle (for every circle has a pole from which all the lines to the circumference are equal), he adds further 'of the points lying within the figure'. It is not without reason that he takes into account the center only, ignoring the pole because he wishes to restrict his consideration to what lies in a plane, and the pole is above the plane assumed."

centered on the earth. This abstract notion could, in turn, be materially represented by mechanical models in metal and wood in order to assist astronomers in their visualisation of celestial motions.[79]

Now, to return to Fine's interpretation of the ontological status of geometrical motion, it is in any case clear that if he admitted the distinction between the motion of geometrical objects and the motion of natural substances from an ontological point of view, the fact that he related them insofar as they can both be expressed according to the category of position confirms that he had a properly spatial understanding of the notion of flow of the point, and of geometrical motion in general.

2.7 Motion and the Composition of the Continuum

Another notable aspect of Fine's treatment of genetic definitions is that, in certain parts of his commentary on the *Elements*, he assimilated the generative flow or motion of the point to an infinite process of multiplication of the point. This characterisation appears in particular in his commentary on Euclid's definition of the point.

> Length, the first of dimensions, is produced from the abstract flow of the point, through the infinite multiplication of the point by itself.[80]

To this relates what he wrote in the commentary on Df. I.3, that the line "starts with a point, is composed of an infinite number of points, and ends with a point."[81]

The fact of describing the flow of the point as a process of infinite multiplication of the point by itself evokes the fourteenth-century debates between divisibilists and indivisibilists (or atomists) concerning the composition and divisibility of the continuum,[82] whether mathematical or physical, which were raised at the occasion of philosophical, theological or logical discussions, notably when considering the possibility of the motion of angels.[83] These debates aimed, more fundamentally, to address Aristotle's discussion of

---

[79] On this topic, see, for instance, Menn (2015).

[80] Fine (1536, p. 1), Df. I.1: "*Ex cuius quidem puncti abstracto defluxu, per infinitam suiipsius multiplicationem, longitudo dimensionum primaria conficitur.*"

[81] Fine (1536, p. 1), Df. I.3: "*Lineae autem limites, sunt puncta. Incipit enim à puncto, & ex infinitis conficitur punctis, in punctumque terminatur.*"

[82] On these debates, see Maier (1949), Murdoch (1982), Murdoch (2009), Robert (2010) and Robert (2017).

[83] Following Aristotle's conception of the relation between the composition of motion, time and magnitude, the fact that angels were considered as deprived of matter and indivisible would make it impossible for them to be endowed with motion. On the context of these discussions and their motivations, see, for instance, Murdoch (1982) and Murdoch (2009). On the topic of angels in the fourteenth-century discussions on the composition of the continuum, see Sylla (2005).

continuity in *Physics* VI,[84] in which the Stagirite rejected atomism, that is, the theory admitting continuous quantity as composed of indivisible and extensionless parts (points or atoms).[85] Aristotle's discourse on this topic showed that the concept of continuous magnitude as produced by the multiplication or successive addition of indivisible elements leads to absurd consequences, given that what is partless cannot, even by multiplying itself an infinite number of times, constitute something that has parts and that is continuous. By appealing to the genetic definition of the line, Eratosthenes (c. 276–c. 194 BC), Theon and Hero[86] precisely intended to guarantee the continuity of the line, since the motion through

---

[84] Aristotle, *Physics* VI.1, 231a20–231b6. See, in particular, Aristotle, *Physics* VI, 231a20–29 (Barnes 1995, I, pp. 390–391): "Now if the terms 'continuous', 'in contact', and 'in succession' are understood as defined above—things being continuous if their extremities are one, in contact if their extremities are together, and in succession if there is nothing of their own kind intermediate between them—nothing that is continuous can be composed of indivisibles: e.g. a line cannot be composed of points, the line being continuous and the point indivisible. For the extremities of two points can neither be *one* (since of an indivisible there can be no extremity as distinct from some other part) nor *together* (since that which has no parts can have no extremity, the extremity and the thing of which it is the extremity being distinct)." See also Ps.-Aristotle, *On indivisible lines*, 971a22–71b4 (Barnes 1995, II, pp. 1533–1534): "Again, a line is a magnitude; but the putting together of points constitutes no magnitude, because several points put together occupy no more space than one. [...] And since the partless has no dimension, it follows that a continuous magnitude cannot be composed of partless items. Hence neither can a line consist of points nor a time of nows."; 972a1–6 (Barnes 1995, II, p. 1535): "Again, if it is absurd for a point to be by a point, or a line by a point, or a plane by a line, what they say is impossible. For if the points form a series, the line will be divided not at either of the points, but between them; whilst if they are in contact, a line will be the place of the single point. And this is impossible."; 972a13–19 (Barnes 1995, II, p. 1535): "It is clear, then, from the above arguments that a line does not consist of points. But neither is it possible to substract a point from a line. For, if a point can be substracted, it can also be added. But if anything is added, that to which it was added will be bigger than it was at first, if that which is added be such as to form one whole with it. Hence a line will be bigger than another line by a point. And this is impossible."

[85] On this issue, see also *supra*, p. 26.

[86] See *supra*, pp. 25–25.

which it is produced itself belongs to continuous quantity. Aristotle, in *Physics* VI,[87] asserted indeed the homology between magnitude and motion, as well as time.[88]

Although Fine did not explicitly refer to any of his predecessors' discussions on this issue, the fact that he compared the flow of the point to a process of infinite multiplication echoes the position defended by some of the contributors to these debates, such as Henry of Harclay (c. 1270–1317), who assumed continua as actually composed of an infinite number of indivisible (and extensionless) elements.[89] Indeed, as suggested by other passages of Fine's geometrical works,[90] the point would multiply itself by continuously leaving a trace of itself in its motion, which would allow us to think that the line is composed of all the individual and successive traces which the point produces of itself in its flow or motion. For Harclay, the fact that the point is deprived of parts would not lead to the absurdities laid out by Aristotle in *Physics* VI, since, to him, extensionless indivisibles may be conceived as touching by existing in different positions.[91] Such a distinction of position between the points of a line could, to a certain extent, be interpreted by Fine to occur in the composition of the line through the flow of the point.[92]

---

[87] Aristotle, *Physics* VI.1, 231b18–23 (1995, I, p. 391): "it is plain that everything continuous is divisible into divisibles that are always divisible; for if it were divisible into indivisibles, we should have an indivisible in contact with an indivisible, since the extremities of things that are continuous with one another are one and are in contact. The same reasoning applies equally to magnitude, to time, and to motion: either all of these are composed of indivisibles and are divisible into indivisibles, or none. This may be made clear as follows. If a magnitude is composed of indivisibles, the motion over that magnitude must be composed of corresponding indivisible motions: e.g. if the magnitude ABC is composed of the indivisibles A, B, C, each corresponding part of the motion DEF of Z over ABC is indivisible."; Aristotle, *Physics* VI.2, 232b20–25 (1995, I, p. 393): "And since every motion is in time and a motion may occupy any time, and the motion of everything that is in motion may be either quicker or slower, both quicker motion and slower motion may occupy any time: and this being so, it necessarily follows that time also is continuous." and 233a11–21 (*ibid.*): "Moreover, the current arguments make it plain that, if time is continuous, magnitude is continuous also, inasmuch as a thing passes over half a given magnitude in half the time, and in general over a less magnitude in less time; for the divisions of time and of magnitude will be the same. And if either is infinite, so is the other, and the one is so in the same way as the other; i.e. if time is infinite in respect of divisibility, length is also infinite in respect of divisibility; and if time is infinite in both respects, magnitude is also infinite in both respects."

[88] On this topic, see Vitrac (2005a, p. 21).

[89] On Henry of Harclay, see Maier (1949, pp. 161–162, 165 and 168–169) and Murdoch (1981). Such as adopted in the seventeenth century by Cavalieri, this position was qualified of semi-atomism by A. Moretto in Moretto (1984, p. 99).

[90] Fine (1532, fol. 51r), I.2: "Ex cuius intelligibili fluxu, *non secus ac si vestigium relinqueret*, linea secundum mathematicos causari describitur: longitudinem dimensionum primariam acquirendo." (My emphasis.) See *supra*, p. 42.

[91] Murdoch (1981, p. 243–245).

[92] It is notable that when asserting that a continuum (here, time) can only be constituted by indivisibles (instants) according to one dimension, Harclay uses the notion of flow (in Harclay, *Quaestio* I, G, Mss. Assisi 172, fol. 151v–152r; Vat. Burgh. 171, fol. 24r-v, quoted in Murdoch

Before we go further in the consideration of this issue, it should be noted that a similar discourse was set forth in Fine's commentary on the definition of the straight line (Df. I.4), where he presented the straight line as the principle of the plane surface. There, Fine stated that the line is imagined to produce the surface by leaving successive traces of itself throughout its transversal flow.

> *A straight line is a line which lies equally between its points.* As is the line which is led from one point to another according to the shortest path, connecting the extreme points to those which are intermediary and evenly positioned. [...] From the imaginary flow of the line, as if it left the trace of lines succeeding to each other, breadth is consequently added to the other dimension and the surface is described.[93]

According to this passage, the surface, which is said to result from the transversal flow of the line, would be constituted by the succession of different traces of the same line. This process would thus be equivalent to a multiplication of the line by itself, according to the way this process was previously described by Fine when dealing with the composition of the line. However, Fine did not actually say here that the surface is generated by an infinite process of multiplication of the line, nor that it is actually composed of lines (or of multiple traces of a line). Moreover, he manifested here a certain caution in handling this notion by saying that the line generates the surface in its flow *as if* it left the trace of lines succeeding to each other ("ac si succedentium ad invicem linearum vestigium relinqueret"). He then also clearly presented this flow of the line as an imaginary process.

This cautious attitude, which was not set forth by Fine as he dealt with the composition of the line, would be coherent with the indivisibilist conception of Henry of Harclay, who only admitted the actual existence of an infinite number of indivisibles within *linear* or unidimensional continua, considering that two- and three-dimensional infinites only exist in a potential manner.[94] Yet, given the homology, in Fine's characterisation of genetic definitions, between the relation of the point to the line and the relation of the line to the surface, it rather seems that these conceptual restrictions were also applied to the composition of the line by the point, though it was not made clear in the above-quoted passages (Df. I.1 and Df. I.3).

This may be confirmed by another passage of his commentary on the *Elements*, which deals with the principles of geometry. In this context, Fine also presented the definition of the line as produced by the flow of a point, but he qualified the assertion that a line is

---

(1981, n. 35, p. 233).): "non est latitudo necessary profunditas in temporis dimensione. Causatur enim *ex fluxu* indivisibilis." (My emphasis.)

[93] Fine (1536, p. 2), Df. I.4: "*Recta linea est, quae ex aequali sua interiacet puncta.* Utpote, quae à puncto in punctum brevissime ducitur, ipsa terminativa puncta intermediis aequali positione connectens. [...] Ex lineae autem imaginario fluxu, ac si succedentium ad invicem linearum vestigium relinqueret, latitudo dimensionum altera respondenter acquiritur, describitùrque superficies".

[94] Murdoch (1981, p. 233).

composed of points by stating that geometrical magnitude, if it may be imaginarily conceived as composed of indivisible parts, is always divisible into divisible parts, conforming to its continuous nature.

> Magnitude is imagined to eventually reach, through its continuous division by itself, a minimal part (although it is by nature always composed of divisible parts), which, in other words, could not be further divided, as if it were entirely deprived of dimension, in a way similar to the unit in discrete quantity. Just as each number is produced by the multiplication of the unit, we imagine that magnitude similarly takes its origin or is produced from such a part, or indivisible mark, by the abstract or translational motion (*abstractum seu transsumptivum motum*)[95] of this little mark.[96]

In this passage, Fine conceded indeed that, although the point is the origin of the line and may be imagined as the remaining element after a successive process of division of the line, the line is not *actually* composed of indivisible parts. At least, these indivisible parts can never be *actually* reached at any moment of the division of the line. Yet, he still admitted, and even advised, to imaginarily conceive or even intellectually assume that magnitude is divisible into indivisible parts, since this is how one obtains the geometrical concept of point as deprived of parts. To a certain extent, this hints at the fictitious status some philosophers, also among the fourteenth-century opponents of indivisibilism, attributed to the concept of point.[97] The notion of point as an elementary part of the line would therefore only serve a heuristic function, to help abstract the notion of partlessness that is crucial to Euclid's definition of the point, and also to allow us to make an analogy between

---

[95] The juxtaposition of *abstractus* and *transumptivus* in this passage shows again the complexity of the term *transumptivus* such as used by Fine, since what he is dealing with here is a mere translation or transport of the point from one place to another. Still, the fact that this motion is abstracted from any matter or substance allows us to consider it as a motion in an improper or metaphorical sense, being called "motion" by analogy with the motion of physical substances, that is, without taking into consideration most of its characteristics as an attribute of physical substances. Yet, if it may be called motion only metaphorically, it is precisely because it is merely translational, being considered only according to the category of position. Hence, if the adjective *abstractus*, used here as a synonym of *transumptivus*, to designate the motion of the point causing the geometrical line points to the fact that this motion is not a motion in the proper sense, it also expresses the fact that this motion is only taken according to the category of position, leaving aside all the other features which express the state of a substance according to the Aristotelian doctrine of categories.

[96] Fine, *Principiorum libri primi interpretatio* (Fine 1536, p. 1): "Fingitur enim magnitudo per continuam suiipsius divisionem (quanquàm in semper divisibilia naturaliter distribuatur) devenire tandem ad partem minimam, quae videlicet amplius dividi non possit, ac si foret omni dimensione privata: instar quidem unitatis in discreta quantitate. Ut quemadmodùm ex unitatis multiplicatione, omnis conficitur numerus: haud dissimiliter ex huiuscemodi parte, vel indivisibili nota, per abstractum seu transsumptivum eiusdem notulae motum, omnem effingamus oriri seu produci magnitudinem".

[97] Aristotle (*Metaphysics* 992a20) attributed this view to Plato. On this issue, see Heath (1956, I, pp. 155–156) and Murdoch (1982, pp. 573–574).

the composition of magnitude from the point through motion and the composition of number from the unit through multiplication.

It nevertheless remains that the conception of the line as resulting from the multiplication of a point, which only appeared in a mathematical context within Fine's work, clearly suggests the admission of indivisibles within the line, at least at a certain level of interpretation. Generally speaking, indivisibilists such as Harclay were quite aware of the logical and mathematical difficulties raised by their conception. And he responded to them through ontological and gnoseological distinctions which enabled him to address these difficulties in a rational manner.[98] Harclay notably appealed, in this regard, to the distinction between the human and the divine apprehension of the relation between continua and indivisibles, since only God would have the power to seize (but also to create) an infinite number of points in a finite line.[99] As we will see, this distinction between the human and divine perspective on the composition of magnitudes was not entirely absent from Fine's philosophy of mathematics, though it was not made explicit in the context of his commentary on Euclid's definitions.

2.8   The Relation Between the Generation of Magnitudes and the Generation of Numbers in Fine's Commentary on Book II

As we have just seen, Fine wrote, in the above-mentioned passage,[100] that the line is always divisible into divisible parts, making a clear distinction between the modes of composition of discrete and continuous quantity. Yet, it remains that the comparison he then made between the flow of the point as the principle of magnitude and the multiplication of the unit as the principle of number echoes his comparison, in other passages of his commentary on the *Elements*, between the flow of the point as the principle of the line and the multiplication of the point. Thus, although Fine conceded the difference between the mode of composition of numbers and the mode of composition of magnitudes, the concept of line as generated by the multiplication of a point suggests the admission of a potential correspondence between the respective modes of composition of numbers and of magnitudes, which would be founded on the analogous roles and properties of the point and of the unit in geometry and arithmetic, respectively. Indeed, the point and the unit may be assimilated to each other not only insofar as they are both indivisible, but also inasmuch as they both correspond to the principle of their respective types of quantity. In other words, just as it is possible, for Fine, to imagine or hypothetically assume that the infinite division of magnitude will result in the identification of an indivisible element, it would be possible

---

[98] Murdoch (1981, pp. 222–230).

[99] Murdoch (1981, pp. 230–232).

[100] See *supra*, p. 63.

to imagine or hypothetically assume that the infinite multiplication of the point will result in the composition of continuous magnitude. In both cases, the infinite character of these processes would guarantee that they are only interpreted by the mathematician as virtually, and not actually, possible.

Now, if Fine did nevertheless demonstrate that he was aware of the difficulties raised by the comparison between the modes of composition of numbers and of magnitudes, he was however able to fully admit the correspondence between the generation of numbers and the generation of magnitudes when dealing with geometrical or figurate numbers. Figurate numbers, according to ancient Pythagorean arithmetic, are numbers that are formed by the alignment of discrete point-units in one or several juxtaposed rows and which form a line or a plane or solid geometrical figure (e.g. a rectangle, a triangle, a cube or a pyramid). These ordered aggregations of discrete point-units were taken to represent, for instance, the addition or multiplication of the unit a given number of times (producing a linear number), or the multiplication of a number (of points) by itself or by one or two different numbers (producing either plane or solid numbers). The concept of geometrical number, which was expounded in the *Arithmetical introduction* of Nicomachus of Gerasa (ca. 60–ca. 120),[101] was commonly known and taught in medieval and Renaissance Europe through the *Arithmetical institution* of Boethius (c. 480–524), which paraphrased Nicomachus' treatise.[102] The Pythagorean notion of geometrical number was also taken up by Euclid, in Df. 16 to 19 of Book VII of the *Elements*.[103]

Iamblichus, in his commentary on Nicomachus' *Arithmetical introduction*, wrote, to explain the generation of linear numbers, that "as the point, whose flow ($\dot{\rho}\acute{v}\sigma\iota\nu$) geometers consider the line to be, is the principle and element of length, the arithmetical unit will, by similitude, fulfil the function of the sign and of the point as the principle of discrete quantity ($\pi o \sigma o \tilde{v}$); indeed, as if it were flowing ($\dot{\rho}\upsilon\epsilon\tilde{\iota}\sigma\alpha$) from itself and extending itself ($\delta\iota\alpha\sigma\tau\tilde{\alpha}\sigma\alpha$) in one dimension according to its own magnitude, it will first bring about a length".[104] In

---

[101] Nicomachus of Gerasa, *Introduction to arithmetic* II.6–18 (Hoche 1866, pp. 82–114).

[102] Boethius, *De institutione arithmetica* II.5–19 for linear and plane numbers and II.20–25 for solid numbers (Friedlein 1867, pp. 90–115).

[103] Euclid (Heiberg 1884, p. 186) and (Heath 1956, II, p. 278), Df. VII.16: "And, when two numbers having multiplied one another make some number, the number so produced is called plane, and its sides are the numbers which have multiplied one another"; Heiberg (1884, p. 186) and (Heath 1956, II, p. 278), Df. VII.17: "And, when three numbers having multiplied one another make some number, the number so produced is solid, and its sides are the numbers which have multiplied one another"; Df. VII.18: "A square number is equal multiplied by equal, or a number which is contained by two equal numbers"; Df. VII.19: "And a cube is equal multiplied by equal and again by equal, or a number which is contained by three equal numbers."

[104] Iamblichus (Pistelli 1894, pp. 57, 6–12), IV, § 4 (my translation from the Greek and the French translation by Vinel, in Iamblichus (Vinel 2014, p. 127)). On the meaning of this passage, see also Iamblichus (Vinel 2014, pp. 35–36). It may be noted that Nicomachus' text somewhat already depicted the progression of numbers as a kinematic process, linear numbers being said to "advance ($\pi\rho o\chi\omega\rho o\tilde{v}\nu\tau\epsilon\varsigma$) by the addition of 1 in one and the same dimension" (Hoche 1866, pp. 86–87) and (transl. D'Ooge 1960, pp. 239–240, II.7). This is less obvious in Boethius' *De institutione*

this passage, which intends to explain and provide a foundation for the Pythagorean notion of geometrical number as theorised by Nicomachus, it is admittedly not the generation of the line that is compared to the arithmetical process of multiplication of the unit, as in Fine's commentary on Euclid. But it is rather the generation of numbers that is compared to the generation of the geometrical line from the flow of the point, presenting it as a spatial process, that is, as the extension of an indivisible unit "in one dimension according to its own magnitude". This comparison allowed him to establish a correspondence between the modes of generation of magnitudes and of numbers through the notion of flow of the point.[105] Iamblichus wrote indeed that the unit produces linear numbers (and from there plane and solid numbers), *as if* (ὡσανεί) it were flowing and extending itself, in the manner of the point in geometry. What he meant by this is that the unit deploys itself into a series of juxtaposed discrete points, which spatially translates the successive addition or multiplication of the unit. The process of addition or multiplication here is certainly not conceived as infinite, nor is the resulting number of point-units that form linear numbers, unlike the multiplication of points that would compose the geometrical line in Fine's commentary on the *Elements*. It sets forth however the connection, rather than the opposition, between the modes of generation of numbers and magnitudes.

The analogy between the kinematic generation of magnitudes and the spatially-comprehended mode of generation of numbers through the multiplication of point-units arranged in lines, surfaces and volumes, was regularly taken up in early modern commentaries on the *Elements*, in particular when commenting on Book II.[106] Indeed, as this book established relations of equivalence between quadrangular areas, commentators were tempted to interpret this book as expressing a set of arithmetical and algebraic operations.[107] In Fine's commentary on Euclid's Df. II.1, a definition which states that "any rectangular parallelogram is said to be contained by the two straight lines containing the right angle",[108] the generation of a rectangular surface by the flow of a line-segment

---

arithmetica, II.5.1, where geometrical numbers are rather presented as obtained through a mere aggregation of points (Friedlein 1867, p. 90) and (transl. Masi 1983, p. 131): "A linear number is one beginning from two with unity added on. To one same and continuous line you add the accumulated quantities".

[105] It should nevertheless be noted that, although Iamblichus claimed, in order to explain the concept and mode of production of linear numbers, the similarity rather than the distinction between the mode of production of number from the unit and the mode of production of the line from the point (comparing the arithmetical unit to a point and the generation of numbers to the flowing of the unit from itself, thereby producing a length), he also asserted that the relation of numbers to the unit differs from the relation of magnitudes to the point insofar as the point is not a part of the line as the unit is a part of number (Iamblichus, *In Nicomachi arithmeticam*, IV § 4–8, in Vinel 2014, pp. 126–127; see *supra*, n. 115, p. 26). See also N. Vinel's analysis of this passage and of the discussions on the topic in ancient philosophy, in Iamblichus (Vinel 2014, pp. 36–41), as well as Dye and Vitrac (2009, pp. 183–189).

[106] On this topic, see Malet (2006). For earlier examples, see Corry (2013).

[107] On this issue, see *supra*, n. 79, p. 16.

[108] Euclid (Heath 1956, I, p. 370), Df. II.1.

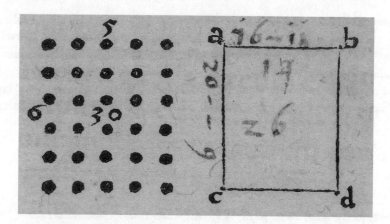

Fig. 2.1 Oronce Fine, *In sex prioris libros geometricorum elementorum Euclidis Megarensis demonstrationes*, 1544, p. 38, Df. II.1. Diagram illustrating the correspondence between the generation of rectangular numbers by the multiplication of two numbers and the generation of rectangular parallelograms by the flow of one line-segment along and according to another at right angles. Courtesy Max Planck Institute for the History of Science, Berlin

along and according to the length of another at right angles is thus explicitly compared to the generation of a rectangular number by the multiplication of two different numbers.

> We must imagine that the straight line AB flows according to a direct path in C, and that point B describes the side BD, or that the straight line AC reaches B by a straight flow, and that point C produces the side CD, just as are abstractly described rectangle parallelograms. In a similar manner, a number multiplied by any number different from it will produce a plane and rectangular number.[109]

In the margin of Prop. II.1, Fine wrote furthermore that "the multiplication of numbers transmitted by arithmeticians is understood through this proposition,"[110] suggesting the didactic dependence of the arithmetical concept of multiplication on the geometrical notions and operations set forth in Book II.

The correspondence between the mode of generation of rectangular numbers and the mode of generation of rectangular parallelograms is also made clear through the diagram that accompanies Fine's commentary on Df. II.1. This diagram sets the geometrical rectangle ABCD side-by-side with a rectangular number obtained from the multiplication of a row of 5 point-units by 6 (Fig. 2.1).

---

[109] Fine (1536, p. 43), Df. II.1: "Imaginanda est igitur *ab* recta, fluere directa via in *c*: et punctum *b* describere latus *bd* vel *ac* rectam, venire recto fluxu in *b*: atque punctum *c* efficere latus *cd*. Ita enim abstractivè describuntur parallelogramma rectangula. Ad quorum similitudinem, numerus per alium quenvis numerum multiplicatus, planum atque rectangulum efficit numerum".

[110] Fine (1536, p. 44), II.1: "Ex hac propositione, numerorum ab Arithmeticis tradita colligitur multiplicatio."

This geometrical understanding of numbers is also very likely what allowed the authors of Latin medieval arithmetical works (in particular in the tradition of Boethius' *De institutione arithmetica*), to designate the arithmetical process of multiplication (of two or three numbers) by the terms *ductus* or *ductio*,[111] whereby the result of a multiplication was to be called a product (*productus* or *productum*).[112] Hence, in Campanus' commentary on Euclid, for instance, while the verb *duco* was used to designate the generation of

---

[111] See Boethius (Friedlein 1867, p. 21), I.9: "IIII enim tricies et bis, vel quarter XXXII *ducti* CXXVIII" or (*ibid.*, 24) I.10, 13: "In ea enim dispositione quae est par, ut idem reddunt II per XVI multiplicati quod IIII per octonarium *ducti*: utroque enim modo XXXII fient", where this term designates the multiplication of a number by another and II.5, 1 (*ibid.*, p. 90): "Linearis numerus est a duobus inchoans adiecta semper unitate in unum eumdemque *ductum* quantitatis explicata congeries", where it designates the production of linear numbers by the successive addition of the unit to each number from two, and II.6, 2 (*ibid.*, 91): "Quadratum enim ita *ductae* lineae in quattuor, pentagonum in quinque triangulos, exagonum in sex et ceteros in suorum angulorum modo mensuraque per triangulos partiuntur", where it designates the mode of generation of polygons (and first of all the square) from the line in geometry. See also, for example, Sacrobosco, *De arte numerandi*, VI (Halliwell 1839, p. 12): "In multiplicatione duo sunt numeri necessarii, scilicet, numerus multiplicandus et numerus multiplicans. Numerus multiplicans adverbialiter nuncupatur. Numerus vero multiplicandus nominalem recipit appellationem: potest et jam tertius numerus assignari qui productus dicitur, perveniens ex *ductione* unius in alterum." Here the term used is that of *ductio*, which is a synonym of *ductus*, as shown by the following passage, which deals with the production of geometrical numbers: IX, (*ibid.*, p. 20): "Numerus superficialis est qui resultat ex *ductu* numeri in numerum, et dicitur superficialis quia habet duos numeros denotantes sive mensurantes ipsum, sicut superficies duas habet divisiones; scilicet, longitudinem et latitudinem. [...] Sciendum quod si ducatur in se semel, fit numerus quadratus [...]. Si *ducatur* in alium, fit numerus superficialis et non quadratus, ut binarius *ductus* in ternarium constituit senarium, numerum superficialem, et non quadratum. [...] Radix autem numeri quadrati est ille numerus qui ita *ducitur* in se, ut bis duo sunt quatuor." See also Johannes Regiomontanus' (1436–1476) *De triangulis omnimodis* (as this text will be explicitly referred to in a similar context by Commandino and Clavius), in Regiomontanus (1533, p. 18), I.16: "Resumpta figuratione prima, numerus *k* in numerum *l ductus*, efficiat numerum *lm*." (All emphases mine.)

[112] This term appears in its arithmetical sense in commentaries on Euclid's *Elements*, such as that of Campanus, in Df. VII.10 (Lefèvre 1516, fol. 93r : "*Productus* vero dicitur: qui ex eorum multiplicatione crescit"). In Luca Pacioli's commentary on Euclid, which was based on the text of Campanus, the notion of *productum*, defined as what results from the *ductus* of one line on the other in the context of Book II, is presented as synonymous to the notions of *multiplicatio*, of rectangular surface, of rectangle and of rectangular parallelogram (Pacioli 1509a, fol. 12v: "*Productum*. Quod fit ex ductu unius in alterum. Superficies rectangula. Rectangulum. Multiplicatio. Parallelogramum rectangulum. Nomina sinonima"). See also the above-quoted passage from Sacrobosco's *De arte numerandi* VI (Halliwell 1839, p. 12): "[...] potest et iam tertius *numerus* assignari qui *productus* dicitur, perveniens ex ductione unius in alterum." where the result of the multiplication is designated as the *numerus productus*. (All emphases mine.)

lines,[113] the arithmetical term *productus* was used in Book VII to designate the result of a multiplication.[114]

The correspondence between the use of the term *ductus* and its derivatives in the context of geometry and the use of the same terms in the context of arithmetic is set forth by Fine in his commentary on Euclid's Df. I.30 (the definition of the square). In this context, the line transversally led at right angles according to its own length (*in seipsam ducta*) in order to produce a square is indeed assimilated to any number multiplied by itself or "led" on or in itself (*in seipsum ductus*) to produce a square number and which, in relation to the latter, corresponds to its root.

> Each side of any square may be indifferently called its root. Indeed, the square results from a given line which is abstractly led on itself according to a straight path, just as the number led on itself produces a square number.[115]

In his commentary on Df. II.1, Fine did not use the term *ductus* to designate both the mode of generation of numbers and that of magnitudes, as he did in Df. I.30. But it remains that, by associating the geometrical notion of *fluxus* with the arithmetical process of multiplication, Fine also aimed to describe a mode of generation that would be common *in principle* to all mathematical quantities. The only reservation to this conception is that, for him, the flow of the point would compose the line only through an infinite multiplication of the point by itself, so as to guarantee the infinite divisibility of the line.

Generally speaking, even if Fine was aware of the difficulties raised by the comparison between the generation of numbers and the generation of magnitudes, it seems that he was eager to make this comparison possible, notably for its ulterior application to practical geometry, in which arithmetical operations were applied to the measurement of lengths, areas and volumes. Indeed, in the prologue of the second book of the *Geometria*, he

---

[113] See, for instance, Euclid's first postulate (Campanus, in Lefèvre 1516, fol. 3v): "A quolibet puncto in quemlibet punctum, rectam lineam *ducere*". See also the translation of Bartolomeo Zamberti (Lefèvre 1516, fol. 5r): "Ab omni signo in omne signum: rectam lineam *ducere*." (My emphasis.) In the Greek text of Euclid's *Elements*, the verb corresponding to *duco* is ἀγάγω (Euclid, Heiberg 1883, p. 8, Post. 1: "Ηἰτήσθω ἀπὸ παντὸς σημείου ἐπὶ πᾶν σημεῖον εὐθεῖαν γραμμὴν ἀγαγεῖν."), which, as seen above, was chiefly used in a geometrical context to designate the drawing of straight lines. Indeed, as established by Mugler (1958–1959, I, pp. 40–41), the term *ductus* was generally used in a mathematical context to translate the Greek term ἀγωγή, employed by ancient geometers within constructions to designate the process by which a line is led from one point to another.

[114] Campanus (Lefèvre 1516, fol. 93r): "*Productus* vero dicitur: qui ex eorum multiplicatione crescit." (My emphasis.)

[115] Fine (1536, p. 7), Df. I. 30: "Omnis itaque quadrati unumquodque latus, radix eiusdem indifferenter adpellatur. Fit enim quadratum, ex data linea recta abstractivè in seipsam rectissimè *ducta*: quemadmodum numerus in seipsum *ductus*, quadratum efficit numerum".

presented practical geometry as the fruit to be obtained from the study of Euclidean geometry.[116]

2.9   The Metaphysical Interpretation of the Generation of Quantities

As previously mentioned, Fine did not refer to any metaphysical interpretation of the generation of mathematical objects in his geometrical teaching. Yet, his association between the generation of numbers and the generation of magnitudes could, to a certain extent, be understood as a literal interpretation of the ancient correspondence between the arithmetical and geometrical notions of ῥύσις. This would be, in any case, consistent with the reference he made to the Pythagorean cosmogonic doctrine in the preface of his *Arithmetica practica*.

> For the unit, which is the root and origin of all numbers, remains in itself, from itself and around itself indivisible, although each number arises and is generated by its accumulation, and is eventually resolved into it. In the same manner, all discrete and composed things that may be contemplated in the universe have been divided and gathered by the eminent creator of things in a determined number, into which they must eventually be resolved.[117]

According to this passage, the relation between numbers and magnitudes, as for their mode of generation and of composition, would be justified on a metaphysical level by the admission of a correspondence between numbers and magnitudes in God's mind and

---

[116]Fine (1532, fol. 64r), *Geometria*, II.1: "Duo sunt, optime lector, quae in omni disciplina, studiosis omnibus solent esse non iniucunda. unum est, facilis in disciplinam introductio: qua & via doctrinae, & sensus eiusdem universus aperitur. Reliquum esse videtur, *collectus ex ipsa disciplina fructus, susceptorum laborum compensator gratissimus*. Praemissis itaque generalibus ipsius Geometriae rudimentis, ad elementorum Euclidis, & succedentium nostrorum operum intelligentiam isagogicis: consequens nobis visum fuit, universam *Geometriae* subnectere *praxim*, hoc est, linearum, superficierum, & corporum, ex demonstratis Euclidis elementis, ostendere mensuram. Ea potissimum intentione, ut succedentium & geometricorum, & coelestium instrumentorum usum (quae non poterant his, sine iactura carere) redderemus faciliorem: & ijs etiam satis pro nostra virili parte faceremus, quos eiuscemodi practicis geometricarum subtilitatum exercitamentis, novimus plaerumque delectari". (My emphasis.)

[117]*Arithmetica practica*, in Fine (1532, fol. 1r): "Nam unitas omnium numerorum radix & origo, in se, à se, ac circum seipsam unica vel impartibilis permanet: ex cuius tamen coacervatione, omnis consurgit & generatur, omnisque tandem in eam resolvitur numerus. Quemadmodum cuncta quae seu discreta, sive composita inspectentur universo, à summo rerum conditore in definitum digesta, redactave sunt, & demum resolvenda numerum." On this text and its significance for Fine's conception of mathematics, see Axworthy (2016, pp. 50–56).

through the arithmetical principle of the divine Creation. As stated here, all things and all quantities would have been organised and determined by God according to the model of numbers. This conception of Pythagorean origin, which was set forth in Boethius' *Arithmetical institution*,[118] was defended and popularised in France from the end of the fifteenth century by Jacques Lefèvre d'Étaples (c. 1455–1536) and his disciples, Charles de Bovelles (1479–ca. 1567) and Josse Clichtove (ca. 1472–1543).[119]

In a different manner, Fine's notion of flow of the point as a multiplicative process would more directly evoke the philosophical thought of Robert Grosseteste (c. 1168–1253) (and, after him, of Henry of Harclay, among others).[120] Grosseteste considered indeed that the universe was created by God through the infinite multiplication of points or atoms of light from a single source in all directions and that the cosmos therefore consists of an infinite number of points filling the totality of its finite sphere, whereby the point would represent the unit of measurement of all beings.[121] Grosseteste considered moreover that the constitution of bodies occurred by the flow of points.[122] In this framework, only God would be able to know the quantity of indivisibles (or light-atoms) in the universe, as well as the number of points that compose a line. If this intellectual stance should be fully representative of Fine's thought in this regard—it is unclear that it does—it could help us understand why he admitted the conception of the line as generated by the multiplication of a point while denying the mathematician the hope to ever actually reach an indivisible point through the successive division of the line, except as a useful fiction to understand the geometrical notion of point. The actual generation of the line by the infinite multiplication of the point, to which Fine identified its generation through the flow of the point, would indeed reflect a state of magnitudes only known to God.

* * *

Throughout his two major geometrical works, the *Geometria* and his commentary on the first six books of Euclid's *Elements*, Fine provided numerous examples of the function he attributed to genetic definitions and to generative motion in geometry. The motion that was attributed to geometrical objects in this context is presented as the cause of each of the three dimensions, but also of the immense variety of geometrical figures with their own specificities, a process which is shown to start with the generation of the line from the point and to end with the constitution of the most complex figures. In this framework, it is

---

[118] Boethius (Masi 1983, pp. 75–76) I, 2, 1: "From the beginning, all things whatever which have been created may be seen by the nature of things to be formed by reason of numbers. Numbers was the principal exemplar in the mind of the creator."

[119] On the circle of Lefèvre d'Étaples and its importance for the early sixteenth-century French mathematical and philosophical culture, see Oosterhoff (2018).

[120] On the relation between Grosseteste and Harclay, see Murdoch (1981, pp. 240–241).

[121] McEvoy (1978), McEvoy (2000, pp. 87–95), Lewis (2005) and Lewis (2013, pp. 239–242). See also Lindberg (1976, pp. 96–99).

[122] Grosseteste (Panti 2011, pp. 79 and 119–126).

not only the motion of the point, of the line and of the surface that is attributed a generative function, but also that of solid figures. Although the motion of the solid is unable to cause any magnitude of higher dimension, it would be able to generate solid figures of different configurations and sizes through the extension of their sides.

The motion that is attributed thereby to points, lines and geometrical figures by the mathematician is presented by Fine as the condition of the synthetic deployment of the science of geometry as a whole, which starts with the definition of the point and ends with the construction and demonstration of the properties of the most complex figures. It also enables the mathematician to ultimately show the dependence of the latter on the former through a reversed process of analysis. Genetic definitions thus constitute, for Fine, a necessary instrument of geometrical knowledge and teaching, explaining the spatial properties of geometrical objects while revealing their mutual connections and dependencies. They also offer a foundation to Euclid's first three postulates, since the genetic definitions of the straight and circular lines then allow to legitimate the construction procedures authorised in the framework of problems and theorems by displaying their dependence on the definitions by property of the straight line and of the circle.

In spite of the epistemic importance attributed to genetic definitions in his geometrical teaching, Fine did not go as far as to consider these as definitions in the proper sense, reserving this status to definitions by property. He thus designated genetic definitions as *descriptiones* rather than *definitiones*, following the older terminology found in medieval commentaries on Sacrobosco's *Sphaera* when comparing Euclid's kinematic definition of the sphere with Theodosius' definition of the sphere by property. As the authors of those commentaries on the *Sphaera*, he did not consider motion as an essential attribute of geometrical objects.

Although Fine took care to specify the abstract and intelligible nature of geometrical motion, and to clearly distinguish it from physical motion, he did not go as far as to relate it to a metaphysical process, at least not in the context of geometry, where the flow of the point, the line or the surface are straightforwardly presented as local processes.

Yet, in a philosophical context, he may have related it to an ontologically higher process, that is, when describing the divine creation of the world as founded on the model of numbers in the preface of his *Arithmetica practica*. This reading would be confirmed by the tensions that appear in Fine's discourse on the composition of the continuum, where he interpreted the flow of the point as an infinite multiplication of the point, establishing thereby a connection between numbers and magnitudes as for their mode of generation. This connection was also set forth in his commentary on Euclid's Df. II.1, where he related the generation of magnitude through the flow of a point, a line or a surface to the multiplication of numbers, taking as a basis the Pythagorean doctrine of geometrical or figurate numbers. It remains that, while the flow of the unit had a clearly metaphysical meaning, beside its mathematical interpretation, in the context of works such as Iamblichus' commentary on Nicomachus' *Arithmetical introduction*, this metaphysical meaning is not apparent in Fine's commentary on Euclid, where it is merely presented as relevant to the arithmetical understanding of Book II of the *Elements*.

Generally speaking, although Fine was aware of the difficulties which genetic definitions could raise with regard to the assessment of the essential properties of geometrical objects, both with respect to the mereological relation of the point to continuous magnitude and with regard to the ontological status of motion in general, he did not wish in any way to avoid introducing motion within his commentary on the *Elements*. He actually made an extensive use of it in this context, both as a pedagogical tool and as an epistemic foundation for the whole constitution of geometry. The generative motion of geometrical objects also displayed the relation between arithmetic and geometry, to the point that motion then appeared as a unifying principle for the whole of mathematical knowledge, which Fine considered to embrace all four disciplines of the *quadrivium* and to extend, by a relation of subalternation, to geography and optics.[123]

Fine's general attitude toward motion and genetic definitions in his commentary on the *Elements* and in the *Geometria* seem to have been greatly determined by his role as Royal lecturer in mathematics, which required of him to offer his audience and readership an accessible and intuitive approach to Euclidean geometry. Indeed, although the institution of the Royal lecturers placed an emphasis at first on teachings useful to the humanistic study of ancient texts, the lectures of these royal professors did not lead to any degree and were in principle open to anyone, that is, also to an audience less familiar with the sources and concepts taught within the university.[124] Fine's *Protomathesis* offered as such a hybrid representation of mathematics, which was partly determined by the curricular model of the Faculty of the Arts and which diverged from it by the prominent place it attributed to practical mathematics. In this context, the study of Euclidean geometry was presented as an introduction to practical geometry, which itself ultimately aimed to prepare students and readers for the study and application of the principles of cosmography, taught immediately after the *Geometria* in the *Cosmographia, sive sphaera mundi*.[125] The *Cosmographia* was

---

[123] See, for instance, the preface to the French translation of his *Cosmographia*, Fine (1551a, sig. aa2r): "Du jugement de tous ceux qui sont de sain esprit & bonne volunté, Sire, il n'est chose plus agreable entre les humains, & digne de plusgrande louenge, qu'en postposant les accidens & vanitez de fortune, communiquer aux autres dons & graces que l'on a receu du createur: & restituer principallement les bonnes sciences en leur integrité [...] comme sont les nobles & divines mathematiques, c'est à sçavoir *arithmetique, geometrie, musique & astronomie, avec leurs subalternes geographie & perspective*." (My emphasis.) On Fine's division of mathematics, see Axworthy (2016, pp. 301–303).

[124] Pantin (2006).

[125] Axworthy (2020, pp. 226–227). In the preface of the *Geometria*, Fine asserts the necessity of his *Arithmetica practica* and of his *Geometria* for the astronomical teaching that follows, in the succession of treatises which compose the *Protomathesis* and which is provided by the *Cosmographia*. Fine (1532, fol. 50r): "Non incommodum iudicavimus, studiose lector, post Arithmeticae praxim, insignora Geometriae tradere rudimenta. utpote, quae non modo succedentibus nostris geographicis vel astronomicis operibus, passim sese offerunt accommoda: verumetiam universo mathematicarum studio videntur admodum necessaria."

not only Fine's masterpiece and most popular work,[126] but it displayed his teaching on astronomy (along that of geography), which represented, to him, the aim for which one should study arithmetic and geometry,[127] conforming to what Regiomontanus held in the inaugural oration to his lessons at the University of Padua from 1464.[128] Hence, his promotion of a kinematic understanding of geometry, despite the ontological restrictions he placed on the nature of geometrical motion when considered from a philosophical point of view, may also have been intended to prepare his students to visualise planetary motions,[129] while helping them grasp more easily the abstract notions of point, line, surface and solid, such as defined by Euclid. To this adds his properly constructive approach to geometry, whereby the generative motion of points, lines, surfaces and solids is presented as essential to the constitution of geometrical knowledge.

Fine's diverse considerations on motion in geometry and on the relation between numbers and magnitudes, within mathematics and in nature, also display a philosophical eclecticism proper to Renaissance humanism. Within his treatment of motion in geometry, the philosophical conceptions on mathematics of Aristotle, Proclus and of the Pythagoreans seem indeed to converge, in addition to the medieval authors who discussed the composition of the continuum in mathematics, in nature and in God's mind.

---

[126] The *Cosmographia* was reprinted twice in 1542 (in an unabridged and abridged version), in 1551, 1552, 1555 and, in French, in 1551 and 1552. On the international reception and teaching of Fine's *Cosmographia*, see Axworthy (2020, p. 193).

[127] Fine (1542, sig. *2r–v): "Quanquam enim ipsae Mathematicae, omne philosophandi genus adaperiant, et in universum cunctis opitulentur artibus: eò tamen omnes tendere videntur, ut Caeli suscipiendi peculiarem sortitae sint curam. Quam beatissimam contemplationem, Astronomiam vocant."

[128] Regiomontanus (2008, p. 137): "Inter omnes autem hasce disciplinas astronomia instar margaritae non modo sorores suas, reliquas inquam scientias medias, verum etiam omnium disciplinarum matres geometriam et arithmeticam longe antecellit; cuius ortum prae vetustate nimia haud satis comperimus ita ut aeternam aut mundo concreatam non inique putaveris."

[129] On learning how to visualise celestial motions in the premodern period, see Barker and Crowther (2013).

Jacques Peletier

3

3.1   The Life and Work of Jacques Peletier and the Significance of His Commentary on the *Elements*

Jacques Peletier (Le Mans, 1517–Paris, 1582)[1] was a mathematician, a poet,[2] a reformer and promoter of French as a scientific language.[3] He practiced medicine and occupied positions as a professor and director of pedagogical institutions.[4] As a mathematician, some

---

The content of this chapter is partly based on a preliminary study of Peletier's treatment of the notion of flow of the point that was published in Axworthy (2017).

[1] On Peletier's life and career, see Jugé (1907) and Arnaud (2005, pp. 19–21).

[2] Peletier was a member of the literary circle of the Pléiade, alongside Pierre de Ronsard (1524–1585) and Joachim du Bellay (ca. 1522–1560). Peletier wrote, as such, several poems that display the nature and perfection of mathematics, in particular the *Louange de la Sciance* published in 1581 (Peletier 1581, fol. 40r–62r) and *A ceulx qui blâment les mathématiques* first published in 1547 (Peletier 1547, 1904, pp. 104–105). He evolved in other poetic and humanistic circles, such as the literary circle of Marguerite de Navarre (1492–1549) in Lyon in the years 1550, which were his most prolific years in terms of mathematical and literary production.

[3] He worked to reform the spelling of the French language, in his *Dialogue de l'ortografe e prononciation françoese* (Peletier 1550), and promoted the use of the French language for the diffusion of scientific theories and for teaching.

[4] Beside his multiple travels, he occupied a position as a director of the Collège de Bayeux in Paris from 1543 to 1547, at the Collège d'Aquitaine in Bordeaux in 1572, at the Collège du Mans in Paris around 1578, and as a professor of mathematics, for instance, at the university of Poitiers in 1579, for which position he wrote an inaugural discourse entitled *Oratio Pictavii habita in praelectiones Mathematicas* (Peletier 1904). He held most of these positions only for a short while. One of his first known professional positions was the one he held, around 1540, as secretary of René du Bellay (1500–1546), the cousin of the poet Joachim du Bellay (Jugé 1907, p. 23).

---

of his most important contributions were in the domain of algebra. [5] He also significantly contributed to the Euclidean tradition through his commentary on the first six books on the *Elements* published in 1557 in Lyon by Jean de Tournes (1504–1564).[6] In this commentary, Peletier did not hesitate to challenge Euclid's definitions and propositions, discussing their philosophical implications as well as their mathematical validity.[7] In this regard, he became an important actor of the sixteenth-century debates on Euclidean geometrical concepts and practices, in particular concerning the status of the angle of contact[8] and the use of superposition within congruence theorems.[9] He was mostly involved in these debates as an opponent of Clavius.

Peletier is important to consider here not only because he appealed to genetic definitions in his commentary on Euclid, but also because he rejected the use of superposition in geometry, which has been taken to presuppose the motion of geometrical figures. This procedure, which was used in classical geometry to demonstrate the congruence of geometrical figures, was regarded by Peletier as mechanical rather than properly geometrical.[10] Peletier's treatment of superposition will not be considered in detail here,[11] as the present study focuses on generative motion. Yet, the difference in Peletier's attitude toward the non-generative type of motion implied by the method of superposition and the generative motion referred to within genetic definitions is not to be entirely dismissed, as it raises the question of the nature of the motion that he considered admissible in geometry. I will therefore briefly touch upon this issue in the conclusion of this chapter.

As Peletier also introduced genetic definitions in his scientific poetry, and in particular in one of his poems entitled *Louange de la Sciance* (published in 1581 in his *Oeuvres intitulées Louanges*),[12] where he displayed in a clearer manner the ontological significance he attributed to such definitions, I will consider this text in parallel to his commentary on Euclid.

---

[5] Peletier (1554). On Peletier's contribution to the development of sixteenth-century algebra, see Cifoletti (1992).

[6] Peletier (1557). This work was reedited in 1610 and translated in French in 1611 and 1628 (Peletier 1610, 1611, 1628).

[7] Axworthy (2013).

[8] That is, the angle comprised between the circumference of a circle and a straight line tangent to it, which is dealt with by Euclid in Prop. III.16. On this debate, and its ulterior developments, see Maierù (1991), Loget (2000, pp. 165–280), Loget (2002) and Rommevaux (2006). See also Dear (1995b), pp. 53–55).

[9] In Euclid's *Elements*, the propositions that appeal to superposition are Props. I.4, I.8 and III.24. On this debate, see Mancosu (1996, pp. 29–31), Loget (2000, pp. 171–177), Palmieri (2009, pp. 474–476) and Axworthy (2018).

[10] Peletier (1557, p. 16), Prop. I.4: "Figuras Figuris superponere, Mechanicum quippiam esse: intelligere verò, id demùm esse Mathematicum." See also Axworthy (2018, pp. 13–25).

[11] This topic was the focus of Axworthy (2018). See also *infra*, pp. 96–97.

[12] *Louange de la Sciance*, in Peletier (1581, fol. 40r–62r).

After considering Peletier's terminology and use of genetic definitions in his commentary on Euclid, I will examine the ontological status he attributed to geometrical motion, first, in the *Louanges* and, then, in the *Elements*. I will also analyse the connections, in Peletier's thought, between geometrical objects and natural phenomena pertaining to light, as well as his position on the issue of the composition of the continuum. The last sections of this chapter will be devoted to the relation between the generation of magnitudes and the arithmetical operation of multiplication, as well as to the limits of human knowledge faced with the geneses of geometrical objects.

As will be shown here, Peletier's attitude toward generative motion in geometry was quite different from Fine's approach insofar as he made a more restricted use of the term *fluxus* to describe the process through which magnitudes are understood to be generated, reserving it to the generation of the line only. This resulted, for a great part, from Peletier's acknowledgment of the metaphysical implication of the ancient notion of ῥύσις σημείου. It nevertheless remained that Peletier treated the flow of the point as a spatial process in his commentary on Euclid. Peletier's case thus offers an exemplary representation of the underlying tensions between these two levels of interpretation, spatial and non-spatial, mathematical and metaphysical, of the ancient notion of ῥύσις σημείου.

3.2 Formulation and Distribution of Genetic Definitions in Peletier's Commentary on Euclid

In his commentary on the *Elements*, Peletier appealed to genetic definitions on very few occasions, two of which are his commentary on the definition of the point (Df. I.1) and his commentary on the definition of the straight line (Df. I.4):[13]

> The line is conceived as generated by the uninterrupted longitudinal flow of the point.[14]

> And just as the line springs from the continuous flow of the point, the surface originates from the transversal leading of the line.[15]

In the commentary on Df. I.1, the definition of the line as the flow of the point is mentioned among different definitions of the point, which is then defined as the principle or origin of the line. In Df. I.4, Peletier considers the line itself (represented by the straight line only), which is then defined as derived from the flow of the point and as the origin of the surface. In both quotations, the genetic definition is presented as complementary to Euclid's definition.

---

[13] Df. I.2 (Peletier 1557, p. 2) is the other occurrence (see *infra*, p. 89).

[14] Peletier (1557, p. 2), Df. I.1: "Ex Puncti fluxu perpetuo in longum, gigni intelligitur Linea".

[15] Peletier (1557, p. 3), Df. I.4: "Ac quemadmodum ex Puncti fluxu in continuum, exit Linea: ita ex Lineae in transversum ductu, oritur Superficies."

These are practically the only passages in which Peletier referred to the notion of *fluxus* in his commentary on the *Elements*. However, he referred in Df. I.4 and in other passages to other generative processes, for which he used other designations. In fact, while he used the term *fluxus* to describe the generation of the line from the point, Peletier used the term *ductus* to describe the generation of the surface and of the solid from the line and the surface, respectively. This was the case in the above-quoted passage of Df. I.4, where he also presented the generation of the surface from the line: "the surface originates from the transversal leading (*ductus*) of the line".

As was mentioned before, the term *ductus* (corresponding to the Greek term ἀγωγή) was commonly employed in geometrical constructions in the middle ages and in the Renaissance.[16] Indeed, although *ductus* and *duco* were also used to translate φορά and its related verbal forms,[17] these terms were chiefly used to express the procedure through which the geometer is said to construct lines in the first postulate of the *Elements* ("Ab omni puncto in omne punctum rectam lineam *ducere*")[18] and thereby in the framework of Euclid's constructions. In this context, the term *ductus* thus possessed a specifically operative meaning, implicitly pointing to the intervention of the geometer. On the other hand, the term *fluxus* (ῥύσις) was attributed, in philosophical contexts, a metaphysical meaning, pointing to the creative power of the divine principle in ancient Pythagoreanism and Neoplatonism.[19]

Therefore, by reserving *fluxus* for the generation of the line from the point and *ductus* for the generation of the surface from the line, Peletier seems to have attributed a different status and different ontological implications to these two processes in the context of Euclid's geometry.

The term *motus* is also indirectly used in Peletier's commentary on Df. I.15 and I.16 (i.e. the definitions of the circle and of its centre). In this context, Peletier started by presenting Euclid's definition of the circle, to compare it afterwards with its genetic definition, which he related to Euclid's genetic definition of the sphere (Df. XI.14).[20] As

---

[16] See *supra* p. 68.

[17] For instance, φέρεσθαι and περιφέρεσθαι were translated as *circumduci* in Campanus' Latin translations of Euclid's Df. XI.14. Compare Heiberg (1885, p. 4): "Σφαῖρά ἐστιν, ὅταν ἡμικυκλίου μενούσης τῆς διαμέτρου περιενεχθὲν τὸ ἡμικύκλιον εἰς τὸ αὐτὸ πάλιν ἀποκατασταθῇ, ὅθεν ἤρξατο φέρεσθαι, τὸ περιληφθὲν σχῆμα"; Campanus (Lefèvre 1516, fol. 189v): "Super quamlibet lineam semicirculo descripto, si linea illa fixa semicirculus tota revolutione *circunducatur*, corpus quod describitur, sphaera nominatur. Cuius centrum: constat esse centrum semicirculi *circunducti*." See also Zamberti, in (Lefèvre 1516, fol. 190v): "Sphaera, est quando semicirculi manente dimetiente *circunductus* semicirculus in seipsum rursus revolvitur unde incoepit, circum assumpta figura." (My emphasis.)

[18] Peletier (1557, p. 9), Post. 1. As noted above, the Greek term translated here by *ducere* is indeed ἀγαγεῖν. See *supra*, n. 92, p. 22.

[19] See *supra*, p. 27.

[20] Peletier (1557, p. 6), Df. I.16: "Siquis verò factionem seu creationem Circuli sibi exponi petat, instar Definitionis Sphaerae quam Euclides libro undecimo daturus est".

he put it, the genetic definition of the circle would teach the production or creation of the circle (*circuli factio sive creatio*), instead of its disposition or attribute (*circuli affectio sive passio*).

> *The circle is a plane figure contained by one line called the periphery, which is such that all the lines extended toward it from a single point that exists within it are equal.* Now, this point is called the centre of the circle. This is the most common definition of the circle, which explains its disposition or, as they say, its attribute. However, if someone wished to be taught the production or creation of the circle, as does the definition of the sphere that will be given in Euclid's Book XI, it will be as such: *The circle is the trace of the straight line carried around (circumduci) in a plane, one of its extremities remaining fixed, until it reaches the point where it started to be moved (duci).* For if the line AB started to be moved from point B around point A, through points C, D and E, until it became the same line AB again, the circle BCDE would be described. And through this description is perfectly expressed the whole property of the circle. Indeed, the fixed point A will be called the centre, while the circumscribing trace drawn by the mobile point (*mobile punctum*) B will be called the periphery.[21]

The term *motus* is thus not directly used in this passage, but it is suggested by the indication of the mobility of the point B (*punctum b mobile*) in the example provided to illustrate the genetic definition of the circle. This mobile point then corresponds to the moving extremity of the line-segment AB rotated around its other extremity A remained fixed (Fig. 3.1). This motion of the point is shown to result from the rotation (*circumductio*) of a line-segment, the latter therefore properly corresponding to the mobile element, as was the line generating the surface in Df. I.4. This confirms that Peletier equated *ductus* and *motus* in his commentary on Euclid. In turn, it confirms that the *ductus* of the line generating the surface in Df. I.4, which is etymologically and semantically related to the notion of *circumductio*, is straightforwardly understood as a spatial process and as a local motion (*motus*) in the proper sense. If Peletier distinguished *ductus* and *fluxus* as for their ontological implications, he then also distinguished *motus* and *fluxus* in the same manner. What this passage shows furthermore is that Peletier clearly admitted motion in a straightforward sense (as local transport) within geometrical definitions and that he considered genetic definitions as proper definitions.

---

[21] Peletier (1557, p. 6), Df. I.15–16: "*Circulus, est Figura plana, una linea contenta quae Peripheria appellatur: ad quam ab uno puncto introrsùm existente omnes porrectae lineae sunt aequales. Punctum autem illud, Centrum Circuli vocatur.* Haec Circuli definitio notissima est: quae ipsius affectionem, seu, ut dicunt, passionem explicat. Siquis verò factionem seu creationem Circuli sibi exponi petat, instar Definitionis Sphaerae quam Euclides libro undecimo daturus est: ea erit huiusmodi. *Circulus, est vestigium lineae rectae in plano circunductae, altero extremorum manente fixo, donec ipsa unde duci coepit, redierit.* Ut, si linea *ab* super *a* puncto duci incipiat in orbem à puncto *b*, per *c*, *d*, & *e* puncta, donec ipsa rursus *ab* facta sit: descriptus erit Circulus *bcde*. Atque ex hac descriptione, graphicè exprimitur tota Circuli proprietas. Punctum enim illud fixum *a*, Centrum dicetur: vestigium verò à puncto *b* mobili circunscriptum, Peripheria."

Fig. 3.1 Jacques Peletier,
Euclidis Elementa Geometrica
Demonstrationum Libri sex,
1557, p. 6, Df. I.16. Diagram
illustrating the generation of the
circle by the rotation of a line-
segment around one of its fixed
extremities. Courtesy Max
Planck Institute for the History
of Science, Berlin

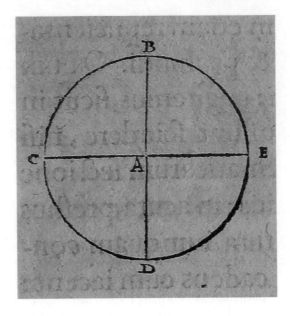

Moreover, with regard to the generative motions of the line and of the surface (which are termed *ductus* or *ductio*), Peletier did not establish, in his commentary on Euclid, a clear distinction between the generative processes referred to within definitions and the operations through which geometers produce their objects in the context of constructions. Indeed, the fact that, in the above-quoted passage, the motion supposedly performed by the line-segment, and by which the circle is said to be created, is expressed in the passive mode (*linea recta circumducta*; *duci*) easily allows one to conceive this process as the result of an external intervention. This would also be confirmed by the operative connotation of the term *descriptio*, as it is used here (which is not, as in Fine, used then by opposition to *definitio*).

On the other hand, when referring to the generation of the line through the flow of a point, no external intervention is ever suggested, even indirectly, the point being presented as the sole agent of the generative process.

Whether it is understood as intrinsic or as externally-induced, the motion attributed to geometrical objects in the context of definitions would itself play, for Peletier (as it was for Fine), an important role in conveying the essential properties of geometrical figures. In the case of the circle, for instance, it is through the intuition of the motion of the line-segment's mobile extremity, while it is thought to rotate about its other fixed extremity, that one apprehends the essential properties of the circle ("ex hac descriptione, graphicè exprimitur *tota Circuli proprietas*").

Now, in spite of the terminological differences between the genetic definition of the line and the genetic definitions of other magnitudes, nothing appears to distinguish these as for their logical or epistemic functions in Peletier's commentary on the *Elements*. For that matter, both the genetic definition of the line and that of the circle (as those of the surface and plane figures in general) would aim to teach and account for the spatial properties of the defined magnitudes by displaying their mode of generation, as would Euclid's definition of the sphere. This is marked in particular by the homology established by Peletier in his commentary on Df. I.4 between the generation of the line by the flow of a point and that of the surface by the transversal leading of the line ("quemadmodum ex Puncti fluxu in continuum, exit Linea: ita ex Lineae in transversum ductu, oritur Superficies").[22]

The homology established by Peletier between the genetic definitions of the line and of the surface also confirms that both aimed to convey properly spatial processes in this context, since the line, in Df. I.1 and in Df. I.4,[23] is said to be generated by a longitudinal (*in longum*) flow of the point, just as the surface, in Df. I.4, is said to be generated by a latitudinal or transversal (*in transversum*) *ductus* or leading of the line. As in Hero's *Definitiones* and in Fine's *Geometria* and commentary on Euclid, the flow of the point is presented as the origin of a generative process which takes place in several steps and which ends with the derivation of solid figures. At each step, the spatial properties (configuration, proportions, dimensions and position) of the given or previously-obtained geometrical object determine the spatial properties of the magnitude that is derived from it, even if these properties, since they apply to abstract and universal notions in the context of definitions, are not metrically determined.

3.3   The Flow of the Point in Peletier's *Louange de la Sciance*

If Peletier, contrary to Fine, used a different terminology to designate the generation of the line from the point and the generation of the surface and of the solid from the line and from the surface, respectively, it would be due, as was previously suggested, to the fact that the generation of the first dimension from the point, *and it only*, would ultimately refer to a metaphysical process.

Before looking at the hints left in this direction by Peletier in his commentary on Euclid, let us first look at his *Louange de la Sciance*, where the metaphysical origin of this notion of flow of the point is clearly set forth and which clarifies the terminological distinction between *fluxus* and *ductus* found in the commentary on the *Elements*. In this text, the flow of the point generating the line, from which all plane and solid figures ultimately take their origin, is indeed assimilated to the instantaneous emanation of the universe from a divine unitarian principle.

---

[22] Peletier (1557, p. 3). See *supra*, p. 77.

[23] See *supra*, p. 77.

> For this whole universe gained its form together, all numbers have been and are at the same
> time as the unit; matter, and form, and all things, had no principle; points, which extended at
> once and instantaneously, produced an infinite number of lines, areas, and bodies.[24]

The production of magnitude from the flow of the point—Peletier even says here from the
extension of the point—would occur simultaneously and by the same principle as that
through which all numbers, and all material and formal multiplicity, were initially produced from the divine One.[25] The point itself is thus presented as the indivisible principle
of the universe and of all the things it contains, according to a cosmogonic vision evocative
of the Pythagorean, Platonic and Neoplatonic doctrines.[26]

> The point, entirely simple and one, but of immense extension, demonstrates that, with it,
> Nature starts everything. [...] Thus, infinite points flowing in themselves produce the finite
> line between its two extremes, and the line led on itself, as much as it is long, produces the
> square, and finally the square, which is led on the line equal to its side, demarcates the cube.[27]

This passage shows that Peletier not only knew the metaphysical interpretation of the flow
of the point as a divine principle, but also that he maintained, in his depiction of the
primordial process of the universe's creation, the terminological distinction between the
flow of the point ("points s'ecoulans") and the leading of the line and of the surface ("la
ligne se condùit"; "le quarre se meine") to generate the surface and the body. This would
retrospectively confirm that this terminological distinction between flowing and leading
already intended to convey an ontological difference between the generation of the line and
those of the surface and of the solid in his 1557 commentary on Euclid.

  This passage of the *Louanges* also implicitly points to the correspondence between the
composition of continuous magnitudes from the point and the composition of numbers
from the arithmetical unit, which we have previously seen with Fine. Indeed, the fact that
the line, which is presented as finite ("Ligne finie antre ses deus extremes"), is also
described as produced by an infinite number of points flowing *in themselves* ("s'ccoulans

---

[24] Peletier (1581, fol. 57r): "Car tout cet Univers, à pris sa Forme ansamble:/Tous Nombres ont eté,
e sont, aussi tòt qu'Un:/Matiere, e Forme, e Tout, n'urent principe aucun:/A coup, e an l'instant les
Poins, qui s'estandiret, Lignes, Eres, e Cors an l'Infini randiret." I would like to thank Alain Legros
for his help in interpreting this passage.

[25] As shown by K. Banks (2007), Peletier resorted to a metaphysical and geometrical notion of
outward expansion of magnitude from a single origin in another poetic work, *L'Amour des Amours*,
where the divine light emanating from the poet's beloved lady, along with the desire of the poet and
the divinity of love itself, are presented as expanding in all directions to stretch toward the limits of
the sphere of the universe.

[26] Vinel (2010).

[27] Peletier (1581, fol. 56r–v): "Le Point tout simple e un, mes d'etandue immanse,/Demontre, que par
lui Nature tout commance./[...] Einsi infiniz Poins s'ecoulans an soemémes,/Font la Ligne finie antre
ses deus extremes:/E an soe par-apres la Ligne se condùit,/Autant comme el ét longue, & le Quarre
prodùit:/Puis an fin le Quarre, qui se meine an la Ligne/Egale a son cote, e le Cube designe".

an soemémes"), evokes Fine's comparison of the flow of the point to an infinite multiplication of the point. The notion of points flowing in themselves also echoes, to a certain
extent, the above-quoted passage of Iamblichus' commentary on Nicomachus concerning
the generation of geometrical numbers, insofar as the point-unit was then said to flow from
itself and to extend itself to produce figurate numbers.[28] Moreover, the transversal leading
of the line-segment along and according to its own length is said to produce the square, and
the transversal motion of a square along and according to one of its side, to generate the
cube. This evokes quite distinctly the composition of square and cubic numbers by the
arithmetical *ductus* of a number on itself once or twice, this number led on itself
corresponding to the root of the square or cubic number.

Nevertheless, it is important to note that Peletier did not use here the arithmetical term of
multiplication to describe the mode of composition of the line from the point, as did Fine,
but precisely that of flow or flowing, which he only considered as proper to the generation
of the geometrical line.

It may also be noted that the two above-quoted passages of the *Louange de la Sciance*
seem to present two different generative processes originating from the point. One is an
extension or expansion of the point ("les Poins, qui s'estandiret"[29]), through which Nature
is said to have produced all things ("Lignes, Eres, e Cors an l'Infini randiret"). The other is
a flowing of the point ("Poins s'ecoulans an soemémes"[30]), which is then specifically
presented as the principle of the line ("Font la Ligne finie"), and not of other types
of magnitude, nor of numbers or physical bodies. In a sense, the notion of expansion of
the point, when it is used to designate the generation of the entire universe, would seem to
bear a stronger metaphysical connotation than the notion of flow of the point. The latter
would as such rather evoke its mathematical and spatial understanding, since the point is
only said to generate the geometrical line.

However, in both Iamblichus and Proclus, we find an assimilation of the extension of the
point to its flow when referring to its generative function. Indeed, in his commentary on
Df. I.6, Proclus wrote that, "in the region of immaterial forms the partless idea of the point
has prior existence. As it goes forth from that region, this very first of all ideas *expands*
itself ($\delta\iota\alpha\sigma\tau\dot{\eta}\sigma\alpha\varsigma$), *moves* ($\kappa\iota\nu\omicron\dot{\upsilon}\mu\epsilon\nu\omicron\varsigma$),[31] and *flows* ($\dot{\rho}\dot{\epsilon}\omega\nu$) towards infinity".[32] In

---

[28] See *supra*, p. 65.

[29] Peletier (1581, fol. 57r). See *supra*, n. 24, p. 81. See also Peletier (1581, fol. 56r–v): "Le Point tout
simple e un, mes d'etandue immanse".

[30] Peletier (1581, fol. 56r–v). See *supra*, n. 27, p. 82.

[31] The fact that the notion of expansion and of flow ($\dot{\rho}\dot{\upsilon}\sigma\iota\varsigma$) of the point are not here distinguished from
its motion ($\kappa\dot{\iota}\nu\eta\sigma\iota\varsigma$), while it appears to be in the commentary on the postulates (see *supra*, p. 48),
would be due to the fact that these processes are here understood as taking place on an intelligible
level and do not have the status of the motion produced by the mathematician in his imagination when
studying the properties of extended magnitudes.

[32] Proclus (Friedlein 1873, p. 101) and (Morrow 1992, p. 83). On this passage, see also *infra*,
pp. 106–107.

Iamblichus, these two processes—expansion and flow—are both attributed to the point-unit when describing the generation of geometrical numbers.[33] Hence, it does not seem that Peletier (who knew Proclus' commentary on the *Elements*[34]) admitted a difference between these two processes on a metaphysical level, especially as, in this context, the geometrical point is not distinguished from the primordial One from which all things have emanated. This is confirmed by the passage of the *Louanges* where Peletier wrote that the extension of points is what produced all magnitudes (lines, areas, bodies) instantaneously and infinitely.[35]

3.4   Peletier's Metaphysical Discourse in the Commentary on Euclid

In his commentary on the *Elements*, Peletier mainly aimed to adopt a mathematician's perspective on the point and on its mathematical relation to the line or to continuous magnitude in general, using the notion of the flow of the point in order to express the spatial properties of the line. Nevertheless, he also presented certain elements of this metaphysical interpretation of the generation of the line. In his commentary on Df. I.1, for instance, he quoted a passage from Proclus' commentary on Euclid,[36] which itself refers to Book 10 of Plato's *Republic*[37] and in which the genesis of the universe is represented as starting with the flow of the axis of the world from a primordial indivisible element. This passage of Peletier's Euclid also compares points to Epicurean atoms in order to show their primordial function in the constitution of the universe.

> Plato calls the hypostasis or the substance of points "adamantine", for it is eternal, stable, incorruptible, always identical to itself. [He says that] the universe revolves around them and is moved all around in a dance. They are the atoms of Epicurus, seeds of all things.[38]

---

[33] See *supra*, p. 65.

[34] On Proclus' influence on Peletier, see Axworthy (2013) and *supra*, pp. 84 and 128–130.

[35] Peletier (1581, fol. 57r): "Poins, qui s'estandiret, Lignes, Eres, e Cors an l'Infini randiret."

[36] Proclus (Friedlein 1873, p. 90).

[37] Plato, *Rep.* X, 616c.

[38] Peletier (1557, p. 2), Df. I.1: "Plato Punctorum hypostasin seu subsistentiam vocat adamantinam: nempe aeternam, stabilem, incorruptibilem, quaeque eodem semper modo habeat: Universum circa ipsa converti, ac circunquaque in plausum moveri. Eae sunt Epicuri Atomi, omnium rerum semina". *Cf.* Proclus (Friedlein 1873, p. 90) and (Morrow 1992, p. 74): "This is why Plato declares that the substance of these axes is as hard as adamant, thus indicating their irreversible, everlasting, steadfast, unchangeable being. The whole 'spindle,' he says, moves about these axes, celebrating their unity in a dance." *Cf.* Plato, *Rep.* X, 616c (transl. Emlyn-Jones and Preddy 2013, pp. 470–471): "Stretching down from either end was the spindle of Necessity by means of which all the circles turn. Both its shaft and the hook were made of adamant." See also Plato, *Tim.* 40c (transl. Bury 1929, pp. 84–87): "the choric dances of these same stars and their crossings one of another".

The present assimilation of the points to the atoms of Epicurean physics is also asserted in the commentary on Df. I.5 through the correspondence between the geometrical notions of point, line and surface and the physical notions of atom, matter and form. Both conceptual triads are then said to give rise to the body.

> The points are atoms; lines are matter; surfaces are form. And from these originates the body, which is long, broad and deep.[39]

The body is only described as a tridimensional object, but it would be logical to suppose that it is also taken here to represent the natural body, insofar as it would stem from the synthesis of atoms, matter and form, just as the geometrical solid stems from the synthesis of points, lines and surfaces according to the system of correspondences proposed here by Peletier. Through these correspondences, the procession of the body from the point, the line and the surface, is to a certain extent mirrored by the constitution of physical substances from atoms, matter and form. And both deployments would express in different manners the mode according to which the universe would have sprung from the primordial One. This is confirmed by the fact that the identification of the geometrical solid to the natural body, composed of atoms, matter and form was again presented in the *Louange de la Sciance*. In this context, this correspondence between the composition of geometrical and natural bodies was intended to show the compatibility between ancient atomism and the Aristotelian theory of nature.

> As for he who is in the School of high reputation, who with great difficulty refutes atoms, may he receive his sentence, leave authority aside and, by appealing to reason, give precedence to truth, by whomever it be pronounced or written, be it Epicurus or Democritus. If I do not however want to strip the form of its value, much less do I want to judge matter as fruitless. May the atom be the point; the line, matter; the area, form; the body, the entire mass, so that the debates of natural philosophers be settled and the School be made consistent with the facts of Nature.[40]

This passage confirms the primordial connection Peletier conceived between the constitution of geometrical objects and of physical substances and which was founded on the primordial identity of the point and of the atom (as expressions of the divine One) at the first stage of the universe's creation.

---

[39] Peletier (1557, p. 5), Df. I.5: "Puncta igitur, atomi sunt: Lineae materia: Superficies forma. Atque ex his Corpus, quod longum, latum, & crassum est."

[40] Peletier (1581, fol. 56v): "Quant à celui, que tant an l'Ecole on repute,/Qui si peniblemant les Atomes refute,/Reçoeve sa santance: arriere Autorite,/E aveq la Reson, valle la Verite,/De quiconque ele soèt prononcee, ou ecrite:/Soèt l'auteur Epicure, ou l'auteur Democrite. Si ne veu-je pourtant, la Forme deprimer:/Ancor' moins la Matiere inutile estimer/*L'Atome soèt le Point, la Ligne, la Matiere:/L'Ere, la Forme soèt: le Cors, la masse antiere,*/Pour des Phisiciens les debaz acorder,/E aus fez Naturez l'Ecole acommoder." (My emphasis.)

3.5   The Motion of the Point and the Propagation of Light

In Peletier's commentary on Euclid, the passage comparing the points to atoms, lines to matter, and so on, was preceded by a discussion on the ontological status of geometrical objects and, in particular, on the distinction and relation between the different types of geometrical objects (point, line, surface, solid) with physical substances. In this context, Peletier stated both the impossibility for us of finding geometrical points, lines and surfaces in nature and the possibility of reaching their intuition through the perception of certain material objects, and in particular through dust particles moving in the light of the Sun, light rays, as well as shadows and colored surfaces.

> Although it seems that the point, the line and the surface can only be seized by the intellect, that they do not exist in reality and that they cannot be shown, each of these things can however be represented by something in nature. Points are indeed assimilated to the indivisible corpuscles which play together in the rays of the Sun, lines, to those very rays, the surfaces, to shadows, since they can never go underground, or even colours, according to the opinion of the Pythagoreans.[41]

This passage, which appears to be partly inspired by Hero of Alexandria's *Definitiones* (Df. 8, i.e. the definition of the surface)[42] and partly by Proclus' commentary on Df. I.5,[43] displays the privileged relation that was established, in Peletier's thought[44] and in the larger context of premodern optics, between geometrical objects and natural phenomena pertaining to light, and thereby between geometry and optics more generally. In ancient and medieval optics, this privileged relation was founded on the subordination of optics to geometry in Aristotle's *Posterior analytics*[45] and by the ontological foundation of the object of optics on the object of geometry in the *Physics*.[46]

---

[41] Peletier (1557, p. 3), Df. I.5: "Ac tametsi, Punctum, Linea, & Superficies intellectu tantùm capi videantur, non re existere, neque ostendi posse: habet tamen unumquodque horum in rerum natura quo repraesentetur. Puncta enim corpusculis insecabilibus assimilantur, quae in radijs Solis colludunt: Lineae radijs ipsis: Superficies umbris, ut quae terram nunquam subeant: seu etiam coloribus, ex Pythagoreorum sententia."

[42] Hero (Heiberg 1974, p. 20): "And every shadow and every colour may be understood as a surface, wherefore the Pythagoreans also called surfaces colours" (my translation from the Greek and from the German translation by Heiberg).

[43] Proclus (Friedlein 1873, pp. 114–115) and (Morrow 1992, p. 93): "we get some perception of it [the surface] when we look at shadows. These are without depth, since they cannot go under the ground, and have only breadth and length".

[44] On the place of light in Peletier's scientific poetry, see Pantin (1984), Pantin (1995, pp. 214–224) and Banks (2007). On this quotation in the context of Peletier's commentary on Euclid's *Elements*, see Axworthy (2013).

[45] Aristotle, *Posterior Analytics* I.13, 78b37–38 and 79b10–13.

[46] Aristotle, *Physics* II.2, 194a8–12.

In relation to the kinematic understanding of geometrical objects, and notably of the line as generated by the motion of a point, it may be noted that, in the medieval optical tradition derived from Ibn al-Haytham and developed by the thirteenth-century Perspectivists (Roger Bacon (ca. 1220–ca. 1292), Witelo (fl. c. 1270–1285), John Pecham (ca. 1230–1992)),[47] light propagation had been explained in mechanistic terms, as it was sometimes attributed motion and speed and compared to the motion of material bodies.[48] This mode of explanation was taken up and further developed by Descartes in his 1637 *Dioptrique*,[49] where the action of light particles was presented as the cause of the generation of rectilinear rays of light and where the behaviour of light was illustrated by the motion of a ball or a stone projected in the air.[50] As was explained by M. Smith, no motion *per se* was then attributed to light, as it was not considered as material in the same way as other physical substances, but this kinematic understanding of light was considered useful to explain the phenomena of reflection and refraction.[51] Hence, in this framework, the geometrical definition of the line as caused by the flow or motion of a point would provide a perfect expression of this mechanistic explanation of the propagation of light in a straight line from its point of origin, all the more as neither processes were considered as motions in the strict sense.

There is no direct reference, in Peletier's commentary on Euclid, to any mechanistic representation of light, nor to any optical doctrine. Nevertheless, it seems possible to draw a relation between the two frameworks, geometrical and optical, from his appeal to the genetic definition of the line and the privileged connection he established between the geometrical line and light rays, where the flow of the geometrical point could be physically represented by the propagation of light along a rectilinear path. This would be corroborated by the fact that Peletier also introduced the genetic definition of the line in his commentary

---

[47] Lindberg (1967).

[48] Lindberg (1967, p. 335) and Smith (1987, pp. 44–53).

[49] On Descartes' debt to the medieval Perspectivists, see Smith (1987, pp. 7–12, 45–48 and 64–65). It is important however to keep in mind that Descartes rejected the Perspectivists adhesion to the theory of the multiplication of species. See Descartes (1637, *Dioptrique*, 5): "Et par ce moyen vostre esprit sera delivré de toutes ces petites images voltigeantes par l'air, nommées des *especes intentionelles*, qui travaillent tant l'imagination des Philosophes."

[50] Descartes (1637, *Dioptrique*, p. 8): "Au reste, ces rayons doivent bien estre ainsy toujours imaginés exactement droits, lors qu'ils ne passent que par un seul corps transparent, qui est par tout esgal a soy-mesme: mais, lors qu'ils rencontrent quelques autres cors, ils sont sujets a estre détournés par eux, ou amortis, en mesme façon que l'est le mouvement d'une balle, ou d'une pierre jettée dans l'air, par ceux qu'elle rencontre. Car il est bien aysé a croire que l'action ou inclination a se mouvoir, que j'ay dit devoir estre prise pour la lumière, doit suivre en cecy les mesmes loys que le mouvement."

[51] Smith (1987, pp. 29–31, 46 and 67–69). As Smith explained (Smith 1987, p. 69), the failure to consider the propagation of light as a motion in the proper sense is precisely what would prevent Descartes, as his medieval predecessors, from achieving a fully adequate explanation of light. On Descartes' optical work in general and the place held by motion in this context, see, for instance, Shea (1991, pp. 227–249).

on the definition of the straight line (Df. I.4). Moreover, this relation would be ontologically founded in Peletier's thought, since this process would be in both cases regarded as
distinct from the physical motion of material bodies. In the above-mentioned optical
tradition, the propagation of light, just as the flow of the point, was indeed regarded as
deprived of motion and spatiality in the strict sense, being rather conceived under the mode
of instantaneous emanation.

This privileged connection between the propagation of light and the flow of the point in
Peletier's thought would be confirmed again by his scientific poetry, in which light holds an
important place, as was demonstrated by I. Pantin and K. Banks. In this context, light was
mostly understood in a metaphysical sense, which evokes the metaphysics of light developed by Robert Grosseteste and his followers.[52] This medieval theory, which was
influenced by Neoplatonic conceptions, presents light as the principle of the cosmos and
of all it contains,[53] representing the first emanation of God and of the divine archetypes in
the physical world.[54] In this framework, as was shown in the previous chapter,[55] the
universe would have been produced by light through the infinite and instantaneous
multiplication or self-generation of an indivisible primordial source of light[56] in all
directions.

The role attributed to the flow of the point in Peletier's *Louange de la Sciance* clearly
evokes this doctrine, as it is then presented as the principle of the universe, which would
have created all things, all matter, form and quantity, through its instantaneous expansion
and flow in all directions.[57] The fact that, in this text, the universe is said to have proceeded
from points infinitely "flowing in themselves"[58] is furthermore evocative of the process by
which the primordial light-atom would have multiplied itself infinitely in all directions to
form the cosmos in the medieval metaphysics of light. In Grosseteste, and later in Roger
Bacon's optics, this process was applied to the explanation of optical phenomena, and in
particular of the propagation of light. The *species* or formal properties of light were then
said to reach the eye by radiating in a sphere through an instantaneous process of replica-
tion or multiplication of themselves in a straight line in all directions.[59] The atomistic
implications of such a theory were however not interpreted literally in this context, since

---

[52] This theory was presented in the previous chapter.

[53] See *supra*, n. 121, p. 71.

[54] McEvoy (1978).

[55] See *supra*, p. 71.

[56] J. McEvoy (McEvoy 1978, p. 130) actually described the source of this process as "a primordial
point". See also Lindberg (1976, pp. 96–99) and Lewis (2005).

[57] See *supra*, pp. 81–82.

[58] See *supra*, n. 27, p. 82: "Le Point tout simple e un, mes d'etandue immanse,/Demontre, que par lui
Nature tout commance./[...] Einsi infiniz Poins s'ecoulans an soemémes [...]." (My emphasis.)

[59] Lindberg (1976, pp. 98–99 and 113–115) and Smith (1981). See also Smith (1987, pp. 40–46,
52 and 75–78).

the propagation of light along straight lines was regarded as intrinsically continuous,[60] just as the flow of the point in Peletier's definition of the line in the context of geometry (as will be shown in the following section). Also it is important to reiterate that this process was not considered as a mechanical or kinematic process *per se*, as there would be no local motion of bodies,[61] but rather an instantaneous outward radiation, diffusion or emanation of a form or *species*. In the same way, the line in Peletier is not strictly defined as a motion, but as a flow, even if it has certain properties that enables us to interpret it as a local motion in a geometrical context. In any case, the Neoplatonic influences common to this theory of light[62] and to many aspects of Peletier's philosophy of mathematics allows for their comparison in this regard, both on the metaphysical and on the mathematical level.

3.6   The Flow of the Point and the Composition of the Continuum

The comparison between the commentary on the *Elements* and the *Louange de la Sciance* thus revealed certain tensions in Peletier's notion of flow of the point, which were easily resolved through the distinction between its mathematical and its philosophical treatment. Now, a similar situation occurs when considering Peletier's interpretation of the relation between discrete and continuous quantity in both texts. Indeed, in the *Louange de la Sciance*, the generation of the line is compared to the flowing of an infinite number of points in themselves, which evokes Fine's comparison of the flow of the point to an infinite multiplication of the point, as well as Iamblichus' description of the generation of geometrical numbers by the flow of a unit. Yet, in the commentary on Euclid, the flow of the point is explicitly presented as a means of distinguishing the modes of generation of discrete and continuous quantity.

> Unlike number, which arises from the accumulation of units, the line does not come from the addition of points, but rather from their continuous flow. And, in this, the continuous differs from the discrete, because the continuous can be infinitely divided without the point ever being reached as the unit is reached in discrete quantity.[63]

Peletier states here that magnitude differs from number insofar as the unit produces number by addition, while the point produces the line through its continuous flow. Now, this is the very argument that was set forth by Hero, Eratosthenes and Theon of Smyrna[64] to explain

---

[60] Lindberg (1967, p. 341); Smith (1981, pp. 578–580, 1987, pp. 41–44).

[61] Lindberg (1967, p. 337) and Smith (1987, p. 46).

[62] Lindberg (1967, pp. 335–336).

[63] Peletier (1557, p. 2), Df. I.2: "Non igitur sicut ex accumulatis Unitatibus fit Numerus, sic ex additis Punctis fit Linea: sed ex ipsorum fluxu continuo. Atque in hoc differt Continuum à Discreto, quòd Continuum infinitè dividatur: neque ad Punctum unquam deveniatur, ut in Discretis ad Unitatem".

[64] See *supra*, pp. 25–26.

how magnitude and continuous quantity can originate from the point without being conceived as resulting from an addition or juxtaposition of indivisible parts,[65] as are, on the contrary, numbers or discrete quantity. Hence, although the *Louanges* will, much later, present the generation of the line as resulting from a process that is comparable to a multiplication of points (even if the word multiplication is not used then), Peletier clearly considered, in his commentary on the *Elements*, the mode of generation of magnitudes as absolutely distinct by nature from the mode of generation of numbers. This shows that Peletier was aware of the logical and mathematical aporias raised by the notion of magnitude as caused by the point in a geometrical context. He therefore made sure, in the commentary on Euclid, to avoid interpreting the generation of the line from the flow of the point as an addition or multiplication of points.

However, this was only the case when Peletier intended to offer a mathematical account of the relation between the point and the line. Indeed, in the passage where he evoked the connection between the modes of composition of physical and geometrical bodies, which proposed a philosophical rather than a mathematical perspective on the issue, the mode of composition of geometrical magnitudes was compared to the mode of composition of material substances as interpreted in ancient atomistic doctrines. Thus, if, on a philosophical level, the generative motion of the point may be related to the mode of composition of discrete quantities, this would not be the case within a properly mathematical discourse. In this context, the continuity of the flow of a unique point is taken as the guarantee of the continuity of the resulting line, which cannot then be considered as composed of indivisible parts, even in an infinite number.

3.7   The *ductus* of the Line and the Multiplication of Numbers

Peletier's commentary on Euclid's Book II provides a representation of the relation between discrete and continuous quantity that raises further questions concerning the mode of composition of magnitudes. In his commentary on Df. II.1, Peletier, like Fine and other authors considered here, explicitly compared the generation of the rectangular parallelogram by the transversal motion of a line-segment along another at right angles (for which he used the term *ductus*) to the multiplication of two numbers. This comparison was meant to explain why rectangular parallelograms that are contained by equal sides have equal areas, whichever side is considered as moving along the other, according to the arithmetical principle of commutativity.

> A rectangular parallelogram is produced by the leading (*ex ductu*) of a straight line on a straight line, in the manner of numbers, as we have said for the sake of ease. It is therefore necessary to

---

[65] On this issue, see *supra*, pp. 60–61.

remember that to lead a greater line on a smaller one is the same as leading a smaller line on a greater one, just as 3 by 4 and 4 by 3 produce the same result.[66]

The fact that Peletier here relates the mode of composition of the rectangular parallelogram to the multiplication of two numbers does not mean that he is contradicting the position he adopted in Book I regarding the essential differences between the modes of composition of numbers and of magnitudes, but it certainly qualifies it. This comparison is, generally speaking, representative of an attitude he demonstrated elsewhere in his commentary on Euclid, where he defended the kinship and similitude *in principle* between numbers and magnitudes. For instance, in his commentary on Euclid's Prop. III.16, he stated that "numbers are the mirrors of all things".[67] He also asserted, in the French version of his treatise of practical geometry,[68] and more specifically in a proposition that offers an arithmetical treatment of Euclid's Prop. I.22, that he has "taught this proposition by the means of numbers, as an example, to show that numbers fraternise nearly everywhere with geometrical quantities, for which they are as their exponents".[69]

---

[66]Peletier (1557, p. 52), Df. II.1: "Parallelogrammum Rectangulum fit *ex ductu* lineae rectae in lineam rectam, idque instar Numerorum: facilitatis, ut diximus, gratia. Ubi etiam meminisse oportet. *Ducere* maiorem lineam in minorem, idem esse ac si minorem in maiorem *ducamus*: sicut 3 in 4, & 4 in 3, idem efficiunt." (My emphasis.)

[67]Peletier asserted this in order to prove that the angle of contingence is not an angle since it is defined by Euclid as smaller than any rectilinear angle and does not, as such, obey the Axiom of Archimedes (as expressed in Euclid's *Elements*, Prop. X.1 and taken up then by Peletier), which guarantees the mutual comparability of quantities of the same kind (Peletier 1557, pp. 74–75). "*Si à maiori duarum Quantitatum auferatur maius quàm dimidium, ac rursus ex reliquo maius quàm dimidium, idque continuò fiat: relinquetur tandem magnitudo, minor magnitudine minore posita* Verbi causa, Sint duo anguli, *a* quidem rectilineus, & *bcd* angulus (si modò sit angulus) contactus: Vult Prima Decimi: ut, si auferatur ab angulo *a*, maius quàm dimidium, ac rursus à reliqua parte maius quàm dimidium: sicque continuò ex residuis partibus maius quàm dimidium: tandem relinquatur minor angulus quàm *bcd*. Cuius demonstrationem hic non appono, quum ex sequentibus pendeat. Nulla tamen in tota Geometria Propositio est, quae (ut sic dicam) magis naturaliter vera sit. Quod ex *Numeris* (*in quibus rerum omnium imagines*) luce clarius evadit. Quis enim non videt, propositis duobus Numeris 8 & 2, quum ab octonarius maius quàm dimidium abstuleris, ut quinarium: tum à ternario residuo, maius quàm dimidium, ut binarium: relinqui unitatem, posito binario minorem?" (My emphasis.)

[68]*De l'usage de geometrie* (Peletier 1573), which corresponds to the French translation and adaptation of his *De usu geometriae* (Peletier 1572).

[69]Peletier (1573, p. 35): "Ce Probleme icy est proposé generalement en lignes pures par Euclide, en la vintedeuziesme Proposition du premier livre: *laquele nous avons icy declairée par nombres, comme en maniere d'exemple, pour monstrer que les nombres fraternisent quasi par tout avec les quantitez Geometriques: Desqueles ils sont comme les exposans.*" (My emphasis.) The term "exposant" used here designates in French the algebraic notion of exponent, which was known at the time of Peletier (see Nicolas Chuquet's *Triparty en la science des nombres*, in Marre 1881, p. 152) and was used by Peletier in his *Algèbre* (Peletier 1554, p. 8) and before that in the *Arithmetica integra* of Michael Stifel (1487–1567) (Stifel 1544, fol. 236r). Although Peletier's main goal was to assert the fraternity of numbers and magnitudes in general, this notion specifically evokes the algebraic treatment of

It remains that, even in the passages where he appealed to numbers to explain certain properties of geometrical objects, Peletier remained cautious when comparing numbers and magnitudes,[70] for which he specified, in the above-quoted commentary on Df. II.1, that the arithmetical interpretation of this definition only intends to make it more comprehensible to the reader (*facilitatis gratia*). It may also be noted that, at the beginning of his commentary on Book V, which deals with the theory of ratios and proportions as applied to magnitudes specifically,[71] Peletier explained that, while he intended to appeal to numbers when dealing with the *definitions* of this book for the sake of clarity, he would generally avoid doing so when dealing with the *propositions* in view of the differences between number and magnitude.[72] Yet, he also added that he would sometimes speak of quantity in general terms (in place of magnitude) when the taught notions are applicable to both numbers and magnitudes.

3.8   Genetic Definitions and the Limits of Human Knowledge

Hence, if Peletier's position on the notion of flow of the point in geometry, and on its relation to the mode of composition of numbers, may seem somewhat contradictory from one text to another, this contradiction is only apparent. For, as he wrote in the French version of his *De usu geometriae*, the true relation between numbers and magnitudes,

---

geometrical problems. *Exposans* thus means here both the algebraic *exponents* and numbers as the principles of magnitudes.

[70] See, for instance, his commentary on Prop. I.47 (Peletier 1557, p. 49): "Nunc autem hoc Theorema quonam pacto ad Numeros accommodetur, obiter ostendemus. In Numeris itaque locum praecipuè habet, quum maximus ad medium fuerit ut 5 ad 4: scilicet in proportione, quam vocant, sesquiquarta: & medius ad minimum ut 4 ad 3: hoc est, in proportione sesquitertia. [...] Isoscelia verò Rectangula ex Numeris non conficiuntur. In quo id dignum consideratione est, quod in Geometricis est evidentius & Demonstrationi promptius, *id in Numeris veritatem non habere*. Nunquam enim duo Quadrati Numeri aequales Quadratum Numerum componunt. Atque eam ob causam, lateris Quadrati ad Diametrum proportio incognita. Est enim Diameter radix seu latus duorum Quadratorum aequalium in unum Quadratum iunctorum: ob id, irrationale: hoc est, ut in Arithmeticis dicitur, radix Surda." (My emphasis.)

[71] The theory of ratios and proportions as applied to numbers is presented separately, within the arithmetical books of the *Elements*.

[72] Peletier (1557, p. 112), Df. V.1: "In explicandis huius Quinti Definitionibus, Numeros nobis accommodabimus. Id enim disciplinae gratia in Principiorum ostensione licet: In demonstrationibus autem Propositionum, Geometrica dignitas servanda est. Alia quippe ratio & natura Continuorum, atque alia Discretorum. In quibusdam tamen adeò religiosi non erimus: nempè dum Quantitatis vocabulum pro Magnitudine usurpabimus. Quanvis enim hoc peculiarius, illud generalius sit utrunque tamen Geometria parvo discrimine sibi vendicat." It may be noted additionally that, in this context, Peletier directly applied the verb *ductus* to the multiplication of numbers (Peletier 1557, p. 116, Df. V.10): "Huius enim numeri singularis proprietas est, quòd tantum efficiat duplicatus, quantum in se *ductus*." (My emphasis.)

which primordially correspond to divine principles, cannot be fully reached by the human intellect.[73] Along the same line, he wrote, in the *Louange de la Sciance*, that the causal relation between points, lines, surfaces and solids, and even between bodies and atoms, just as their true geneses, is only accessible to God.

> To seek a greater or smaller number of surfaces within the body, of lines within surfaces, and of points within lines, or why these are constituted in such and such way, is like wanting to shatter the closet of Nature, to unveil the treasure of this great immortal, who only is in himself infinite and only knows himself to be so.[74]

We find a similar observation in the commentary on Df. I.16, where Peletier presented the genetic definition of the circle. He asserted then that due to the divine origin of geometrical objects and due to the finiteness of the human intellect faced with the transcendent perfection of divine entities, it is practically impossible to know the true mode of generation of magnitudes or, in other words, the true genesis and order of existence between points, lines, surfaces and solids, as well as between the rectilinear and the circular. Thus, the causality established by the geometer between these objects through genetic definitions would not enable us to establish the truth concerning these questions.

> And nobody should exert himself to find out which is first between the straight and the circular. But if someone was forced to give an opinion, he would judge appropriately and as a philosopher, if he declared that both exist simultaneously. For the circle turned about in the plane produces the straight line. Indeed, to the mind, nothing will come before or after. The understanding can barely seize points as having existed before lines, or lines before surfaces, or finally surfaces before bodies, just as, among philosophers, to say that the universe once was not exceeds the capacity of our souls and to say that it always has been is above all admiration. But, as for us, our aim is to establish all things in their proper order and to reduce them to art to the measure of our understanding, and attempt to decide, as much as possible, what is probable and the least false. [. . .] This diversity of things exercises us, in which it is enough, for us, to accommodate the conjecture to practice. For what do we think we can accomplish by the means of art in the things which Nature has so skilfully made? Or what can we understand regarding the things which have emanated in a divine manner, when we judge them in a human manner?[75]

---

[73] Peletier (1573, pp. 30–33), XVII: "*Trouver l'aire d'un triangle donné, par le moyen des nombres.* [. . .] Icy ne se contentera le Geometrien, allegant qu'en tels triangles non orthogones, la perpendiculaire *ef*, est inconnue: & partant ne se peut autrement avoir l'aire du triangle. Et avec bonne raison. [. . .] *Car la vraye estimation est inconnue à l'artifice: & est connue à seule nature.*" (My emphasis.)

[74] Peletier (1581, fol. 56v): "Mes chercher dans le Cors, les Eres, plus ou moins,/E les Lignes an l'Ere, an la Ligne les Poins,/Ni pourquoe il an vient tele, ou tele facture,/C'et vouloer defonser l'armoere de Nature,/Pour comter le trezor de ce grand Immortel,/An soe seul infini, e seul se sachant tel".

[75] Peletier (1557, p. 6), Df. I.16: "Neque est quòd quisquam se fatiget inquirendo, utrum sit prius Rectum an Rotundum. Sed si quis sententiam ferre cogatur: ut Philosophus, rectè iudicabit, si utrunque simul esse pronuntiaverit. Nam & Circulus in plano rotatus, Rectum procreat. Menti quippè

To Peletier, the fact of supposing that all things, including magnitudes and their parts, came to exist instantaneously and simultaneously certainly transcends the capacities of the human understanding, but it would be the most reasonable assumption on a philosophical level. It would notably allow us to avoid the problems raised by the admission of an infinite chain of causes in the universe. In the context of a commentary on the definition of the circle, this echoes Proclus' assertion, in his own commentary on Euclid, that "in the circle the center, the distances, and the outer circumference all exist at the same time, so also in the paradigms there are no parts that are earlier in time and others that come to be later, but all together at once".[76]

Peletier's assertion of the divine nature of geometrical objects and of the limits of the human mind faced with the geneses of figures indirectly evokes the discourse of Nicholas of Cusa (1401–1464), in his *De coniecturis*, where it is stated that the geometer, through his mental constitution and study of geometrical objects, recreates in his own manner and according to his own intellectual faculties the world created by God.[77] Yet, because of the infinite disproportion between the human understanding and the divine mind, the notions brought forth by the geometer concerning his objects, their properties and mutual relations, are merely conjectures, through which the human understanding attempts to participate in the divine intuition of the universe.

Peletier's description of the humility of the mathematician, faced with the conjectural character of his knowledge concerning the geneses and mutual relations of his objects, also resonates with the reaction later displayed by Michel de Montaigne (1533–1592)—albeit to defend a different position about geometry—concerning the geometrical notions of parallel lines and asymptotes, of which he notably learned from Peletier himself (in person as well as in writing). These complex and somewhat contradictory notions led him to see in certain abstract geometrical concepts, in particular those that implied the notion of infinite, and,

---

nihil prius neque posterius. Immò puncta ante lineas: aut lineas ante Superficies: aut denique superficies ante corpora fuisse, vix cogitatio ipsa complecti potest. Sicut apud Philosophos, Universum aliquando non fuisse, captum animorum excedit: semper fuisse, supra omnem admirationem est. Nos autem, quantum cogitatione assequimur, omnia suo ordine statuere, atque ad artem reducere: conamur iudicio quoad eius fieri potest, probabili, minimeque fallaci. [...]. Sed nos haec rerum varietas exercet: in qua satis nobis est coniecturam ad usum accommodare. Quid enim nos efficere posse putamus arte, in ijs quae Natura tam affabrè fecit? aut quid ingenio consequi, quum de his quae divinitùs emanarunt, humanitùs iudicamus". On this passage and its ontological and epistemological implications, see also Axworthy (2013) and *infra*, p. 124.

[76] Proclus (Friedlein 1873, p. 153) and (Morrow 1992, p. 122).

[77] Cusa (Koch and Bormann 1972, p. 7), *De coniecturis*, I.1.5: "Coniecturas a mente nostra, uti realis mundus a divina infinita ratione, prodire oportet. Dum enim humana mens, alta dei similitudo, fecunditatem creatricis naturae, ut potest, participat, ex se ipsa, ut imagine omnipotentis formae, in realium entium similitudine rationalia exserit. Coniecturalis itaque mundi humana mens forma exstitit ut realis divina." On Cusanus' influence on Peletier, see Staub (1967, pp. 20–21) and Arnaud (2005, pp. 50 and 73–77).

from there, in geometry in general, a dangerous attempt at rivalling theology and at corrupting human morality by overthrowing intuition and worldly experience.[78]

Thus, for Peletier, the distinctions and the causalities the geometer would establish between the different types of magnitudes, by appealing to motion or to any other conceptual means (such as suppression of dimensions), are just conjectures or artificial means to help us understand the properties of these objects in a manner appropriate to our understanding and to adapt this knowledge to our needs. So if, on a philosophical level, the notion of flow of the point referred, for him, to the primordial procession of the multiple from the divine One, which is not fully accessible to our understanding, on a mathematical and properly scientific level, the flow of the point would only correspond to an imaginary and spatialised representation of the relation between the point and the line, which is then regarded as absolutely distinct from the arithmetical relation of the number to the unit.

* * *

In his commentary on the *Elements* and in his *Louange de la Sciance*, Peletier showed that, like Fine, he considered genetic definitions as relevant to geometry. In this context, these would have the function of teaching the mode of production (or description) of the defined magnitude while accounting for its mathematical properties.

Among the various terms Peletier used to describe this process, the term *fluxus* was exclusively used for the generation of the line, while he employed the terms *ductus* (as well as the term *motus*, indirectly) for the generation of the surface and of the body. This distinction, which was maintained in both works, have different ontological implications, which were acknowledged by Peletier. Indeed, while *ductus* designated a properly mathematical, spatial and operative process, *fluxus* possessed, in addition to a spatial understanding, a metaphysical meaning. On the philosophical or metaphysical level, the flow of the point generating the line pointed to the primordial and instantaneous emanation of the multiple from the divine One, which was also represented by the unitarian principles of numbers and physical bodies, namely, the unit and the atom.

In the context of geometry, the flow of the point generating the line had, for Peletier, the same status as the leading or motion generating plane and solid figures, such as the parallelogram, the circle and the sphere. However, Peletier did not go as far as to synonymously designate the flow of the point by the terms *motus* or *ductus*, nor did he apply this notion of flow of the point to the explanation of Euclid's two first postulates, as had done Fine. It is very likely his awareness of the metaphysical origin of the notion of flow of the point, and his faithfulness to the Platonic ontology of mathematics,[79] that led him to avoid using the terms *motus* or *ductus* to describe the generation of the line by the point.

---

[78] On Montaigne's views on the geometry of parallel lines and asymptotes, see Dear (1995b, pp. 52–53) and Calhoun (2017).

[79] See, for instance, Peletier (1557, sig. 4v): "De cuius initijs huc nihil afferre constitui. Non ab Aegyptiis, non à Chaldaeis, non à Phoenicibus, illius originem requiram. Scientias quippè aeternas esse semper existimavi: atque ut in Mente divina, ab aeterno infixam fuisse Mundi constitutionem: sic

In Peletier's conception, the metaphysical meaning of the notion of flow or emanation of quantity from an indivisible element offers an ontological foundation for the geometer's appeal to genetic definitions in geometry. It legitimates also the conception of the point as the origin of magnitude, in spite of the logical difficulties raised by the notion of continuous quantity as caused by a partless principle. With regard to this, Peletier was careful to avoid assimilating the generation of the line by the flow of the point to the infinite multiplication of the point in a geometrical context, although he considered it admissible on a metaphysical level.

Moreover, although Peletier did compare in Book II the generation of rectangular parallelograms by the *ductus* of a line to the multiplication of a number by another and therefore related the modes of composition of numbers and of magnitudes, this comparison would only be valid, to him, as a means of illustrating and teaching certain properties and relations of geometrical objects. It was in particular useful to explain the quantitative relation between the two sides of a rectangle parallelogram and its area. Nevertheless, a perfect understanding of the relation between discrete and continuous quantity would remain beyond our grasp, since, as Peletier wrote when dealing with irrational quantities, "the true estimation is unknown to the practitioner, and is only known to nature".[80] Hence, even if numbers and magnitudes, as well as all created beings, have the same origin and mode of composition in God's mind and in the constitution of the universe, their true connection would be inaccessible to man, just as the spatial relation between point, lines, surfaces and solids. Indeed, as Peletier explained in his commentary on Df. I.1, lines, surfaces and solids are regarded by the geometer as heterogeneous by nature, the parts of lines, surfaces and solids being apprehended as lines, surfaces and solids, respectively.[81]

It remains that, in Peletier's commentary on Euclid, no difficulties were raised against genetic definitions on account of the essential connection established by Aristotle between motion and physical matter, of which mathematical objects were considered to be deprived. On the contrary, by adopting in a philosophical context the metaphysical interpretation of the flow of the point defended by the Pythagorean and Neoplatonic doctrines, he was free to connect the mode of composition of geometrical solids and of physical bodies from the point and the atom, respectively, in the constitution of nature. Still, on an epistemological level, the full nature of this relation would remain inaccessible to the human mind.

---

disciplinas, caelestia quaedam semina esse: quae in nobis insita, & pro rata cuiusque portione exculta, fructum edunt". This passage describes geometrical notions as innately set within our souls, appealing to the Platonic doctrine of innate ideas (as set forth in *Meno* 81b–86a and in *Phaedo* 75c–77a) and of the primordial role of mathematics in the universe created by the Demiurge (as set forth in the *Timaeus*). On Peletier's Platonic influences, see Demerson (1975), Pantin (1984) and Axworthy (2013).

[80] Peletier (1573, 33): "Car la vraye estimation est inconnue à l'artifice: & connue à seule nature." See also *supra*, n. 73, p. 92.

[81] Peletier (1557, p. 2), Df. I.1: "Sed quia Magnitudinum partes, naturam totius denominationemque retinent, partes enim Linearum, lineae sunt: Superficierum, superficies: & Corporum, corpora: alioqui vaga & confusa esset rerum substantia". On this aspect, see also Axworthy (2013).

Now, if Peletier admitted the use of genetic definitions in a geometrical context, he nevertheless expressed clear reservations concerning the admissibility of the non-generative kinematic process that was associated with geometrical superposition on the basis of its allegedly mechanical and non-geometrical character. As I have shown elsewhere,[82] Peletier did not reject this procedure in view of its appeal to motion, but because this non-generative motion would involve the free or non-regulated transport of a figure from one place to another and would therefore not be able to demonstrate the equality of the figures by a rational deduction, but only through empirical judgment. On the other hand, the motion involved in genetic definitions, such as that of the circle, would show how a figure may be generated bearing all the attributes stated in the definition by property. This motion would then rationally establish and display the quantitative relation between the whole and the part within the generated figure.

Hence, although Peletier found it useful to introduce genetic definitions in his exposition of Euclid's geometrical books in order to teach the spatial properties of geometrical figures, as he did also in the context of an arithmetical understanding of the definitions of Book II, he was attentive to the philosophical as well as to the mathematical implications of motion and genetic definitions in a geometrical framework. For this reason, he took care in his commentary on Euclid to convey the conjectural nature of kinematic notions, in addition to their utility for the constitution of geometry.

---

[82] Axworthy (2018).

François de Foix-Candale

4

4.1 The Life and Work of François de Foix-Candale and the Significance of His Commentary on the *Elements*

François de Foix, count of Candale (Bordeaux, 1512—Bordeaux, 1594),[1] was a French humanist, mathematician,[2] engineer[3] and alchemist.[4] He was born in a powerful family of the nobility of Guyenne and of the city of Bordeaux and occupied as such several ecclesiastical functions, notably as Bishop of Aire-sur-Adour, in the region of Bordeaux (from 1576 to 1594).[5] He also inherited a lordship as "Captal" of Buch, in the province of Gascony. He was a relative of the King Henri IV (Henri de Navarre),[6] as well as a friend of Michel de Montaigne and of the mathematician Élic Vinet (1509–1587).[7]

---

[1] On Foix-Candale's life, education and work, see Dagens (1951), Harrie (1975, pp. 9–15 and 27–63) and Faivre (2008).

[2] As a mathematician, Foix-Candale notably participated to the Pope's reform of the calendar and would have entered in a disagreement with Clavius on this issue (Harrie 1975, pp. 39–41).

[3] Foix-Candale would have invented mechanical devices and measuring instruments (Harrie 1975, pp. 35–37).

[4] On his activities in alchemy, engineering and applied mathematics, see Harrie (1975, pp. 35–37, 42–44, 122–127 and 175–185).

[5] On these various ecclesiastical positions, see Harrie (1975, pp. 55–61).

[6] On Foix-Candale's kinship to Henri de Navarre, see Berger de Xivrey (1843, pp. 77–78) and Harrie (1975, p. 36).

[7] Harrie (1975, pp. 48–49); Limbrick (1981).

© The Author(s), under exclusive license to Springer Nature Switzerland AG 2021
A. Axworthy, *Motion and Genetic Definitions in the Sixteenth-Century Euclidean Tradition*, Frontiers in the History of Science,
https://doi.org/10.1007/978-3-030-95817-6_4

Candale was a promoter of the mathematical sciences. He founded a chair of mathematics at the Collège de Guyenne in Bordeaux in 1591[8] and held a "grand salon littéraire" at his castle of Puy-Paulin, where he gathered mathematicians, humanists and other scholars from the region of Bordeaux and beyond.[9] He published in 1566 a Latin edition and commentary of the fifteen books of Euclid's *Elements* (that is, including the apocryphal Books XIV[10] and XV),[11] to which he added a sixteenth book on the regular polyhedra,[12] as well as a short treatise on mixed and composed regular solids.[13] Several of his successors (among whom Billingsley, Clavius and Isaac Barrow) drew in part or in totality from these additional books for their edition of the *Elements*.[14] Foix-Candale's commentary was reprinted in 1578. He added at this occasion a seventeenth and an eighteenth book, where he pursued his investigation of the properties and relations of the regular polyhedra.[15] Jean Bodin (ca. 1530–1596) described Foix-Candale as the "French Archimedes" (*l'Archimede des François*)[16] and Guy Lefèvre de la Boderie (1541–1598) wrote of him that he "surpassed all scholars in the princely and royal art of mathematics."[17]

---

[8] Harrie (1975, p. 61).

[9] Harrie (1975, pp. 47–48).

[10] This book was authored by Hypsicles (c. 190–c. 120).

[11] Foix-Candale (1566 and 1578).

[12] Foix-Candale (1566, fol. 192r–101v): *Francisci Flussatis Candallae Elementorum geometricorum Liber decimussextus.*

[13] Foix-Candale (1566, fol. 102r–104r): *De mixtis et compositis regularibus solidis.*

[14] Billingsley (1570, fol. 445v–458r): *The sixtenth booke of the Elementes of Geometrie added by Flussas* and Billingsley (1570, fol. 458v–463r): *A briefe treatise, added by Flussas, of mixt and composed regular solides*; Clavius (1611–1612, fol. 610–637): *Elementum decimumsextum, Quo variae solidorum regularium sibi mutuo inscriptorum, & laterum eorundem comparationes explicantur, a Francisco Flussate Candalla adiectum, & de quinque corporibus*; Barrow (1751, p. 325): "A Brief Treatise (Added by Flussas) of Regular Solids." On the reception of Foix-Candale's additional books, see Harrie (1975, pp. 174–175).

[15] Foix-Candale (1578, pp. 507–536): *Francisci Flussatis Candallae Elementorum geometricorum liber XVII, Qui et solidorum regularium compositorum primus* and Foix-Candale (1578, pp. 537–575): *Francisci Flussatis Candallae Elementorum geometricorum liber XVIII, Qui et solidorum regularium compositorum secundus.*

[16] Bodin (1597, p. 367); Harrie (1975, p. 27). This expression was taken up by several of his contemporaries (Harrie, 1975, p. 35, n. 38 and n. 40). It is to be noted that Jean Bodin authored one of the prefaces of Foix-Candale's commentary on Euclid (Foix-Candale 1566, sig. A4r). On the relation between Foix-Candale and Bodin, see Harrie (1975, pp. 52–53).

[17] Le Fèvre de la Boderie (1578a, fol. 32r): "Toy De Candale ardent plus clair que la chandelle/Qui le monde illumine, & chasse la nuict d'elle,/Qui es Prince de nom, de vertu, & de sang,/Et *entre les sçavans qui tiens le premier rang/En l'Art prince & royal de la Mathematique*/Dont tu sçais rapporter les regles en pratique." (My emphasis.) See Harrie (1975, p. 34).

Foix-Candale also published in 1574 a Greek-Latin edition,[18] as well as a Greek-French translation the same year,[19] of Hermes Trismegistus' *Poimandres* (or *Pimander*) and an extensive commentary in French of the same work in 1579.[20] These were described by A. Faivre as one of the most representative works of the Neo-Alexandrian Hermetic tradition of the second half of the sixteenth century.[21]

Foix-Candale's commentary on Euclid was the first extensive commentary on the entire text of the *Elements* that was written and printed in France.[22] Its additional books on the regular polyhedra, which were taken up in various later commentaries on Euclid, including those of Billingsley and Clavius, earned it an international reputation within the early modern Euclidean tradition. These books, and their treatment of the regular polydra, were notably said to have influenced Johannes Kepler (1571–1630).[23] Within the tradition analysed here, and within its French branch in particular, Foix-Candale's commentary on Euclid is related to that of Peletier both because of its underlying Platonic and Neoplatonic influences and because Foix-Candale followed Peletier in rejecting superposition as an admissible procedure in geometry.[24]

---

[18] Foix-Candale, *Mercurii Trismegisti Pimandras utraque lingua restitutus, D. Francisci Flussatis Candallae industria*. Bordeaux: Simon Millanges, 1574.

[19] *Le Pimandre de Mercure Trismegiste nouvellement traduict de l'exemplaire grec restitué en la langue françoyse par François Monsieur de Foys de la famille de Candalle*. Bordeaux: Simon Millanges, 1574.

[20] *Le Pimandre de Mercure Trismegiste de la philosophie chretienne, cognoissance du Verbe divin, et de l'excellence des œuvres de Dieu traduict de l'exemplaire grec, avec la collation de commentaires, par François Monsieur de Foix*. Bordeaux: Simon Millanges, 1579. On this work, its context of production and publication, as well as its reception, see Dagens (1951), Dagens (1961), Yates (1964, pp. 173, 179, 182 and 406), Limbrick (1981), Faivre (2008), Moreschini (2009) and Giacomotto-Charra (2012).

[21] Faivre (2008, p. 377). See also Dagens (1951), Harrie (1975) and Limbrick (1981). On the reception of this work in the early modern period, see Harrie (1975, p. 133–135).

[22] Prior works that were published in France dealt with all fifteen books, but these consisted mainly in editions without added commentary. This was the case of J. Lefèvre's compared edition of Campanus and Zamberti (Lefèvre 1516), or of the editions of Petrus Ramus (Pierre de la Ramée) (1515–1572) (only enunciations) (Ramus 1545) or of Stephanus Gracilis (fl. 1550–1580) (which also contains the commentary on the tenth book by Pierre de Montdoré (ca. 1505–1570)) (Gracilis 1557).

[23] Kepler (1993, I, p. 46). See Westman (1972) and Mehl (2003).

[24] Foix-Candale (1566, fol. 5v), Prop. I.4: "Alteram demonstrationem huic quartae exhibere cogimur, ne praebeatur aditus, quo *ulla mechanicorum usuum instrumenta in demonstrationes incidant*. Nam Campanus ac Theon hanc demonstrantes, triangulum triangulo superponunt, angulumque angulo, sive latus lateri, demonstrationem potius instrumento palpantes, quàm ratione firmantes: quod tanquam prorsus alienum à vero disciplinarum cultu reijcientes, aliam demonstrationem absque figurae, anguli seu lineae transpositione, protulimus ratione elucidatam". Cf. Foix-Candale (1566, fol. 6v), Prop. I.8: "Huius alteram demonstrationis partem resecavimus eò quòd trianguli transpositione uteretur, quod quidem *mæchanicum spectat negotium à vera mathesi alienum*, posita anguli qui ad z hypothesi ex quarta huius sumpta." and Foix-Candale, (1566, fol. 27r), Prop. III.24:

Beside the commentary on the *Elements*, the *Pimandre*, that is, Foix-Candale's 1579 commentary on the *Poimandres*, represents his most significant work and a crucial contribution to Renaissance Hermeticism. It was, for him, the occasion to display his thought on a variety of issues and domains, from theology, to natural philosophy and mathematics, and therefore constitutes a valuable complementary source to approach his philosophy of mathematics. For this reason, this text will also be considered here insofar as it will allow us to better understand his discourse on the status of geometrical objects in the commentary on Euclid.

After presenting Foix-Candale's treatment of genetic definitions in his commentary on the *Elements*, this chapter will thus look at the connections between this work and the *Pimandre* with respect to the issue at stake. This will be the occasion to explore the relationship between the conceptions of Peletier and Foix-Candale concerning the geometrical notions of sphere and circle and the epistemic status of geometrical knowledge. As in the previous chapters, I will briefly look at the use that was made of geometrical motion in the commentary on Df. II.1.

As we will see, Foix-Candale's treatment of genetic definitions is very different from those of the other authors considered here, even in comparison to Peletier, given that he did not take up genetic definitions that convey the translation or rotation of a point or a magnitude of lower dimension, even when defined in terms of flow. Similarly to Peletier, his treatment of genetic definitions reveals both the convergence and the frictions that existed, in this framework, between the mathematical and the metaphysical understanding of geometrical objects and of their geneses.

"Quoniam Theon & Campanus hanc demonstrare conati sunt, aut hi à quibus demonstrationes sumpserunt, *instrumento ferè mechanico, nempe coaptata figura supra figuram*, quod indignum traditione mathematica supramodum existimatur. [...] Campanus verò unam sectionem per puram alterius *superpositionem, tanquam instrumento mechanico metitur ut aequalem probet, quod esse argumentum verè mechanicum*, patet. [...] Quare non intelligit figuras superponendas figuris ut aequales aut inaequales percipiantur, sed figurarum aut aliorum quorumvis subiectorum quantitates, ratiocinante argumento convenire cognitae, adinvicem sibimetipsis illae quantitates aequales dicentur, non autem quae experimento congruere palpantur, illae aequales dici debeant. Mathesis enim ex praeassumptis certis necessariò concludit, non autem ex sensibus externis praxim operantibus saepius fallacem". See also Foix-Candale (1566, sig. e3r): "Propositiones demùm à principiis genitae, hoc praescripto disponantur, ut subsequentes à prioribus, non autem à posterioribus demonstrandae sint, ac earum demonstrationes solis disciplinæ legibus construendae tanta religione decorentur, *ne ullum in eis demonstrandis intercedat mecanici instrumenti iuvamen*." (My emphasis.)

4.2 Formulation and Distribution of Genetic Definitions in Foix-Candale's Commentary on Euclid

The treatment reserved to genetic definitions by Foix-Candale, in his commentary on the *Elements*, stands out among those of other commentators considered here by the fact that he did not describe the generation of the line or of the surface through the flow or motion of a point or of a line, but rather described this process as a forward motion, a stretching or an expansion of the line or of the surface itself. Nevertheless, he still presented the point as the efficient principle of magnitude ("punctum quantitatem agens") in his commentary on Df. I.1.[25]

This particular mode of generation of magnitude, for which he used the term *progressus*, is set forth in the commentary on Df. I.3 and I.6 of the *Elements*, which respectively define points as the limits of the line and lines as the limits of the surface.

> *The limits of the line are points.* Points are called the limits of lines, since to limit is to stop the progress (*progressus*) of any quantity, because the length of a line, since it is the only dimension in the line, assumes as a limit the extremity or demarcating point, which surely is also used as such by the other dimensions—these are, of course, breadth and height –, although it is none of them. That is why the progress of the line is stopped or limited by the point.[26]

---

[25] Foix-Candale (1566, fol. 1r), Df. I.1: "signum esse intelligat, quantitatem agens". If the text defines quantity and not magnitude specifically as the object of geometry, it is because, as will be shown later, Foix-Candale understood geometry and arithmetic as sharing a common origin (see *infra*, n. 159, p. 132). He also considered that geometry has a larger scope than arithmetic, given that it concerns both commensurable and incommensurable quantities, although its principles cannot be correctly understood by the human intellect without the help of arithmetic because of the "confused" nature of continuous quantity: (Foix-Candale, 1566, sig. e3r): "diximus enim Geometriam & Geometricarum & Arithmeticarum magnitudinum comprehendere respectus, non tamen eam facilitatem qua per discretionem illae exprimuntur habitudines" (on this issue, see *infra* pp. 135–136). Hence, geometry is defined in this context as the science of quantity: "Quia geometriam, agimus cuius unicum obiectum est quantitas". A similar assertion appears in the preface (Foix-Candale, 1566, sig. e2v): "Geometriam igitur nihil aliud esse sentiemus, quàm artem qua maioris, minori, & aequalis patefit natura, eius verò Geometriae unicum esse obiectum quantitatem." This is also corroborated by the title of Foix-Candale commentary on the *Elements*, which presented Euclid's *Elements* (all fifteen books) as a geometrical treatise, and not as a treatise of arithmetic and geometry: *Elementa Geometrica, Libris XV. Ad Germanam Geometriae Intelligentiam è diversis lapsibus temporis iniuria contractis restituta, ad impletis praeter maiorum spem, quae hactenus deerant, solidorum regularium conferentiis ac inscriptionibus.*

[26] Foix-Candale (1566, fol. 1r), Df. I.3: "*Lineae autem limites sunt, signa.* Limites lineae signa vocat, eò quòd limitare sit alicuius quantitatis *progressum* terminare, quia verò longitudo lineae cùm sit sola in linea dimensio, sumit sibi terminum limitem, sive designatorem signum, quod quidem aeque ut illa utitur reliquis dimensionibus, latitudine scilicet et altitudine, hoc est nulla. Terminatur itaque aut limitatur lineae *progressus* signo." (My emphasis.)

> *The extremities of the surface are lines.* We will say that the progress of the surface is stopped or limited by the line for the same reason we said that the progress of the line is stopped or limited by the point.[27]

This process of expansion, which is applied here to both the generation of the line and the generation of the surface, is also indirectly referred to in Df. I.13, which defines the boundary. In this context, Foix-Candale conveyed this notion through the idea of a figure stretching from its centre to its periphery, a process which is then termed *motus*.

> *A boundary is the extremity of something.* Since the extremities of any quantity enclose it by its boundaries, we say that the boundaries are these extremities which constrain magnitude. Some certainly call the boundary "end" because, by the motion which is assumed in the displayed magnitude, which goes from the smallest part to the extremities, we reach by this motion any extremity once we have found the end of this motion.[28]

To a certain extent, the idea of a figure caused by the internal stretching of its magnitude brings on a quasi-physical notion of motion and expansion, that is, as temporally-determined, since the boundary of the figure would appear at the point in time when the motion (i.e. that by which the figure is understood to unravel) ends. However, for Foix-Candale, the temporal factor would not be equivalent to that which is considered in physical substances and motions, as this expansion would only be held to take place in the mind and imagination of the mathematician, where the generation of the defined figure takes place without any determination of quantity, place or velocity. Moreover, as will be shown further, this process also evokes an ontologically superior state of magnitudes, which may be physically represented by the propagation of light, as suggested by the role Foix-Candale attributed to light in the divine creation and communication of intelligible forms in his *Pimandre*.[29]

---

[27] Foix-Candale (1566, fol. 1v), Df. I.6: "*Superficiei extrema, sunt lineae.* Qua ratione diximus lineae *progressum* signo terminari sive limitari, sic dicemus superficiei *progressum* terminari vel limitari linea." (My emphasis.)

[28] Foix-Candale (1566, fol. 2r), Df. I.13: "*Terminus est, quod alicuius est extremum.* Quoniam cuiusvis quantitatis extrema eam concludunt suis terminis: dicemus terminos esse ea extrema quae magnitudinem coarctant: quem quidem terminum aliqui finem vocarunt, eò quòd supposito in oblata magnitudine motu, ab intima parte ad extremas, ubi extremum aliquod eo motu attigimus, ibi huius motus finem reperimus." (My emphasis.)

[29] On this aspect, see *infra*, p. 121.

4.3   *Progressus* **Versus** *Fluxus*

In Foix-Candale's commentary on Euclid, the notion of flow (or motion) of the point, of the line or of the surface to generate the line, surface or solid, respectively, was thus somehow replaced by the idea of a generation of magnitudes through a continuous stretching or expansion of the same type of magnitude within the imagination. In this context, the point, the line or the surface, when they are defined as the extremities or ends of the line, the surface or the body is only what stops the progression of the considered magnitude and not what causes its generation.[30] This is confirmed by the fact that, in order to explain the interpretation of the spatial boundary of a figure as an end—*terminus* expresses both concepts—in the commentary on Df. I.13, Foix-Candale suggested that, in a mathematical context, the end or the boundary (*extremum*) of a figure coincides with the end (*finis*) of the motion through which it is assumed to be generated.[31]

The use of the notion of *progressus* as a means of describing the generation of magnitudes may have aimed at avoiding the difficulties raised by the representation of the extremity of a given magnitude (a point, a line, a surface) as its efficient and material cause, as it was incompatible with the Aristotelian notion of continuum. This is marked by the fact that Foix-Candale wrote, in his commentary on Df. I.1 and in his preface, that the point, although conceived as the cause of quantity, is not a part of it, just as the instant is not a part of time.[32]

However, it is not entirely clear whether Foix-Candale conceived that the extremity or boundary is imposed from outside to stop the stretching of the line or figure, or whether it is itself caused by the cessation of the expansion. As a matter of fact, the point or the line, even when conceived as an extremity, would not correspond to an actually discrete element of the line or figure, just as the centre of a figure would not itself be regarded as an actually separate part of the figure's magnitude.[33] On the other hand, according to the formulation

---

[30] Foix-Candale (1566, fol. 1r), Df. I.3: "Limites lineae signa vocat, eò quòd limitare sit alicuius quantitatis progressum *terminare* [. . .]. *Terminatur* itaque aut *limitatur* lineae progressus signo." and (fol. 1v), Df. I.6: "Qua ratione diximus lineae progressum signo *terminari* sive *limitari*, sic dicemus superficiei progressum *terminari* vel *limitari* linea." (My emphasis.) See also *supra*, n. 26, p. 103 and 27, p. 104.

[31] Foix-Candale (1566, fol. 2r), Df. I.13: "quem quidem terminum aliqui finem vocarunt, eò quòd supposito in oblata magnitudine motu, ab intima parte ad extremas, ubi extremum aliquod eo motu attigimus, ibi huius motus finem reperimus." See *supra*, n. 28, p. 104.

[32] See, on this issue, Foix-Candale's commentary on Df. I.1, in Foix-Candale (1566, fol. 1r), Df. I.1: "signum esse intelligat, quantitatem agens: Nullam autem partem, sed tantùm animi conceptum: Cuiuslibet quantitatis partem optatam, vel locum limitantem, dividentem, aut designantem. Quod equidem im temporum quantitate instans, veluti im pondere, momentum, idem esse dicemus, quae nihil metiuntur sed tantùm limitant." See also, in Foix-Candale's preface, Foix-Candale (1566, sig. e3r): "nec signum magnitudinis, nec instans temporis, neque momentum ponderis, quidpiam in se habeant, sed tantùm principia, termini, sive limites sunt, à quibus eorum nascitur intelligentia."

[33] See the previous note.

of the commentary on Df. I.3 ("the progress of the line is *stopped or limited by the point*"), the notion of *progressus* clearly suggests an exterior intervention of the point or extremity.

In this regard, it is important to note that the discourse Foix-Candale held on these matters evokes that of Proclus, in his own commentary on Df. I.3, as he wrote: "when Euclid says that the line is limited by points, he is clearly making the line as such unlimited, as not having any limit because of its own forthgoing (πρόοδος)".[34] Indeed, Foix-Candale's notion of *progressus* seems to have a similar meaning to Proclus' πρόοδος or forthgoing of the line, for which the point represents the agent of its interruption. Now, in Proclus' commentary, which Foix-Candale very likely knew[35] and positively received in view of his strong adhesion to Platonic and Neoplatonic ideas, the notion of an indefinitely progressing line interrupted by a point is founded on the admission that all beings, and all mathematical objects for that matter, participate in the two metaphysical principles of the Unlimited (ἄπειρον or ἀπειρία) and the Limit (πέρας).[36] While the Unlimited provides infinite potentiality to the universe and to all the things it contains, the Limit endows them with a definite essence and unity, allowing them to participate in the perfection of the divine One.[37] Thus, in geometry, the Unlimited provides magnitudes with the capacity to extend indefinitely, while the Limit bounds this progression and endows it with determination and unity.

In Proclus' commentary, the operation of these principles is however expressed differently depending on the ontological level at which geometrical objects are taken. Indeed, as he wrote in his commentary on Df. I.3, "in imagined and perceived objects the very points that are in the line limit it, but in the region of immaterial forms the partless idea of the point

---

[34] Proclus (Friedlein 1873, p. 101) and (Morrow 1992, p. 82). On the meaning of this notion in Proclus' thought, see *infra*, pp. 128–129.

[35] I have not found any mention of Proclus in Foix-Candale's works, but it would be unlikely that he did not know his commentary on Euclid, either in the Greek version published by Grynaeus in 1533 or in the Latin translation by Francesco Barozzi (1537–1604) published in 1560 (Barozzi 1560). Generally speaking, Foix-Candale rarely quoted other authors in the context of his commentary on Euclid. Notable exceptions are Theon of Alexandria, Campanus and Bartolomeo Zamberti, to whose translations or additions he regularly referred.

[36] Proclus (Friedlein 1873, p. 5) and (Morrow 1992, p. 4): "To find the principles of mathematical being as a whole, we must ascend to those all-pervading principles that generate everything from themselves: namely, the Limit and the Unlimited. For these, the two highest principles after the indescribable and utterly incomprehensible causation of the One, give rise to everything else, including mathematical beings."

[37] On the Limit and the Unlimited and their role as cosmogonic principles in Proclus' thought, see also the *Elements of Theology*, Prop. 89, in Proclus (transl. Dodds 1933, p. 83): "*All true Being is composed of limit and infinite.* For if it have infinite potency, it is manifestly infinite, and in this way has the infinite as an element. And if it be indivisible and unitary, in this way it shares in limit; for what participates unity is finite. But it is at once indivisible and of infinite potency. Therefore all true Being is composed of limit and infinite." (Emphasis proper to the original edition.)

has prior existence".[38] This means that, in the realm of imaginary objects with which the geometer deals and to which local motion may be attributed, magnitude and its limit are distinct and the point intervenes as the agent of the interruption of the indefinite growth of the line. But, in its intelligible and indivisible form, the point is both what produces magnitude and what limits it,[39] conveying the intelligible notion of flow of the point which the imaginary motion of the point intends to imitate, according to Proclus' commentary on the postulates.[40]

As we will see further, both these meanings may be applied to Foix-Candale's conception of the generation of geometrical objects, given that he distinguished a mathematical and a metaphysical state of magnitudes, as had Peletier.[41] Hence, it seems that, in the context of these Euclidean definitions, Foix-Candale conceived the terminating point or the boundary as agents of the interruption of the forthgoing of magnitudes. These would have a proper existence only in the geometer's imagination, where such imaginary representations are used to study the relation between magnitudes and their parts.

However, Foix-Candale did not go as far to admit, in this context, that the point itself is the agent of the generation of the line.[42] He may have intended thereby to avoid the problems raised by this notion with regard to the composition of the continuum. For, if one should leave aside the issue of the composition of magnitude, the conception of the generation of magnitudes as occurring according to the mode of a *progressus* is not incompatible, on a mathematical level, with the concept of line or surface as produced by the flow of a point or of a line, respectively.[43] These two perspectives are actually found in Proclus' commentary,[44] as well as in other sixteenth-century commentaries on Euclid's *Elements*, as those of Fine, Peletier and Billingsley. Indeed, these authors all explicitly defined the line as resulting from the flow of the point (in Df. I.1 and I.2), but also hinted at the generation of the line in their commentary on Df. I.3 in a way which suggests that the

---

[38] Proclus (Friedlein 1873, p. 101) and (Morrow 1992, p. 83).

[39] Proclus (Friedlein 1873, p. 101) and (Morrow 1992, p. 83): "As it [the point] goes forth from that region, this very first of all ideas expands itself, moves, and flows towards infinity and, imitating the indefinite dyad, is mastered by its own principle, unified by it, and constrained on all sides. Thus it is at once unlimited and limited—in its own forthgoing unlimited, but limited by virtue of its participation in its limitlike cause."

[40] See *supra*, p. 48.

[41] See *supra*, p. 83.

[42] He actually presented it as the principle of quantity in general (Foix-Candale 1566, fol. 1r: "quantitatem agens").

[43] Although Foix-Candale discussed genetic definitions in Book XI (the first book of the *Elements* pertaining to solid figures), as will be shown, he did not present a genetic definition of the solid analog to those he provided for the line and the surface in Df. I.3 and I.6.

[44] See, for instance, Proclus (Friedlein 1873, p. 97) and (Morrow 1992, p. 79): "*A line is length without breadth.* [...] The line has also been defined in other ways. Some define it as the 'flowing of a point', others as 'magnitude extended in one direction'."

line is the active principle of its own generation ("The line *starts* from a point and ends in a point") and that the two extreme points determine the beginning and end of this process.[45] Some commentators even presented the connection between the two perspectives, as did Clavius in his commentary on Euclid's first postulate, where he defined the line as progressing from one point to another through a direct flow ("linea recta fluxus directo omnino itinere progrediens").[46]

Nevertheless, Foix-Candale's approach on the matter is relatively different from that of the other commentators considered here. In fact, most of those who aimed at the time to account for the generation of the line when commenting on the *Elements* followed what was philosophically considered as the natural order of causation of substances (from the simplest to the most complex), which is also the order followed by Euclid for the definitions of Book I. Therefore, these other commentators generally placed the genetic definition of the line in their commentary on Df. I.1 or Df. I.2, showing how the lines derives from the point. But Foix-Candale insisted rather on the relation between the generation of the line (conceived in terms of *progressus*) and the cause of its interruption or limitation (which he called the demarcating point) and thus presented a genetic definition of the line only in his commentary on Df. I.3. The situation is the same in Df. I.6 and in Df. I.13, in which the line and the boundary are defined as the limit of the surface and of the figure, respectively.[47]

The contrast with Fine is, in this regard, compelling. Beside Foix-Candale, Fine was the only other commentator of Euclid, among those considered here, who connected generative process and the notion of boundary in his commentary on Df. I.13. Now, the generative process to which he referred in this context, as well as in his commentary on Df. I.3 and I.6, is the description of the line, the surface and the solid by the flow or motion of a point, a line and a surface, respectively.[48] Also, if Fine, like Foix-Candale, presented the line as the active principle of its generation in his commentary on Df. I.3, he also presented it as constituted by the infinite multiplication of the point,[49] which precisely raised the problem which Foix-Candale would have aimed to avoid by appealing to the notion of *progressus* of the line instead of that of flow or motion of the point.

---

[45] Fine (1536, 1), Df. I.3: "[Linea] incipit enim à puncto [...] in punctumque terminatur"; Peletier (1557, p. 2), Df. I.3: "A puncto Linea oritur, & in idem desinit"; and Billingsley (1570, fol. 1v), Df. I.3: "For a line hath his beginning from a point, and likewise endeth in a point".

[46] Clavius (1611–1612, I, p. 22), Post. 1.

[47] Foix-Candale (1566, p. 1v), Df. I.6 and (1566, p. 2r), Df. I.13 (see *supra*, n. 27 and 28, p. 104).

[48] *Cf.* Fine, (1536, p. 2), Df. I.6: "*Superficiei extrema sunt lineae.* Porrò cùm linea, ad descriptionem mota superficiei, recta fuerit, atque in longum lineae rectae uniformiter, brevissimeque traducta: fit superficies, quae plana dicitur"; and (*ibid.*, p. 4), Df. I.13: "*Terminus est, quod cuius finis est.* Utpote, punctum ipsius lineae, linea superficiei, superficies denique solidi: quemadmodùm ex eorundem abstractiva descriptione facilè colligitur." See *supra*, n. 21, p. 42.

[49] Fine (1536, p. 1), Df. I.3: "Incipit enim à puncto, & *ex infinitis conficitur punctis*, in punctumque terminatur." (My emphasis.) See *supra*, n. 22, p. 42.

4.4   The Two Definitions of the Sphere: Euclid Versus Theodosius

As we have seen, the will to guarantee the homogeneity between magnitude and its parts may have been an incentive for Foix-Candale's avoidance of genetic definitions which present the line and surface as generated by the motion of a point or a line. But he may also have avoided such generative processes because they were too close to the mode of generation of the sphere in Euclid's Df. XI.14, which he did not consider as a proper definition. Indeed, in his exposition of this definition, Foix-Candale placed the definition of the sphere proposed by Theodosius in the *Spherics*[50] before Euclid's definition and justified this by stating that the definition of the sphere commonly ascribed to Euclid did not teach the essence of the sphere, but merely its mode of generation. As in Fine's *Geometria*, the mode of generation of the sphere is then designated as a *descriptio*, in order to distinguish it from a *definitio*.

> The Sphere is a solid figure, enclosed by one surface, toward which all the straight lines led *from one point within it are equal. However, the description of the sphere is the circumduction of a semicircle around its fixed diameter until it returns to the place where it started to move.* Since the sphere is the most perfect of all solids, we do not judge unworthy to present its two expositions. And truly, through the first, we express the true definition of its substance, which is fully convertible with the defined term. However, through the second, we have defined its description, clearly demonstrating the rule according to which it should be described. Because Theon and Campanus only expressed the description of the sphere, but not the proper nature of its substance, we placed this definition, which we support by the most powerful discourse of geometry, before the description. Indeed, the perfection of the nature of this solid, by the equality of the lines proceeding from a unique point and, furthermore, by the flexion, everywhere uniform, of its unique and admirable configuration, encloses the solid so skilfully that the regularity of its perfection cannot be corrupted by any split of angles or sides, but sets forth a certain image of its ineffable eternal essence deprived of beginning and of end.[51]

---

[50] See *supra*, n. 141, p. 33.

[51] Foix-Candale (1566, fol. 140r–v), Df. XI. 12: "*Sphaera est figura solida, una superficie comprehensa, ad quam ab uno signo intus existente, omnes rectae lineae ductae sunt aequales. Sphaerae autem descriptio est, circunductio semicirculi manente eius demetiente, quoad unde coepit redeat.* Quoniam omnium solidorum perfectissimum est Sphaera, ipsis duplicem non dedignamur conferre expositionem: priori etenim veram eius substantiae diffinitionem expressimus, cùm diffinito admodum convertibilem. Posteriori verò eius descriptionem diffinivimus, qua lege describenda sit planè demonstrantem: quia Theon ac Campanus Sphaerae solam descriptionem, non autem propriam substantiae naturam depinxerunt, descriptioni eam quam ex potissima geometriae sententia suscepimus diffinitionem praefecimus, quae quidem naturae huius solidi perfectionem, ea linearum ab unico signo prodeuntium aequalitate, ac insuper unicae illius admirabilis faciei aequa undique flexione, tanta solertia solidum conclusit, ut perfectionis eius regularitas, nulla angulorum sive laterum fractura maculanda sit, sed ineffabilis illius aeternae essentiae initio exitúque carentis, aliquam prae se ferat imaginem".

Although Euclid's definition of the sphere was traditionally distinguished from Theodosius' definition insofar as it involved motion, it could be said here that, for Foix-Candale, both definitions did in a certain sense imply motion. For, in Theodosius' definition, the fact that the lines joining the centre of the sphere to the circumference are said to go or advance toward (*prodeuntium*) its periphery clearly resonates with the notion of *progressus* introduced by Foix-Candale to describe the generation of the line or of the surface in Df. I.3 and I.6, and most of all in Df. I.13 through the notion of outward expansion of the figure.

Admittedly, Foix-Candale did not formulate this in this manner in his commentary on Df. XI.14. Moreover, the classical formulation of the Theodosian definition of the sphere, as that of the Euclidean definition of the circle, generally contained a reference to lines *going* or *drawn* from the centre to the circumference, without this having been interpreted as a generative process which would be the cause of the sphere or the circle.

Nevertheless, based on Foix-Candale's commentary on Df. I.13, where the figure in general is conceived as generated by the outward motion or expansion of the magnitude of the figure from its centre until this process is stopped by its boundary, Theodosius' definition of the sphere may be interpreted as signifying the outward motion or expansion of lines which uniformly proceed (*prodeunt*) in all directions from the centre of the sphere to its periphery. Hence, according to this conception, one would be led to assume through Theodosius' definition of the sphere a motion starting from the centre or smallest and innermost part of the figure to its extremities ("supposito in oblata magnitudine motu, ab intima parte ad extremas"). Through this motion, the tridimensional magnitude of the sphere would expand in all directions until this expansion is stopped by its bounding surface.

Admittedly, what is said to proceed or expand from the centre of the figure, according to the definition of Theodosius, is not the tridimensional magnitude of the spherical body, but merely the lines joining the centre of the sphere to its bounding surface. This could be held as problematic with regard to the composition of the continuum, as the lines would then be thought to constitute the whole volume of the sphere. This definition would therefore not be better than a definition of the line as generated by the motion of a point. However, the conflict between the terms used in the Theodosian definition and the mode of generation of the sphere interpreted according to Foix-Candale's commentary on Df. I.13 is only apparent. For what this definition aims to state in the first place (as Foix-Candale would have acknowledged) are the quantitative and spatial properties of the sphere, which are determined by the equality of all the lines situated between the centre of the sphere and its boundary.

But although Foix-Candale referred to this process of expansion of magnitudes in a geometrical context, where it would be held as mathematically relevant, it remains that it would also bear a philosophical meaning, relating to the ontological status of geometrical objects, even if only underlyingly expressed. Indeed, as stated at the end of the above-quoted passage, through the forward motion of the lines from the centre to the circumference of the sphere, Theodosius' definition is able to account for the fact that the sphere possesses no angles and is deprived of beginning and end, conveying thereby an image of its ineffable and eternal essence.[52]

By contrast, Euclid's definition, which defines the sphere as generated through the rotation of a semicircular surface on its axis and which Foix-Candale interpreted as a later addition by Theon of Alexandria (ca. 335–ca. 405) taken up by Campanus, would not be able to express this suprasensible essence of the sphere. And this would not be because it appeals to motion, but because the type of motion it involves does not convey the essential mode of generation of the sphere. This definition would rather express an accidental or extrinsic mode of generation, which is not dictated by the true and essential mode of being of the sphere. As Foix-Candale wrote here, the *descriptio* of the sphere expresses the "rule" or precept instructing how the sphere should be produced or, more literally, drawn out (*lex describenda*) by the geometer, and not the mode of generation that would be conform to its essence.

Thus, by distinguishing in this manner the Euclidean and Theodosian definitions of the sphere, Foix-Candale did not only distinguish two modes of definition of geometrical figures, but also two modes of generation of geometrical figures: one that would be intrinsic and essential to the figure and one that would come about only through the will and action of the geometer, be it carried out instrumentally or in the imagination. In other words, Foix-Candale's motivation for privileging a mode of generation of geometrical objects through the expansion of magnitude in one, two or three dimensions, rather than through the translation of a magnitude of lower dimension, would also be related to the instrumental, extrinsic and non-essential character of the latter, as it would be unsuited to the ontologically higher status and origin of geometrical objects. In the case of Euclid's definition of the sphere, the figure would indeed correspond to a solid generated by the translation, or for that matter the rotation, of the surface of a plane figure around one of its sides. This position would be confirmed by Foix-Candale's rejection of geometrical superposition in view of its alleged mechanical character, since he took it to be instrumentally performed and to subvert, because of this, the purely rational and abstract nature of mathematical demonstrations.[53]

4.5   The Commentary on Euclid and the *Pimandre*

In the commentary on the *Elements*, Foix-Candale remained rather laconic on the topic of the ontological status of geometrical figures. Yet, the fact of attributing to the sphere, such as expressed by Theodosius' definition, an "eternal and ineffable essence" resonates with

---

[52] Foix-Candale (1566, fol. 140r–v), Df. XI. 12: "ineffabilis illius aeternae essentiae initio exitúque carentis, aliquam prae se ferat imaginem". See *supra*, p. 109.

[53] See *supra*, n. 24, p. 101.

what he would later write about this figure in the *Pimandre*, that is, his commentary on the *Poimandres*, which corresponds to the first part of the *Corpus hermeticum*.

This commentary was certainly published by Foix-Candale in 1579, that is, thirteen years after the publication of his commentary on the *Elements*, but there are reasons to think that he already adhered to this doctrine, at least in part, when he published his commentary on the *Elements* in 1566, as will be shown further.[54] It is important at least to note that the *Pimandre* was actually written earlier then 1579, since the preface indicated that it had been completed by 1572.[55] Its publication would have been delayed because of the political and social unrest resulting from the massacre of St Bartholomew's day. Moreover, although it is unclear when Foix-Candale started to work on this commentary, it appears to have taken him quite a long time to complete,[56] which means that he may have started working on the *Pimandre* a few years before 1572, that is, during the time when or immediately after he was working on his commentary on the *Elements*. Furthermore, the fact that Foix-Candale published an augmented edition of his commentary on Euclid in 1578, with an additional book on the regular polyhedra, shows that he continued the geometrical work he had started in his 1566 commentary while he was preparing his works pertaining to the Hermetic tradition, confirming the temporal and conceptual continuity between these two aspects of

---

[54] See *infra*, pp. 115–118.

[55] Foix-Candale (1579, sig. A3v): "Ces commentaires furent prests a publier en lan 1572, & portez par nous a Paris, ou arrivantz, le 26 d'Aoust nous trouvames telz obstacles, le temps & personnes si indisposées a leur publication, que nous fumes contrainctz les raporter, n'ayans eu despuis licence tant pour les miseres universelles, que plus pour les particulieres, d'y mettre aucunement l'œil ou pensée jusques a presant." On the conditions in which this work was composed and published, see Harrie (1975, p. 46–47).

[56] The preface indicates that he was incited to write a commentary on the *Poimandres* by his brother, Frédéric de Foix-Candale (d. 1571), and by his sister Jacqueline de Foix (d. 1580). He would have read this work a number of times before starting to work on its interpretation, an endeavour that he was able to complete only when certain difficulties in his life were to cease ("selon que les empeschemens de noz misere l'ont permis") and only during the hours he was able to take from other occupations ("j'ay employé les heures que j'ay peu emprunter à ceste estude"). Foix-Candale (1579, sig. ã4v): "[nous] considererons la doctrine, que ce Philosophe divin nous presente par cestuy cy, [...] que nous avons voulu travailler de retirer l'intelligence de quelques siens propos: & ce à la persuasion de FEDERIC MONSIEUR DE FOIX, nostre frere, Captal de Buch, & conte de Candalle, homme tres-exercité aux sainctes lettres, & de Dame JAQUELINE DE FOIX notre sœur, personne retirée à la cognoissance & contemplation des choses divines. Qui apres la lecture de ce traicté l'ont estimé si excellent en sa brieveté, qu'ilz en ont grandement desiré l'interpretation. Dont nous avons prins occasion & grand desir de le voir, & l'ayant plusieur fois paßé, & reveu, avons trouvé en ce petit volume un si grand nombre, & de si profonds tesmoignages de la volonté, que despuis il a pleu à Dieu nous signifier par Jesus Christ, que le voyant abandonné de si long temps de toute manière d'expositeurs, avons esté conviés, selon que les empeschemens de noz misere l'ont permis, de prendre peine d'esclarcir les propos de ce bon Philosophe [...]. A quoy desirant obeyr tant à eux que autres aymants Dieu, j'ay employé les heures que j'ay peu emprunter à ceste estude, & à leur occasion l'ay mis en langue Françoise pour plus facile intelligence." (The emphasis is proper to the original text.)

his intellectual work. Moreover, when he founded his chair of mathematics in 1591 at the College of Guyenne, he stated that one of the conditions required of the candidates to the position of professor of mathematics was to demonstrate a new proposition on the topic of regular polyhedra.[57] Kepler, who connected the Pythagorean theory of numbers to Hermetic theses,[58] quoted Foix-Candale's work on regular polyhedra in his *Mysterium cosmographicum*,[59] hinting at the fact that he himself saw a continuity between the French humanist's commentary on Euclid and his commitment to Hermeticism.

Although it is difficult to determine the extent to which Foix-Candale adhered to Hermeticism when he wrote his commentary on Euclid, in particular as little is known of Foix-Candale's life before 1570,[60] J. Harrie, in her thesis dedicated to Foix-Candale's *Pimandre*, considered that his interest in Hermeticism held a central role in the whole of his intellectual life. His involvement in alchemy, mathematics and theology, as well as his inclination toward the Platonic, Neoplatonic and Pythagorean doctrines, would therefore be related to this chief interest,[61] as it was for other Renaissance humanists who preceded him and seem to have influenced him in this regard. These were Marsilio Ficino (1433–1499) and Jacques Lefèvre d'Étaples,[62] but one may also count Adrien Turnèbe (1512–1565),[63] as well as John Dee, to whose 1550 Parisian lectures on Euclid Foix-Candale may have attended.[64] Harrie considered furthermore that all of Foix-Candale's works presented common philosophical views, shaped by his Platonic, Neoplatonic, Pythagorean as well as Hermetic influences, and that these conceptions are properly displayed in their complexity and richness within the *Pimandre*.[65]

On a more general note, it is important to bear in mind that the texts belonging to the Hermetic tradition circulated widely from the late fifteenth century thanks to the Latin translation of the *Corpus hermeticum* by Marsilio Ficino, published in 1471,[66] after the recovery by Leonardo da Pistoia (fifteenth c.) of a Greek manuscript containing fourteen

---

[57] Harrie (1975, p. 174, n. 44).

[58] Kepler (1940, pp. 98–100). As noted by F. Yates (Yates, 1964, p. 442, n.1), the 1940 edition of Kepler's *Harmonices mundi* (Kepler, 1940, p. 534) indicates that Kepler would also have known Foix-Candale's 1574 Greek-Latin translation of the *Poimandres*.

[59] Kepler (1993, I, p. 46). On Kepler's relation to Foix-Candale, see Dagens (1961), Westman (1972) and Mehl (2003).

[60] Harrie (1975, p. 27).

[61] Harrie (1975, pp. 8–9, 42 and 168–169). See also Harrie (1978).

[62] Harrie (1975, pp. 12–14, 108–109 and 162–167).

[63] Harrie (1975, p. 54, n. 136).

[64] Harrie (1975, p. 53). On other points of connection between Foix-Candale and Dee, see also Harrie (1975, pp. 122 and 134).

[65] Harrie (1975, pp. 102 and 162).

[66] Ficino (1471).

treatises belonging to the Hermetic tradition.[67] These texts prompted indeed great interest on the part of Renaissance intellectuals, both because of their philosophical and theological content and because of their association with related scientific and esoteric practices in the fields of astrology, medicine, botany, alchemy, magic and divination.[68] The Hermetic doctrine, which was reinterpreted in the light of Christian faith in the West from late Antiquity, was notably held in the Renaissance as both the foundation and synthesis of several ancient philosophical and theological traditions (among which Pythagorism, Platonism, Orphism and Neoplatonism).[69] It was thus considered by many as a means of retrieving what Ficino called a *prisca philosophia*, or a *philosophia perennis*, as Agostino Steuco (ca. 1497–1548) would later call it,[70] representing a means of reconciliation between the various philosophical and theological doctrines developed from Antiquity in the pre-modern West. The treatises of the *Corpus hermetica* were therefore edited and commented on many times up to the seventeenth century.

In France, many scholars contributed to this tradition before Foix-Candale,[71] starting with Jacques Lefèvre d'Étaples, who published in 1494 a commented edition of Ficino's Latin translation,[72] reprinted again in 1505 with the *Asclepius* and the *Crater Hermetis* of Lodovico Lazzarelli (1447–1500).[73] These works were followed in 1507 by the *Liber de quadruplici vita: Theologia Asclepii Hermetis Trismegisti discipuli cum commentariis* by Symphorien Champier (1471–1539)[74]; the *Mercure Trismégiste, de la puissance &*

---

[67] This history is recounted by Foix-Candale in the preface of his *Pimandre* (Foix-Candale, 1579, sig. A2r): "l'exemplaire Grec fut apporté au seigneur Cosme de Medicis, de Macedoine par un religieux venant des païs orientaux nommé Leonard de Pistoie bon & docte. Et lors Marsille Ficin le tourna en Latin, & voua le premier œuvre, qu'il fit sur la langue Grecque a Cosme son Mecenas."

[68] On the Hermetic tradition in the Renaissance and in the early modern era, see, for example, Yates (1964, 1967), Merkel and Debus (1988), Van den Broek and Van Heertum (2000) and Gilly and Van Heertum (2002).

[69] This was notably acknowledged by Foix-Candale in the preface of the *Pimandre* (Foix-Candale, 1579, sig. A2r–v): "Depuis plusieurs gens doctes escrivants des choses grandes ont allegué des passages & sentences de ce petit traicté, & l'alleguant jusques a noz temps, comme le trouvant faict de grand sçavoir, & conforme a l'Ecriture saincte [. . .] Parquoy nous dirons qu'il a cogneu Dieu, comme les autres Philosophes par les œuvres de nature. Il avoit assez de sçavoir pour ce faire, attandu que nous trouvons, que toutes bonnes escolles de Philosophie, comme la Pitagorique, Platonique, Aristotelique, & autres ont prins leur plus beau & meilleur de son escolle." On the attribution of this role to Hermes by Foix-Candale, see also Moreschini (2009).

[70] Agostino Steuco, *De perenni philosophia libri X* (see Steuco, 1540). On the notions of *prisca theologia* and of *philosophia perennis* in the Renaissance, see Schmitt (1970).

[71] On the French sixteenth-century Hermetic tradition, see Walker (1954), Dagens (1961) and Sozzi (1998).

[72] Lefèvre (1494).

[73] Lefèvre (1505).

[74] Champier (1507).

sapience de Dieu by Gabriel du Préau (1511–1588) in 1549[75]; the edition of the Greek text of the *Corpus hermeticum* by Adrien Turnèbe in 1554, which was based on Ficino's manuscript and which included the Ficinian translation in Latin[76]; the *Deux discours de la nature du monde et de ses parties* by Pontus de Tyard (ca. 1521–1605) in 1578[77] and the French translation of the *De Harmonia Mundi totius Cantica tria* of Francesco Giorgio (1466–1540) by Guy Lefèvre de la Boderie.[78] Foix-Candale, as said,[79] published in 1574 his own edition of the Greek text of the *Poimandres* on the basis of Turnèbe's edition[80] corrected by Joseph Scaliger (1540–1609),[81] together with a Latin translation, as well as a Greek-French edition of this text. In 1579, he published in French his comprehensive commentary of the *Poimandres* entitled *Le Pimandre de Mercure Trismegiste de la philosophie Chretienne, Cognoissance du verbe divin, et de l'excellence des œuvres de Dieu*.[82] These works by Foix-Candale fully belong to the above-described tradition, and the *Pimandre* actually came forth, as said, as one of the most representative texts of French Christian Hermeticism.[83]

Looking for possible traces of Hermetic influences in the commentary on Euclid, one may turn to a verse written in the honor of Foix-Candale by Arnaud Pujol (*Arnoldus Puiolius*) from the "Bordeaux Academy" (*in academia Burdegalensi*). In this verse, the author mentions the name of Hermes (*Mercurius*) as the soul's guide through its journey from body to body[84] when evoking the ancient doctrine of the transmigration of the souls, in reference to Pythagoras' alleged incarnations, notably in Aethalides, the son of Hermes, and in Euphorbus.[85] Although the God Hermes, in classical Greek mythology, was already attributed the function of guiding the souls of the dead to the underworld, this

---

[75] Du Préau (1549).

[76] Turnèbe (1554).

[77] Tyard (1578).

[78] Giorgio (1525); Lefèvre de la Boderie (1578b).

[79] See *supra*, p. 101.

[80] Harrie (1975, p. 8, n. 33).

[81] On the relation between Scaliger and Foix-Candale, see Harrie (1975, pp. 40, 45 and 49–52).

[82] This is the commentary from which the passages quoted in n. 105–116, pp. 120–122 were drawn.

[83] Faivre (2008).

[84] *De Francisci Flussatis Candallae in Euclidem commentariis, Arnoldi Puiolij in academia Burdegalensi I. V. D. Carmen*, in Foix-Candale (1566, sig. A5v): "Unum posse sinunt fieri, ut cum corpore vita/Exiit, in corpus sit reditura novum./Sic Rudius vates animam suscepit Homeri,/Sic animam Euforbus induit Athalidae./Quique prius pisces captabat arundine Pirrhus,/Pythagorae in corpus transiit ille novum./Unde dare ut poßis moestis solatia rebus,/Ne sis qui fueras, forma novanda tibi est./En ego Flussatum formabo ex semine corpus,/In quod *Mercurio* te decet ire duce./Et te Franciscum dicet quicunque videbit,/Tu tamen & tali nomine notus eris." (My emphasis.)

[85] On these alleged incarnations of Pythagoras, see Diogenes Laërtius (Hicks 1925, pp. 322–325), § 8.4–5.

function was transferred to the Egyptian-Greek syncretic figure of Hermes Trismegistus.[86] The Hermetic tradition also attributed a non-negligeable place to the theory of the transmigration of the souls.[87] A reference to this theory in association with the name of Hermes could therefore indicate that philosophical and theological ideas stemming from the Hermetic tradition were being circulated in Foix-Candale's intellectual circle at the time when he wrote his commentary on Euclid.[88] This intellectual circle could notably correspond to what Arnaud Pujol referred to as the *academia burdigalensis*.

In the epistle to Charles IX which prefaced his commentary on Euclid, Foix-Candale referred to the important place of geometry, and of mathematical disciplines in general, among the knowledge required of the priests of ancient Egypt,[89] which was immediately followed by an assertion of the importance of geometry and philosophy for the Church fathers.[90] Although these topics are not specific to the Hermetic tradition, since they are first and foremost related to the *topos* of the origins of geometry, in particular as told by Flavius Josephus (ca. 38–100 AD) in the *Jewish Antiquities*, they strongly resonate with the representation of Hermes trismegistus as an ancient Egyptian priest-king or prophet, who invented writing (in the form of hieroglyphs), who received God's revelation and whose teaching represents the common foundation of all Western philosophical and theological doctrines.[91] The alleged history and representation of Hermes Trismegistus as the inventor of mathematical sciences, who would have transmitted this knowledge to Moses and

---

[86] It may also be noted that, in all the dedicatory verses added to the paratext of the *Pimandre* (Foix-Candale, 1579, sig. A6r–v), Hermes Trismegistus is, in the text, always only designated as Hermes or Mercurius.

[87] On reincarnation in the Hermetic doctrine, see Quispel (2000).

[88] On the intellectual circles of Bordeaux and the status of Bordeaux as an intellectual centre in sixteenth-century France, see Harrie (1975, pp. 15–16 and 47–53).

[89] On this passage, see also Harrie (1975, p. 169).

[90] Foix-Candale (1566, sig. A2v): "Haud secus quàm verae & piae religionis suscepta fides stabilis, animam suaptè natura irae seu exitio devinctam, in aeternum salutis aevum divina clementa latam regenerat. Quae singula sui generis novum hominem edere dicuntur. Nec est quod disciplinarum ac philosophiae principia à religionis consortio explodamus. Solabant nanque veteres Aegyptij à philosophorum coetu, sacerdotes ad religionem elicere, à sacerdotum verò societate, Regem, Persae autem ne ullum quenquam in Regem admitti patiebantur, qui non ante disciplinam scientiamque magorum percepisset. Nec omittendum quàm profuerit propagandae catholicae fidei tantis patribus, praeclara disciplinarum ac philosophiae prudentia."

[91] Foix-Candale (1579, sig. ã4r): "Si est-ce que entre autres & par sus tous ceux de qui nous avons memoire, il a esté un Aegyptien nommé Mercure Trismegiste tres-ancien [. . .] Et de tant que par le Pimandre de Mercure nous trouvons, qu'il a esté vray praecurseur annonçant les principaux poincts de la religion Chrestienne". See also (sig. A5r): "Nous disons cecy à propos de ce tres-grand Mercure Aegyptien, lequel, comme plusieurs escrivent, ayant esté du temps, qu'il n'y avoit en son pays aucun usage d'escriture, fut le premier, qui inventa la manière de faire digerer les lettres & syllabes en painture exterieure, pour le secret & subject de sa pensée, que nous nommons escripture. Laquelle fut premierement par lettres, qu'ilz nommoient Hyerogliphiques." On this specific aspect of Foix-Candale's description of Hermes, see Moreschini (2009).

thereby to the rest of humanity,[92] is also recalled in these terms under the authority of Plato, Iamblichus and Josephus[93] in the preface to Foix-Candale's *Pimandre*[94] written by the humanist Jean Puget de Saint Marc (fl. 1579).[95]

There is, admittedly, no explicit statement in the commentary on Euclid that would indicate with certainty that Foix-Candale fully adhered to the philosophical doctrine found in the *Corpus hermeticum* when he wrote this commentary. But it is at least certain that he then already adhered to some of the Pythagorean, Platonic, Neoplatonic ideas that were propounded by the Hermetic doctrine, and which had been held by Renaissance scholars to stem from the teachings of Hermes Trismegistus. This appears in particular through Foix-Candale's references to the Pythagorean and Platonic conception of mathematics in the epistles and prefaces of his commentary on the *Elements*.[96] Indeed, among the doctrinal elements common to Hermetism, Pythagoreanism, Platonism and Neoplatonism, which (as said) many Renaissance philosophers, such as Ficino and Giovanni Pico della

---

[92] On Foix-Candale's representation of Hermes Trismegistus and of his relation to Moses, see Harrie (1975, pp. 104–105 and 110) and Moreschini (2009).

[93] According to Josephus, it was rather Seth, the son of Adam, who was the inventor of mathematics, his children having engraved on two pillars the entire content of their scientific knowledge in order to preserve it from the flood and fire predicted by Adam. This knowledge would have been passed on to the Egyptian people through Abraham. On Josephus' account of the invention of mathematics, see Goulding (2010, p. 3–6). The notion that the Hebrews passed on knowledge, not only mathematical but also spiritual, to the Egyptians, and to Hermes Trismegistus thereby, was however defended in 1582 by the protestant Philippe Duplessis-Mornay (1549–1623) in his *De la vérité de la religion chrétienne* (Duplessis-Mornay, 1585, pp. 125–126), which was considered by J. Dagens as "the most complete exposition of the Hermetic theology of Ficinian tradition" (Dagens 1961, p. 9). A comparison of Foix-Candale and Duplessis-Mornay is also proposed in Harrie (1975, pp. 247–251) and in Harrie (1978).

[94] Puget de Saint Marc (Foix-Candale 1579, sig. A5v): "Platon en son Phaedre faict mention de ce tres-grand Mercure, luy donnant mesmes noms que Saconiaton: & adjouste davantage, qu'il a esté inventeur de l'Astrologie, Geometrie, & Arithmetique. Et Iamblicus grand Philosophe tout au commencement d'un livre qu'il a faict des mysteres des Aegyptiens dict, que Mercure est tenu de tous les anciens inventeur des artz, & sciences. L'escripture aux Actes des Apostres chapitre septieme, tesmoigne que Moise fust instruict, & endoctriné en toutes les sciences des Aegyptiens. Comme aussi le tesmoigne Joseph en son premier livre des Antiquités des Juifz."

[95] On Jean Puget de Saint Marc, see Harrie (1975, p. 104, n. 12).

[96] Foix-Candale (1566, sig. a6v): "Cum itaque omnium ac singularum naturam datis viribus perquirerem, ut demum illas mihi deligerem, quae animi notitiam certo veròque illustrarent, *illudque audirem (de Mathematicis) Ciceronis à Pythagora, à numeris scilicet & Mathematicarum initiis proficisci omnia*, ac in summo apud illos honore Geometriam fuisse: itaque nihil Mathematicis illustrius esse, pluresque philosophorum sententias, *Geometriam caeterasque Matheses summopere commendantes, nonnihil divinitatis his inesse disciplinis arbitratus sum*, ac aliquod numen prae se ferre, eis solis communicandum, qui intelligendi veri, non qui contentionis aut spectatus cupidi fuerint [. . .]." J. Dagens (Dagens 1961) considered this passage as revealing the connection between Hermeticism and Pythagorean arithmosophy in Foix-Candale's commentary on Euclid.

Mirandola (1463–1494), regarded as historically and conceptually connected,[97] are the transcendence of forms or essences, an ontology of participation linking intelligible and sensible realms, as well as the representation of the Sun as a physical image of a divine principle governing the existence and intelligibility of all things. To this adds the representation of God as a demiurge, who created the cosmos on the basis of the intelligible essences eternally present in his mind. In Foix-Candale's commentary on Euclid, such conceptual elements come forth in particular through the assertion of the existence of transcendent intelligible essences in the divine mind, which God would have communicated to the material realm,[98] as well as through his conception of mathematics as a propaedeutic to the contemplation of theological truths.[99] He notably made repeated references to the Platonic theory of reminiscence,[100] which not only presupposed the transcendence of intelligible forms and the participation of sensible beings in the essences of the intelligible realm, but also the immortality and reincarnation of the soul.

---

[97] On Hermetism and its reception in the Renaissance, see for instance Yates (1964).

[98] See, for instance, Foix-Candale (1566, sig. A6r): "quae *sensibus tanquam corpora vel materies* attrectantur, sive quae intelligentiae ratiocinanti *summi opificis imagini*, velut *eius essentiae* communicant." (My emphasis.)

[99] Dagens (1961); Harrie (1975, pp. 162 and 168). See the following note. Harrie (1978, p. 503–504) wrote that: "It was from Hermes that Foix-Candale ultimately drew his sanction of mathematics' important role in the attainment of religious truth and spiritual purity."

[100] Foix-Candale (1566, sig. A2r-v): "Nimirum eorum qui disciplinas perinde scrutantur, ut earum virtutibus & dogmatum dignitate fruituri. Non autem qui primordiis utcunque perceptis, in plebeios ita saeviunt, ut sola praefulgendi causa illas perlustrasse satis superque propalent. Quorum itaque animi cupido eo splendore fuit ornata, quo genuinos *disciplinarum* afflatus (quos *Plato* à natura humanae menti collatos opinatur) *reminisci*, seu intelligentiae saepius conferre studuerint: hi Mathesium simplicitatem à quovis fuco vacuam, non solum candoris animi testem, verum & eius intelligentiae esse auctricem, qua Philosophiae legibus quaeque regenda subeunt, suo commodo captant, Platonis decretum sequuturi: qui philosophis erudiendis suo lumini praescripserat, nullum sine Geometria ingressurum, his vocibus vetans, generosa philosophiae documenta à quoquam geometriae experte scrutanda esse"; (sig. A3r): "Matheses haud equidem discendas quinimò (ex Platonis sententia) *reminiscendas*"; (sig. A6v): "Hanc subsequuturas ratus sum vero certoque illustratas Matheses sive disciplinas, quarum intellectum natura tam mira arte geniturae animae copulavit, ut quicquid in eis quantumvis abstrusum obiicitur animae discendum, id idem non discere sed verius *reminisci* doceat esse Plato: hae nanque disciplinae ex principiis eidem animae familiaribus nullo aliunde conducto famulatu, sua quaeque construunt theoremata. Quae demum non per novam alicuius documenti notitiam, sed ex primordiorum iam notorum sola *reminiscentia*, menti seu animae innotescunt. Principia nimirum si per se cuivis ratiocinanti animae innotescant, cùm menti offeruntur, non ab illa ideo disci, sed sanius *reminisci* dicemus, tanquam prius scita. Ab iis rursus constructa theoremata priora, ac illorum progrediente obsequio progenita, sola arguendi vi (cuique ratiocinanti animae insita) ex assumptis principiis fluunt. Ratiocinantis nanque animae potissimus usus est, ex praescriptis antecedentibus veris, consequentia deligere necessaria. Non aliud itaque à *reminiscentia* disciplinarum adeptionem esse fatebimur." (My emphasis.)

As expected, most of these doctrinal elements are also found in one way or another in Foix-Candale's *Pimandre*.[101] This commentary was indeed more an occasion to set forth his conceptions on a range of issues pertaining to natural philosophy, cosmology and theology, among other domains, than a mere project of exegesis of the first treatise of the *Corpus hermeticum*.[102] According to J. Dagens, one of the chief aims of this treatise would have been to develop a form of natural theology, whereby Foix-Candale intended to show the compatibility and even the coincidence of the conclusions of philosophy and of theology, notably between pagan philosophy and Christian theology, against Paduan Averroism.[103] To J. Harrie, a key thesis of the *Pimandre* was the interpretation of Christian redemption and salvation according to Hermes' doctrine of regeneration. This doctrine prescribed a gnostic process of purification of the soul through knowledge and piety, that is, as a detachment of the spirit from the corruption of matter, which takes up the scalar epistemological model advocated by the Platonists and the Neoplatonists, as by Christian ascetics and mystics, from Origen (ca. 184–ca. 253) to Ficino.[104]

Hence, the *Pimandre* is important to consider here insofar as it offers complementary information to better understand the ontological status Foix-Candale attributed to geometrical objects in general (and to the sphere in particular) and the conceptual system that motivated his dismissal of translational or rotational generations of geometrical objects as part of their definitions. It also provides keys to understand the epistemological status of geometrical definitions, and, more generally, of geometrical knowledge as a whole in his philosophy of mathematics. As will be shown further, it also echoes certain conceptions held by Peletier concerning the epistemological status of geometrical knowledge.

4.6 The Ontological Status of the Sphere in the *Pimandre*

Foix-Candale's statement that Theodosius' definition of the sphere sets forth an image of its "ineffable eternal essence" is clearly coherent with the doctrinal basis common to the Hermetic, Platonic and Neoplatonic traditions, according to which geometrical objects, and most of all the geometrical circle or sphere, correspond to spatialised images of an ontologically higher substance, of divine and eternal essence, and devoid of spatiality and divisions.

---

[101] See *infra*, n. 113, p. 121; n. 153, p. 131; n. 168, p. 136.

[102] J. Harrie (1975, p. 102).

[103] Dagens (1951).

[104] Harrie (1975, pp. 110–115) and Harrie (1978).

In the *Pimandre*, Foix-Candale asserted the divine origin and perfection of the sphere over all figures,[105] notably on account of the fact that it may rotate on its axis while always occupying the same space, for which the universe was given a spherical shape and was made the cause of the motion of all material beings, from the celestial bodies to the elementary substances.[106] The sphere of the universe, which is the most perfect of all created things, is then compared to the human intellectual faculty, which corresponds to the noblest and most divine part of man[107] and which is deprived of local motion while setting all parts of the human body in motion.[108]

Admittedly, in this context, the rotation of the spherical universe on its axis and the circular motion of the celestial bodies would evoke Euclid's definition of the sphere rather than that of Theodosius. Yet, Foix-Candale's aim then was not to describe the mode of generation and essence of the geometrical sphere, but rather to assert the ability of the sphere, which is deprived of any angle and perfectly uniform in all its parts, to remain in the same space while rotating on its axis. It was indeed for this reason, in addition to the fact that it was the most capacious of all geometrical solids, that it was considered the most perfect and divine of all geometrical figures and was therefore used to shape the cosmos, as the material, mobile and finite expression of its intelligible, immobile and omnipotent divine principle.

---

[105] Foix-Candale (1579, p. 369): "[. . .] la convenance, qu'à la perfection circulaire ou sphericque sur toutes choses materieles, d'aprocher plus a la nature des intelligible." *In marg.* "La forme spherique aproche sur toutes autres l'intelligible perfection."

[106] Foix-Candale (1579, p. 369): "Tout ainsi la sphere celeste, laquelle il a dict estre ce chef, donne mouvement, & faict agiter & mouvoir tous les corps celestes & autres par elle contenus d'un lieu a l'autre, sans toutefois qu'elle mouve jamais de sa place. C'est a cause de la perfection circulaire aux figures planieres, & sphericque aux solides, comme a ce propos. La sphere a de sa nature, & composition de sa figure, une telle propriété qu'estant meuë, voire du plus grand effort, qui se puisse penser, a l'entour de son axe par ce mouvement, elle ne remuë ny occupe autre lieu quelconque, tant soit petit, que le sien propre, non plus durant le mouvement, que durant le repos, sur quelque diametre des siens, qu'elle soit meuë, ce qui ne convient a figure quelconque, que a celle là, a cause de sa perfection. Parquoy son mouvement de vray ravist les corps, qui sont dans elle, & toutes ses parties: & les tire a faire leur circuit, & leur donne mouvement, les tirant d'un lieu en autre. Ce qu'elle, qui les y attire & contrainct y venir, ne faict pas en son tout, ains demeure tousjours en son lieu, ne remuant, tant soit peu, ça ny là, quelque action & vertu de mouvement qu'elle envoye aux autres corps."

[107] Foix-Candale (1579, p. 369): "Ceste pensée, de laquelle veritablement toutes parties soient jugement, intelligence, cognoissance, subtilité, mémoire, invention, & infinies autres parties de l'image de Dieu, sont logées dans le corps humain, principalement au chef [. . .] la pensée merite d'estre dicte chef de toute la composition [. . .] à cause de son excellence, qu'ell'a par dessus toutes les parties de l'homme, comme le chef sur les autres membres."

[108] Foix-Candale (1579, p. 369): "La pensée donc estant chef, elle est dicte mouvoir en la maniere de la sphere, laquelle nous avons dict estre chef. Et pour declarer ce mouvement, qu'il donne a la pensée, qui a la verité n'a aucune agitation, ou ce que nous entendons par mouvement, Mercure n'a voulu taiser la subtilité du mouvement de la sphere pour le comparer au mouvement de la pensée. Laquelle estant cause & mouvant tout ce qui se meut en l'homme, & toutefois elle ne meut, ains repose tousjours en son estat. Tout ainsi la sphere celeste, laquelle il a dict estre ce chef, donne mouvement, & faict agiter & mouvoir tous les corps celestes & autres par elle contenus d'un lieu a l'autre, sans toutefois qu'elle mouve jamais de sa place."

Now, in other passages of the *Pimandre*, the notion of sphere as associated with a higher ontological level of being and as caused by a process of expansion from a single source is conveyed by the representation of light as the divine virtue and essence[109] that is most accessible to man's senses.[110] The primordial light, first emanating from God's Holy Word,[111] is then said to fill all things, allowing both material and intelligible entities to exist, as well as to be seen and known.[112] The light of the Sun, which corresponds to the physical manifestation of the divine intelligible light dispensed by Jesus Christ and the Holy Spirit,[113] is said to illuminate all parts of the universe, its rays reaching the whole celestial realm as well as the most intimate parts of the earth.[114]

---

[109] Foix-Candale (1579, p. 698): "[...] lumiere est ceste divine essence & vertu [...]". See *infra*, n. 112.

[110] Foix-Candale (1579, p. 10): "C'est ce sainct verbe luisant, que Mercure disoit reluire des tenebres, qui a illuminé noz pensées, comme le dict sainct Pol, Dieu qui a dict que des tenebres la lumiere luisoit, il a illuminé noz cœurs, il a voulu nommer la lumiere, voire qui illumine tout homme venant en ce monde, comme nous pourrions penser, à cause que toutes les vertus ou essences de Dieu, que nous pensons estre plus familieres à noz sens, c'est la lumiere [...]".

[111] Foix-Candale (1579, p. 10): "C'est ce sainct verbe luisant, que Mercure disoit reluire des tenebres [...]".

[112] Foix-Candale (1579, p. 445): "*toutes choses remplies de lumiere* qui est une des premieres vertus et essences divines [...] c'est la lumiere qui illumine tout homme venant en ce monde et remplist toutes choses tant materielles que intelligibles. Ceste unicque lumiere & vertu divine illuminant donne cognoissance aux sens de toute chose corporelle: de tant que sans lumiere l'œil corporel ne peut apercevoir le subject, ny sans ceste mesme vertu de lumiere l'œil de l'intelligence, qui est la pensée de l'homme, ne peut cognoistre un subject ou cognoissance, ou intelligence. C'est la mesme vertu divine illuminant, qui secourt les deux, et le corporel et l'intelligible. Et toutes fois ceste lumiere ne peut estre veuë de l'œil corporel en son essence: de tant qu'elle est divine, non plus que les autres essences et vertus divines. C'est donc ceste divine lumiere, qui remplist toutes choses." (The emphasis, which is meant to distinguish the original text of the *Poimandres* from the commentary, is proper to the original text).

[113] Foix-Candale (1579, p. 698): "*Il donne continuellement lumiere a toutes*, à cause que ceste divine essence tant excellente luy est donnée pour la necessité des creatures & toutes choses vivantes, car comme nous avons quelquefois cy devant dict, lumiere est ceste divine essence & vertu, par laquelle Dieu secourt sa creature à luy manifester toutes choses, dont les unes sont corporelles. Pour lesquelles il a commis le Soleil, comme dispensateur de la lumiere, qui sert à la manifestation des corps aux sens corporels, soit aux hommes, ou animaux desraisonnables. Les autres sont intelligibles, lesquelles entre tous animaux ne conviennent qu'à l'homme, lequel n'a besoin de la lumiere du Soleil, pour la manifestation & cognoissance qu'il cherche de telles choses: mais il a tres-grand besoin du Soleil de Justice, Jesus Christ Fils de Dieu & son sainct Esprit, auquel seul apartient la dispensation & ministere de ceste lumiere: par laquelle l'homme reçoit en son entendement mesme secours & clarté, pour entendre & cognoistre ou recevoir manifestation de la chose intelligible, que le corps a receu en ses sens pour apercevoir & voir la chose corporelle. Et par ainsi toute manière de lumiere n'est qu'une mesme vertu, mais estant divine elle a puissance sur plusieurs effects, desquels les corporels sont commis au Soleil, et les intelligibles demeurent en la dispensation du sainct Esprit." (Emphasis proper to the original text.)

[114] Foix-Candale (1579, pp. 698–699): "Et ce Soleil faisant sa charge, donne lumiere & jette ses rayons en toutes creatures y produisant ses effects, *de tant que c'est luy, duquel les bons effects ne*

When discussing the position and role of the Sun in the universe, Foix-Candale wrote that it is however not situated in the geometrical centre of the cosmos, since he explicitly rejected the cosmological model provided by Copernicus, although he acknowledged its value as a mathematical model.[115] Following the Ptolemaic system, he asserted that, in relation to the planets, the Sun is situated above the Moon, Mercury and Venus (the inferior planets) and below Mars, Jupiter and Saturn (the superior planets).[116] And in relation to the entire sphere of the universe, it is situated between the earth, as the centre of the cosmos, and the eighth sphere (the Firmament or sphere of the fixed stars), which corresponds to the boundary of the celestial realm. The Sun would thus move between these two extremes according to a perfectly circular motion. As J. Dagens formulated it, this conception represents a form of "mystical heliocentrism", which was further developed and explicitly

---

penetrent seullement dans le ciel et l'air, illuminant tous les corps celestes en leur region sans aucun empeschement d'autre matiere: mais aussi penetrent en la terre solide & materielle sur toutes choses. Dont il sembleroit l'empeschement estre suffisant, pour retenir le passage du rayon, & l'estouper de manière qu'il ne peust passer: toutefois penetrant toute solidité de matiere & empeschement, qu'elle luy sçauroit presenter, il n'est retenu ou empesché par aucun, qu'il n'enfonce jusques au tres-infime fonds, qui est le centre de la terre et ses abysmes, lieu plus bas de toute l'univers, tant est merveilleuse la puissance d'une essence divine, qu'il n'y a matiere ny solidité quelconque, qui luy puisse resister, ou empescher son action & passage, à faire le commandement de son Seigneur et souverain Dieu." (Emphasis proper to the original edition.)

[115] Foix-Candale (1579, p. 701): "Car disant le Soleil estre assis au millieu du monde en respect de sa quantité, il faudroit qu'il fust au centre, ce qu'il n'est pas, & par consequent qu'il n'eust aucun mouvement, de tant que le monde en son univers est immobile, combien que ses parties soyent mobile. [...] Aucuns pourroyent bruncher en cest endroict, qui auroyent estimé l'oeuvre de Nicolas Copernic avoir est bastie serieuse & cathegorique [...] Non que Copernic vueille asseurer la disposition & situation de l'univers estre ainsi à la verité, mais seulement par supposition qui luy serve à ses demonstrations." On Foix-Candale's comments on Copernicus, see also Dagens (1951).

[116] Foix-Candale (1579, pp. 700–701): "Et ces corps reçoyvent tous la lumiere qu'ils ont en eux, du rayon & regard de Soleil, de tant que c'est à luy seul auquel le souverain createur l'a dispensée & communiquée, pour estre par luy administrée à toutes creatures, selon son estat & prescription d'ordre, de tant qu'il est assis au millieu, comme portant coronne au monde. Ce propos est exprimé en Philosophie, regardant les actions & qualitez, & non en Geometrie, qui auroit seulement consideré la quantité ou mouvement [...] Ce n'est donc ainsi que le Soleil est dit estre au milieu, mais il est dict estre au millieu, non de la quantité ou grandeur de la masse du monde, ains au millieu des actions & puissances divines, administrées tant par luy que les autres corps: & ce de tant qu'il est entre la terre, qui est l'une extremité la plus basse de tout le monde, & l'octave sphere, qui est la plus haute, si precisement qu'il a entour soy les autres six planetes disparties si egalement, qu'il en a trois au dessus de luy, qui sont Saturne, Jupiter, & Mars, & si en a trois au dessous, à sçavoir Venus, Mercure & Lune. Tous lesquels sont ainsi egalement departiz à l'entour de soy, pour recevoir de luy plus facilement sa lumiere & autres actions, si aucunes ils en doyvent recevoir. Et en ceste manière il se trouve au millieu de toutes actions, entre les deux extremitez du monde, à sçavoir entre sa circonference tres-haute partie de luy, & son centre tres-infime partie contraire, & tient ce millieu comme portant coronne au monde."

related to the divine status of the Sun in ancient Egypt by Pierre de Bérulle (1575–1629) in the seventeenth century.[117]

In this framework, light (both intelligible and physical) was defined as a divine and primordial principle of God's creation, whereby the figure of the sphere as resulting from a uniform expansion of space from a single source is associated with a higher ontological state. In this regard, it is significant that, in the preface addressed to Charles IX within the commentary on Euclid, Foix-Candale frequently appealed to terms relating to light and illumination when talking about the truth procured by philosophical or geometrical knowledge and their objects. Moreover, in the Platonic representation of truth and knowledge, the image of the Sun and of its light to represent the divine principle (the Idea of the Good) and its primordial role as first cause and source of knowledge of all things held a central place,[118] and Foix-Candale repeatedly referred to this conception in his preface to Charles IX. Now, this conception also clearly relates to the divine status and causal role of light in the *Pimandre* insofar as it allows all things to be known.[119]

4.7   Foix-Candale and Peletier on the Ontological Status of the Sphere and of the Circle

The foundational place and divine status attributed to light in Foix-Candale's *Pimandre*, as well as the description of truth in terms of light and luminosity in the paratext of his commentary on Euclid, also indirectly recalls elements of the discourse held by Peletier in his commentary on Euclid and in his scientific poetry. While, in his commentary on Euclid, geometrical objects are compared to luminous phenomena,[120] in the *Louange de la Sciance*, Peletier offered a metaphysical representation of the primordial point-unit as instantaneously expanding in all directions to create the universe.[121] As we have seen,

---

[117]Bérulle (1644, p. 170–172) and Dagens (1951). On the role of the Sun in Foix-Candale's *Pimandre*, see also Harrie (1975, pp. 134–135 and 187–190).

[118]Plato, *Republic* VI 509b.

[119]Foix-Candale (1566, sig. A2r): "*radius* primae ac verae solius essentiae"; "unico veritatis *splendore*"; (sig. A2v): "qui philosophis erudiendis *suo lumini* praescripserat, nullum sine Geometria ingressurum"; "Qui quidem habitus, veritate, ratione, symmetria, ordine, reliquisque geometriae subtilissimis viribus mentem *illustrans*." (My emphasis.)

[120]It may be interesting to note that, in the *Pimandre*, Foix-Candale designates the dust or powder moving in the light rays as atoms, which evokes the small bodies dancing in the light of solar rays to which Peletier referred in his commentary on Df. I.5: Foix-Candale (1579, p. 10): "aucuns penseroient voir les rayons passants en ce lieu obscur sans object, quand ils voyent dans iceux rayons, les atomes & poudre que l'air remue continuellement là où il se trouve." *Cf.* Peletier (1557, p. 3), Df. I.5: "Puncta enim corpusculis insecabilibus assimilantur, quae in radijs Solis colludunt: Lineae radijs ipsis" (see *supra*, n. 41, p. 86). Foix-Candale did not however compare these to geometrical lines and points, as did Peletier.

[121]See *supra*, p. 88.

this representation evoked, by many aspects, the role of God's primordial light in the medieval metaphysics of light.

These common elements between the philosophical and theological conceptions of Foix-Candale and of Peletier are chiefly related to the fact that both were influenced by the Platonic and the Neoplatonic doctrines.[122] And, in this regard, a further connection may be established between the ontological perfection and suprasensible origin which Foix-Candale attributed to the geometrical sphere, both in the commentary on Euclid and in the *Pimandre*, and the discourse Peletier held concerning the circle in his commentary on Euclid, where he asserted the limits of the human mind faced with the ontological perfection and divine origin of geometrical objects:

> For what can we understand regarding the things which have emanated in a divine manner, when we judge them in a human manner? The circle taking therefore its origin from itself, seems to come from the rectilinear; it is infinite, and however similar to what is finite: it contains all, as it is the most capacious, but appears however to admit something exterior to itself.[123]

This passage, which conspicuously displays the influence of Cusanus on Peletier's thought,[124] sets forth the contradictions inherent to the nature of the circle, as it is simple and uniform, yet able to contain all other figures; infinite, yet similar to what is finite; self-caused and self-sufficient, yet appearing to stem from something exterior to it (i.e. the rectilinear, as the straight line produces the circle by rotation). Such "coincidence of the opposites" within the circle was also expressed by Peletier in his commentary on Prop. III.1, as he said that, "in the circle, affirmation and negation come together, as do action and privation, and generation and corruption within the universe."[125] As such, the circle may be taken as divine and as similar to the universe, and ultimately to God himself, insofar as it concentrates the properties of all things. It may thus only be defined through a series of oppositions (as described by Cusanus through his concept of *coincidentia oppositorum*), which place it beyond the grasp of human discursive thought.[126]

---

[122] See *supra*, p. 84. A similar conceptual pattern may be found in texts of other sixteenth-century French authors, as was shown in Sozzi (1998).

[123] Peletier (1557, p. 6), Df. I.16: "[. . .] quid ingenio consequi, quum de his quae divinitùs emanarunt, humanitùs iudicamus. Circulus igitur ex se ipse ortus, ex Recto provenire videtur: infinitus, ac finito similis: omnia continens, ut capacissimus, & tamen aliquid extrà se in speciem admittens". *Cf. ibid.* (sig. 4r): "Quid Puncto simplicius? quid Circulo absolutius? at ex illo omnia emanant, in hoc omnia concluduntur. Ut ne Puncto quidem desit infinitatis admiratio. Quid enim tam mirabile, quàm à medij Circuli puncto, quod Centrum". See *supra*, p. 93 and Axworthy (2013).

[124] On Cusanus' influence on Peletier, see *supra* n. 77, p. 94.

[125] Peletier (1557, p. 66), III.1: "Ut in Circulo Affirmatio Negatioque conveniant: sicut in Universo Actio & Privatio, Generatio & Corruptio."

[126] On the role of mathematics in Cusanus' epistemology and metaphysics, see, for instance, Counet (2005).

These oppositions inherent to the nature of the geometrical circle, and its similarity with the divine Creator and with the universe as its material image, are made clearer in a passage of the *De usu geometriae*, Peletier's treatise of practical geometry dating from 1572:

> The excellence of the circle is such that it may be rightfully regarded as the first and the last of the figures.[127] The first because it is enclosed by a single line. And for this reason it is the simplest and most beautiful of all figures. The last because it is the most capacious and largest of all, enclosing all figures in itself, the triangle, the square, the pentagon up to the infinite number of remaining figures, to which it provides rule, measure and proportion, as if all were carved out and cut off from it. And although it appears to have no angles, nor sides, it can however be said [to be composed] of an innumerable number of angles and sides, as the line may be said [to consist in] an infinite number of points, and the surface, in an infinite number of lines, in the manner we imagine God to be, infinite and immense, containing and governing all things.[128]

This passage explains in particular that, if the circle is infinite and similar to God, it is because its circumference may be regarded as composed of an infinite number of angles and because an infinite number of lines join its centre to its circumference, just as the line may be held as composed of an infinite number of points, and the surface of an infinite number of lines. As was shown in the previous chapter,[129] the connection between the discrete and the continuous was only admitted by Peletier on a metaphysical level, since the mathematician may only admit the line as composed of lines and the surface as composed of surfaces. Hence, this coincidence of opposites within the figure of the circle, which allows Peletier to compare the essence of the circle to the nature of God, is to be situated on a metaphysical level. For, in a mathematical context, as Peletier added in the corresponding passage of the French translation of his *De usu geometriae* (published in 1573 as *De l'usage de geometrie*), this infinity, as expressed by the infinite number of lines joining the centre to the circumference of the circle, is only potential or virtual (*infiny en puissance*).[130]

---

[127] Whereas Peletier, in the Latin version (1572, p. 4), uses the term "forms" ("formae"), in the French translation (Peletier, 1573, p. 9), he uses the expression "geometrical figures" ("figures géométriques").

[128] Peletier (1572, p. 4): "Circuli verò ea praestantia est, ut meritò prima & ultima Formarum dici possit: Prima, quòd unica linea sit clausus: Ob id Forma omnium, simplicissima & speciosissima. Ultima, quòd omnium sit capacissima & amplissima, omnes Formas in se includens, Triangulum, Quadratum, Pentagonum, caeteràsque in infinitum. Quibus omnibus regulam, mensuram rationémque praefinit: quasi omnes è Circulo resectae & abscissae sint. Quùmque Circulus nullis angulis, nullísque lateribus contineri videatur, tamen innumerabilium angulorum & laterum ita dici potest, ut Linea innumerabilium punctorum, & Area innumerabilium linearum. Ad cuius species, Deum infinitum, & immensum cogitamus, omnia continentem & gubernantem."

[129] See *supra*, pp. 89–90.

[130] Peletier (1573, p. 10): "Et mesmement le Centre du Cercle est à considerer, comme admirable en sa sorte. Lequel comme il soit posé au vray milieu, & que, comme Point, il semble n'avoir aucunes parties, toutesfois par puissance il est capacissime. Car les lignes innumerables qui se terminent à la

When dealing with the angle of contact, in the *De contactu linearum* from 1563, Peletier asserted again the divine nature of the circle. This assertion was then based on the fact that, within the circle, opposites coincide, but also on the foundational role of the circle in the constitution and understanding of other figures, as well as on its relation to the structure of the universe. In this context, Peletier showed that, if the circle is the first of all figures, it is ultimately because the straight line itself derives from the circle, either by being led from its centre to its circumference or by resulting from the motion of the circle on a plane, and also because the proper motion of the straight line necessarily results in a circle.

> Therefore all the figures are contained and enclosed within the circle, whose circumference is made stable by the perpetual and invariable flow of the points, so that nothing may escape. All other figures have visible angles aspiring to lead to that perfect sum. Within it, the straight lines that go from the centre and that end at the periphery bend back again toward the innermost part. Others are transversal and led crosswise, so that the sight of all actions and operations appear in the greatest capacity, whose points dispense in their infinity as they are dispersed through the constitution of the wheel-shaped figure, their limitless powers being self-sustained from within while sustaining everything. For this reason, the circle is the last figure but also the first, since it is brought about by the revolution of the straight line. Indeed, the straight line cannot create through its own motion any other figure than a circle. And we take it according to this second meaning, so that we may have an art that exercises us. As it happens, the straight line does not come before the circle, since it may be understood to be created by the driving of the circle in a straight line on a plane. And that same plane is once again a circle, that which is God, one and infinite, all embracing, rendered visible by the beautiful orb.[131]

Peletier does not only show here that the straight line finds its foundation in the circle, but also that the straight line and the circle are ultimately one in their mode of generation, mutually causing each other.[132] This is the reason why the circle, yet the simplest of

---

Circonference, sont pareillement toutes tirees du Centre, & icelles mesmes menees de la Circonference, se terminent toutes au Centre: qui le rendent infiny *en puissance*, comme la Circonference." (My emphasis.)

[131] Peletier (1563, p. 43): "Igitur omnes Figurae intra Circulum continentur & concluduntur. Cuius ambitus perpetuo & invariabili Punctorum fluxu sic firmatur, ut nihil quicquam effluere possit: caeterea omnes Figurae angulos habent conspicuos ad summam illam perfectionem iter affectantes. In hoc lineae rectae à Centro ad Peripheriam terminantur, quae rursus ad intima reflectuntur: aliae transversae & decussatim ducuntur, ut in amplissima capacitate appareant actionum & exercitationum omnium species, quas Puncta sua infinitate suppeditant, dum per rotundam fabricam disperguntur: cuius opes inexhaustae foventur intra seipsas, & fovent omnia. Ob id Circulus est figurarum ultima: sed & prima, quum absolvatur à rectae lineae ambitu. Neque enim linea recta suo unius motu aliam figuram creare potest, quàm Circularem. Atque id secundùm sensum accipimus, ut habeamus artem quae nos exerceat. Etenim nec recta linea prior est Circulo: quum & ipsa creari intelligatur ex Circuli ductu in rectum super Plano. Et rursus illud ipsum Planum Circulus est. Ille, ille aeternus Deus infinitus & unus. Omnia comprendens, pulchro spectabilis orbe."

[132] This idea also appears in Peletier's commentary on Euclid's Prop. I.46 (Peletier 1557, p. 45): "Per centrum enim lineas duci, atque in ambitum, non in latum incedere par est. Punctum quippe illud foecundissimum, lineas infinitas circunquaque procreat."

figures, may contain all other figures and ressemble both the universe in its absolute capacity and God in its omnipotence and supreme perfection. The coincidence of opposites in the circle is also marked by the comparability of the infinite-sided polygon to the circle, although this identity remains again only virtual.

Although this is not made as clear here as in the commentary on Euclid,[133] this passage also suggests that the proper mode of generation of the straight line, and with it, of all rectilinear figures and even of the circle itself, may not be determined by the rational and discursive thought proper to the human mind. For, in geometry (i.e. the "art that exercises us"[134]), it is equally possible to define the motion of the straight line as the cause of the circle or the rectilinear motion of the circle on a plane as the cause of the straight line.[135]

Foix-Candale did not so much aim, in his commentary on Euclid's definition of the sphere, to display the oppositions inherent to the definition of the sphere, but rather to set forth the qualities that display the divine essence and origin of this figure, that is, its simplicity, its uniformity, as well as the perfect equality of all the lines that proceed from its unique and indivisible centre to its circumference. Yet, both Peletier and Foix-Candale saw in the perfect simplicity and uniformity of the circle and sphere, which is determined by the equality of the infinite number of lines joining the centre to the circumference, the mark of the ontological superiority and divine origin of these figures. Moreover, as will be shown later,[136] Foix-Candale, in his *Pimandre*, asserted the coexistence of all things within God according to the mode of complication, whereby all opposites exist within him in a state of coincidence, and compared God to the unitarian and indivisible principle of quantity, both of which remain beyond the grasp of the human intellect. One finds therefore, in the philosophics of mathematics of Peletier and of Foix-Candale, common doctrinal elements shared by Christian Hermeticism and by the theological tradition of Neoplatonic inspiration stemming from the works of pseudo-Dionysius Areopagitus,[137] and later developed by Cusanus.[138] J. Harrie actually considered that Foix-Candale followed Cusanus, at least

---

[133] Peletier (1557, p. 6). See *supra*, p. 93.

[134] *Cf.* Peletier (1557, p. 6): "Sed nos haec rerum varietas exercet: in qua satis nobis est coniecturam ad usum accommodare."

[135] Peletier (1557, p. 6): "Neque est quòd quisquam se fatiget inquirendo, utrum sit prius Rectum an Rotundum. Sed si quis sententiam ferre cogatur: ut Philosophus, rectè iudicabit, si utrunque simul esse pronuntiaverit. Nam & Circulus in plano rotatus, Rectum procreat. Menti quippè nihil prius neque posterius."

[136] See *infra*, p. 133.

[137] On pseudo-Dionysius and his relation to Neoplatonic sources (notably Proclus), see Perl (2010) and Dillon (2014).

[138] On the connections made between Hermetism and the Dionysian theological doctrine, as with the philosophy of Cusanus, in the early modern French Hermetic tradition, see Dagens (1961), Harrie (1975, pp. 4, 115 and 166–168) and Sozzi (1998).

through the intermediary of Lefèvre d'Étaples,[139] in attributing a key role to mathematics in the theological process of salvation.[140] In this context, the way mathematicians consider the line, the circle or the sphere to have been generated maintains a conjectural character, though some genetic definitions (such as the definition of the sphere by Theodosius according to Foix-Candale) would be more proper than others to hint at the divine nature of these objects.

4.8   Proclus on the Properties and Constitution of the Intelligible Circle

A further connection between Peletier and Foix-Candale on this issue may also be found in the fact that their respective discourses on the circle and on the sphere both resonate with what was written by Proclus, in his own commentary on Euclid, concerning the properties and constitution of the intelligible circle, that is, the suprasensible and indivisible circle of the intellect, as opposed to the divisible circle of the imagination.[141] In Proclus' commentary, the intelligible circle was shown to be both absolutely simple and containing plurality, finite and infinite, caused by an exterior principle (the straight line) and yet self-sufficient. It was also defined as resulting from a process of expansion or procession (πρόοδος) from its centre, which would be interrupted by the urge of the figure to imitate the simplicity and self-identity of its centre, causing thereby the perfect uniformity of the circle. Thus, according to Proclus, once the geometer has studied the properties of the extended and divisible circle of the imagination and has gone beyond its extendedness and spatiality to contemplate its proper essence, what will be discovered is:

> the truly real circle itself—the circle which goes forth (προϊόντα) from itself, bounds itself and acts in relation to itself; which is both one and many; which rests and goes forth and returns to itself; which has its most indivisible and unitary part firmly fixed, but is moving (κινούμενον) away from it in every direction by virtue of the straight line and the Unlimited that it contains, and yet of its own accord wraps itself back into unity, urged by its own similarity and self-identity towards the partless center of its own nature and the One that is hidden there. Once it has embraced this center, it becomes homogeneous with it and with its own plurality as it revolves about it. What turns back imitates what has remained fixed; and the circumference is like a separate center converging upon it, striving to be the center and become one with it and to bring the reversion back to the point from which the procession (πρόοδος) began.[142]

---

[139] On the importance of Cusanus for Lefèvre's conception of mathematics, as well as for those of his disciples, see Oosterhoff (2018, pp. 125–132, 161–179 and 205–211).

[140] Harrie (1975, p. 168).

[141] On Proclus' ontology of mathematical objects, see for instance Nikulin (2008).

[142] Proclus (Friedlein 1873, p. 154) and (Morrow 1992, p. 122).

According to Proclus, the unfolding of the intelligible circle—an unfolding which is neither spatial nor temporal—would take place thanks to the metaphysical principle of the Unlimited and its determination, through the interruption of this process by the complementary principle of the Limit.[143] Through these two principles, the intelligible circle, in which the centre, the surface and the circumference coincide, would expand from itself and bound itself through the circumference's desire to imitate the simplicity and self-identity of the centre, revolving around it and becoming homogenous with it. Hence, in Proclus' description of the properties and of the (non-spatial) generation of the intelligible circle, we find both Peletier's assertion of the coincidence of opposites within the essence of the circle and Foix-Candale's representation of the generation of the geometrical figure as a uniform expansion and bounding of its own quantity. We also find a justification for Foix-Candale's dismissal of Euclid's definition of the sphere in favour of a definition that, to him, properly displays the true cause of the simplicity and uniformity of the sphere, as well as its eternal and divine condition and origin.

Moreover, Peletier's assertion of the similarity between the circle and God himself, just as Foix-Candale's representation of the Theodosian sphere as an image of the ineffable and eternal essence of this figure, resonate with Proclus' words, when he wrote that the circle, as it enfolds on itself, is urged to return back to its indivisible centre and to "the One that is hidden there". According to the Neoplatonic doctrine derived from the philosophical teaching of Plotinus (ca. 204–270)[144] and on the basis of which Proclus developed his metaphysics,[145] the One (τὸ Ἓν) would correspond to the first principle of all things and would produce everything according to a hypostatic and atemporal mode of causation.[146] The Intellect (νοῦς), as the second hypostasis, would proceed or emanate from the One, and the Soul, as the third hypostasis, would proceed from the Intellect, and so forth until all lower degrees of reality have come to existence, each of them ultimately owing their essence and existence to the One.[147] In Proclus' version, this was held to take place at each stage through a three-fold process consisting in rest (μονή), procession or progression (πρόοδος) and return or reversion (ἐπιστροφή),[148] which is applied to the generation of the intelligible circle in the above-quoted passage of his commentary on Euclid.

In Peletier's characterisation of the circle, this three-fold process is therefore expressed through the series of oppositions inherent to the nature of the circle, whereby the simple,

---

[143] On the two principles of the Limit and the Unlimited according to Proclus, see *supra*, p. 106.

[144] On the philosophical doctrine of Plotinus and its influence in late Antiquity, see Armstrong (1995, pp. 236–268) and O'Meara (2010).

[145] On Proclus' theological and philosophical doctrine and its place within Late Antiquity Neoplatonism, see Lloyd (1995, pp. 302–326).

[146] Armstrong (1995, pp. 236–249).

[147] Armstrong (1995, pp. 250–268).

[148] See, for instance, *Elements of Theology*, Prop. 35, in Proclus (transl. Dodds 1933, p. 39): "*Every effect remains (μένει) in its cause, proceeds (πρόεισιν) from it, and reverts (ἐπιστρέφει) upon it.*" (Emphasis proper to the original edition.)

the finite and the self-caused is conjoined with the multiple, infinite and the caused. And in Foix-Candale's commentary on Euclid, it is chiefly expressed through the geometrical notion of *progressus*, as well as through the notion of sphere as generated by the expansion of its magnitude from its centre and by the uniform interruption of this expansion (as based on his notion of figure in his commentary on Df. I.13).

4.9   The Centre of the Sphere as a Divine Principle

As Proclus did for the intelligible circle, both Peletier and Foix-Candale granted a privileged role to the circle and the sphere, respectively, among geometrical figures, given that it corresponds to the first expression of the divine principle. Now, this role was also granted to these figures in view of their ability to express the unity of the centre, as they would both result from its uniform expansion in all directions in the plane or in three dimensions. For this indivisible centre ultimately represents the first image of the unity of an indivisible and all-encompassing God, from which the universe and all it contains has proceeded. Indeed, while Peletier posited, in the *Louange de la Sciance*, a unitarian divine principle from which the universe instantaneously emanated,[149] this conception was expressed in Foix-Candale's *Pimandre* through the representation of the divine Creator as a unitarian principle, from which all things would have proceeded according to the mode of multiplicity, conforming to the Hermetic theological doctrine.[150]

Thus, if Foix-Candale chose to express the mode of generation of magnitudes through the notion of *progressus* rather than through that of *fluxus* in his commentary on Euclid, it is not only because it allowed him to avoid representing the generation of geometrical figures in an instrumental and non-essential manner and because it could to a certain extent solve the issue of the composition of the continuum. But it is also, more fundamentally, because it offered a more adequate representation of the true mode of procession of magnitude from the point, which mirrors the procession of the multiple from the divine One.

Therefore, when Foix-Candale wrote, in his commentary on Euclid's first definition, that the point is the efficient cause of quantity (*signum quantitatem agens*),[151] it would not be because it would have generated the line through its spatial translation and, from there, would have enabled the surface and all other magnitudes to be caused. It was rather because it represented the undivided principle of all quantity, which compares to God and to his

---

[149] See *supra*, pp. 81–82.

[150] Foix-Candale (1579, pp. 185–186): "Il n'est chose parmy nous qui nous represente plus de la nature & essence divine, que ceste unité: laquelle nous avons dict estre commencent un, & seul de toutes choses, qui veritablement n'apartient à autre que au Dieu souverain. [...] *Dieu à esté commencement un & seul de toutes choses*, à cause que c'est le seul, en qui est vraye essence, ferme, & stable, & de laquelle toutes choses, qui ont essence, la doyvent recepvoir, ne prenant son commencement d'ailleurs que de soy mesmes." (My emphasis.)

[151] Foix-Candale (1566, fol. 1r), Df. I.1: "signum esse intelligat, quantitatem agens."

creative operation. Foix-Candale's discourse on this issue should then be taken on a philosophical rather than on a mathematical level. This is again confirmed by what he later wrote in the *Pimandre*, as he asserted that the undivided unit conceived by mathematicians is what allows us to understand the unity of God in the most adequate manner:

> There is nothing among us that shows us the nature and divine essence to a greater extent than this unit, which we have said is the one and only beginning of all things, which truly belongs to none other than to the supreme God. It is indivisible, continuous (as geometers say) as opposed to number, which is discrete or divided. In the same manner, we understand God as one, indivisible and whole in its entire essence, different from all its creatures composed of differents units, all taking their beginning in this sole divine unity [. . .]. Thus, as all numbers and composed things start from the unit, God is the one and unique beginning of all things, since he is the only being whose essence is true, firm and stable, and from which all things that have an essence necessarily receive their essence, as he takes his beginning from nowhere else than from himself.[152]

It is important to note that, in this text, Foix-Candale does not make a clear distinction between the unit as the indivisible principle of magnitudes, that is, the point, and the unit as the principle of numbers. The unit, taken in this sense, is not either clearly distinguished from God as the divine One, nor numbers from created beings, which are themselves said to be divided into a multiplicity of units, setting forth the Pythagorean element of the Hermetic cosmogonic doctrine.[153]

---

[152] Foix-Candale (1579, pp. 185–186). "Il n'est chose parmy nous qui nous represente plus de la nature & essence divine, que ceste unité: laquelle nous avons dict estre commencent un, & seul de toutes choses, qui veritablement n'apartient à autre que au Dieu souverain. Elle est indivisible, continue (comme disent les geometriens) à la differance du nombre, qui est discret, ou departy: de mesme manière *nous entendons Dieu un non divisible ains entier en toute son essence, different de toutes ses creatures composées de diverses unités, toutes prenant leur commencement en ceste seule unité divine* [. . .]. Aussi comme l'unité commence tous nombres & choses composées, *Dieu a esté commencement un & seul de toutes choses*, à cause que c'est le seul, en qui est vraye essence, ferme, & stable, & de laquelle toutes choses, qui ont essence, la doyvent recepvoir, ne prenant son commencement d'ailleurs que de soy mesmes." (My emphasis.) See *supra*, n. 150, p. 106.

[153] See also, in this regard, the parts of the *Pimandre* which aim to display the ontological supremacy of the number 10 over all other numbers and its assimilation to the divine unity, as the proper cause of the world soul and of all beings, allowing all things to be expressed in numerical terms: Foix-Candale (1579, pp. 608–609): "*Car le denaire, O mon fils, est geniteur de l'ame des nombres, & par le moyen duquel les nombres produisent leurs effects*: detant que le denaire en fin se trouvera estre l'unité, non seulement ame des nombres, mais ame de l'univers. A cause que ceste unité ne prend jamais sa perfection que en Dieu, qui est la vraye unité, de laquelle sont produictz & engendrés tous nombres, & laquelle par consequent produict l'ame, donnant effect, efficace & vertu a tous nombres, quelz qu'ils soient, à cause que toutes multitudes sortent de ceste divine unité." This doctrine is also, in this context, explicitly referred to Pythagoras: Foix-Candale (1579, p. 610): "Et ce à cause de ceste grande & secrette vertu, qui se trouve aux nombres, dont la science est passée presque toute en obly, ne nous estant plus resté, que les petites parties de l'Arithmetique pratiquée, pour subvenir à noz necessitez

Now, in the same context, Foix-Candale referred to his previous geometrical discourse, which is undoubtedly his commentary on Euclid's *Elements*,[154] and designated the indivisible principle of quantity as "confused and undetermined".[155] In the last preface of his commentary on Euclid, as he presented Aristotle's concept of quantity, the geometrical point is indeed said to join the parts of continuous magnitude in a "confused manner" (*confusè*), making it one and undivided.[156] Geometrical magnitude is then also often designated, as it also was in the *Pimandre*,[157] as a confused and continuous quantity by opposition to the discrete quantity dealt with in arithmetic.[158]

While this description chiefly referred to the fact that geometrical magnitude cannot be defined as actually composed of discrete indivisible parts, Foix-Candale also asserted, in the same preface to his commentary on Euclid, that the unit corresponds to the divine principle common to both arithmetic and geometry.[159] By saying this, he was indirectly

corporeles: & ne se trouve plus, ou fort peu, qui entendent les secretz des nombres, *que ce grand Pythagoras entendoit*, & autres, desquelz nous en est demeuré la simple histoire, non la doctrine." (My emphasis.)

[154] Foix-Candale (1579, p. 186): "Nous avons quelquefois dict, traitant la Geometrie...". Foix-Candale is not known to have produced another mathematical work.

[155] Foix-Candale (1579, p. 186): "*Nous avons quelquefois dict, traitans la Geometrie, l'unité estre confuse & indeterminée*, a faute de recevoir discretion ou departement. C'est le propre de la divine nature, qui nous est si confuse pour son infinitude, grandeur, multitude, & puissance de vertus infinies, indicible bonté, plenitude de toute intelligence. Que si nous voulons tascher a la comprendre en son unité & integrité, nous nous y trouverons si confus, que nous y perdrons toute cognoissence & jugement: a cause que l'infinitude de toutes vertus & essence ne peut estre comprinse de nous, qui sommes finis." (My emphasis.)

[156] Foix-Candale (1566, sig. e2v): "Continuam igitur dixit eam quantitatem, cuius partes copulantur aliquo communi termino, quo quidem in unicam & indiscretam magnitudinem *confusè* coëunt." (My emphasis.)

[157] Foix-Candale (1579, p. 186).

[158] Foix-Candale (1566, sig. e2v): "Euclidis praecipuum coeptum fuit Geometriae principia edocere, quae cùm res *continuas & ideo confusas* discutiant [...]". See also, when dealing with the arithmetical part of the *Elements*, Foix-Candale (1566, sig. e3r): "Insuper (quod praeponderat) non rectè suscipiuntur Arithmetices obsequia, quibus ratiocinanti lectori *Geometriae confusas quantitates* discernendas tradidit Euclides, numerorum prolatis quibusdam selectis rudimentis: per quae non numerorum praesertim tradendas leges suscipit, sed *quantitatum confusarum (continui causa)* naturam, *numerorum* famulatu discernere conatur, ut postmodum *per discretionem* (quae sola quantitatis certitudinem menti potest inserere) certas ab incertis quantitatibus, & natura & traditione longè divulsas, ac invicem incomparabiles esse proponat, nulli nempe discretioni communicantes eidem." (My emphasis.)

[159] Foix-Candale (1566, sig. e2v): "Arithmeticam autem discernendi esse dicemus scientiam, quæ cum Geometria commune habet divinum illud principium, quod unitatem nommamus."

pointing to the metaphysical understanding of the relation between number and magnitude at an ontologically superior level.

Hence, if it may be striking for the indivisible principle of magnitude to be described as continuous (as it is in the above-quoted passage), since it was defined as deprived of parts in Euclidean geometry, Foix-Candale aimed, through this characterisation, to present the point as both indivisible and as containing all magnitudes virtually. Through this, it could be simultaneously regarded as the proper cause of all magnitudes and as the expression of the oneness and absolute capacity of God, in which all things lay complicated. This omnipotent unit recalls the centre of the circle in Peletier's *De usu geometriae*, which was said to be infinite in the sense that it virtually connects the infinite number of lines proceeding to the circumference.[160] The same may be said of the centre (or innermost part) of figures in Foix-Candale's commentary on the *Elements*, which is conceived as the starting point of the expansion of magnitude.

Therefore, if Foix-Candale related the point to the arithmetical unit, both in his commentary on Euclid and in the *Pimandre*, the condition of the point-unit itself, being characterised as continuous and confused in its oneness in the *Pimandre*, would be closer to the condition of magnitude than to that of number.[161] Now, it is the oneness that is common to both magnitude and to the divine principle of all quantity that allows geometrical objects, and notably the sphere, to express a higher state of being, which is eternal and ineffable.

4.10   Foix-Candale and Peletier on the Epistemological Status of Geometrical Definitions

As was suggested in the previous section, Foix-Candale also joined Peletier concerning the epistemological status of geometrical definitions, and of genetic definitions in particular, by admitting their conjectural character. Indeed, as was shown earlier, Peletier asserted, both in his commentary on the *Elements* and in his *De contactu linearum*,[162] that the causality posited by the geometer between different geometrical objects (the point and the line, or the straight line and the circle) has no relevance on a philosophical level, since it cannot capture the simultaneousness of their causation and their proper essence as entities of divine origin. As such, he advised us to express geometrical concepts in the manner that is most useful for the study and practice of geometry, while keeping in mind their conjectural character.

Foix-Candale first suggested the conjectural character of geometrical definitions in his commentary on Euclid's definition of the sphere, by stating that the essential definition of

---

[160] See *supra*, p. 125.

[161] In the commentary on Euclid, magnitudes are indeed clearly described as one. Foix-Candale (1566, sig. e2v): "Quælibet enim Geometriæ magnitudo una dicitur."

[162] See *supra*, p. 93 and 126.

the sphere (that of Theodosius), even if it is more appropriate than Euclid's definition to express the nature of the sphere, only offers an *image* of its eternal essence. Given the Platonic background of Foix-Candale's representation of mathematical knowledge in the prefaces of his commentary on Euclid, and the ontological conceptions he presented in the commentary on the *Elements* and in the *Pimandre*,[163] the notion of image would entail a distinction between the true state of the geometrical objects within God's mind and the way they are apprehended and defined in the context of geometry. This would also be the case of essential definitions, even if they offer a more faithful expression of the true essence of geometrical objects than mere descriptions, to which belonged Euclid's definition of the sphere.

In particular, the fact for the sphere to be, as any other geometrical figure, represented spatially in the imagination would make it, for Foix-Candale, improper as such to display its intelligible essence, since he wrote in the *Pimandre* that: "ideas that are in themselves a depiction or a figure in the mind only depend on corporeal things, since incorporeal things have no figure or depiction presented to the senses".[164] What is more, within God, the ideas that are spatially represented to the human mind would have no corporeality, as they are only manifested to the divine mind in their intelligibility and deprived of any spatiality.[165] In other words, the process of expansion the geometer unravels in his imagination as he studies the properties of a figure (to which Foix-Candale refers in his commentary on Df. I.13) would itself correspond to an improper conception of a geometrical object's true state. Indeed, even if it is not taken to occur in physical space and time, it nevertheless implies a succession of states of the figure from the beginning to the end of the process, through which the centre, the surface (or volume) and the boundary of the figure may be distinguished.

Hence, while corporeal objects or figures are images of intelligible principles contained in God, one must not, according to Foix-Candale, linger on them and seek within them the

---

[163] Foix-Candale (1566, sig. A6r): "quae sensibus tanquam corpora vel materies attrectantur, sive quae intelligentiae ratiocinanti summi opificis imagini, velut eius essentiae communicant" and Foix-Candale (1579, p. 740): "Ce sont les vrayes idées qui representent, & figurent toutes choses, qui feurent, ou seront jamais faictes qui sont assises eternellement en Dieu, eternel exemplaire de toutes choses, pourtant toute manière de idées & representations figurées."

[164] Foix-Candale (1579, p. 740): "Et ces idées pourtant en soy representation, ou figure en la pensée, ne dependent que des choses corporeles à cause que les incorporeles n'ont aucune figure, ou representation faicte aux sens en presence". See also the following part of the text: "mais sont representées à l'ame par argumentz, ratiocinations, & conclusions sans aucun dessain, ou figure. Parquoy toutes celles, qui se representent soubz quelque pourtraict, delineation, dessain, ou figure a la pensée dependent infalliblement des choses corporeles."

[165] Foix-Candale (1579, p. 740): "Car bien qu'elle soit corporele, ell ne laisse d'estre aussi presente à Dieu, qui est intelligence, sans avoir son corps, que ayant son corps, & ce à cause que Dieu n'usant d'aucun sens corporel, mais de seule partie intelligible n'a besoin que la chose aye corps, pour luy estre manifestée, ains sans la presence de ce corps, ce corps mesme luy est mieux manifesté, qu'il n'est aux sens corporels de l'homme, qui le voit devant soy."

truth to which they refer.[166] In this regard, the properties and mode of generation displayed by the Theodosian definition of the sphere would point, for Foix-Candale, to the essential condition of the sphere without however directly representing the latter in its true state, that is, as an intelligible and eternal idea within God's mind. This is precisely why this eternal essence of the sphere, in the commentary on Euclid's *Elements*, is said to be "ineffable" (*ineffabilis*), that is, beyond all expression.

The conjectural character of geometry is even more clearly asserted by Foix-Candale when dealing with the distinction between numbers and magnitudes, both in the last preface of the commentary on Euclid and in the *Pimandre*, as the nature of magnitude is then said to exceed the human mind because of the indistinction of its parts.[167] For this reason, its properties as quantity may only be understood by the human reason through numbers, which are divisible into units and therefore commensurable. Indeed, for Foix-Candale, the term "confused" as applied to the point-unit and to magnitude is both understood in the mereological and in the epistemic sense, that is, as denoting, on the one hand, mingling, continuity and complication and, on the other hand, intellectual confusion. As such, numbers were regarded by Foix-Candale in both commentaries as a means of unravelling or explicating what remains complicated and undivided, making magnitude, but also the unit common to all quantity, and God himself thereby, more accessible to the human mind. Just as God is apprehended by the human mind through the multiplicity of his creatures, the nature of quantity in general, and of its undivided

---

[166] Foix-Candale (1579, p. 742): "Et ceste veneration, que tu dois faire a ceste imagination, ne se doibt jamais arester sur elles: mais doit passer outre vers la verité representée par elle, comme a la verité elle en porte le nom, *ne proposant que l'ymage ou representation d'une autre chose.*" (My emphasis.) See also, on the status of images in Foix-Candale's *Pimandre*, Harrie (1975, pp. 131–132).

[167] Foix-Candale (1566, sig. e2v): "Quælibet enim Geometriæ magnitudo una dicitur, Arithmetici verò priorem tantùm unam denunciant, eámque insecabilem ac omni divisione sive discretione privatam, quia tamen confusionem saepius parit continuum Geometriae proprium, miscebit Euclides Geometricis principiis ea arithmeticorum principiorum elementa, quae suae arti geometricae clarius edocendae, facere satis apparebunt. Geometriæ etenim est unicus scopus quantitas, quam (absque Arithmetices auxilio) tam confusam urgente discretorum inopia discentibus offert, ut illis quid tantum maius, minus, vel æquale fuerit exponat, quanto quidem maius vel minus, Arithmetices relinquens subsidio. Volens igitur Euclides Geometriæ intelligendæ faciliores perquirere aditus, eiusdem principia per Arithmeticas distinctiones, ubi par erit, explanare non dedignabitur." and Foix-Candale (1579, p. 186): "Nous avons quelquefois dict, traitans la Geometrie, l'unité estre confuse & indeterminée, a faute de recevoir discretion ou departement. C'est le propre de la divine nature, qui nous est si confuse pour son infinitude, grandeur, multitude, & puissance de vertus infinies, indicible bonté, plenitude de toute intelligence. Que si nous voulons tascher a la comprendre en son unité & integrité, nous nous y trouverons si confus, que nous y perdrons toute cognoissance & jugement: a cause que l'infinitude de toutes vertus & essence ne peut estre comprinse de nous, qui sommes finis."

principle, is apprehended through numbers, whose mode of composition may, on the other hand, be fully apprehended and understood by the human intellect.[168]

In other words, if the Theodosian definition of the sphere expressed the true essence and genesis of the sphere more adequately than the Euclidean definition, this definition would remain however, for Foix-Candale, an imperfect expression of an essence and mode of being that is fundamentally beyond the grasp of the human intellect.

Now, the fact that this geometrical representation possesses a conjectural character (whether it is or not more suitable to represent the ontologically superior state of the sphere) should allow us, in principle, to place the motion implied by the *progressus* of the line or the expansion of figures on the same ontological and epistemological level as the genetic definition of the line as resulting from the flow of the point. This latter definition was indeed regarded by Peletier, in the *Louange*, as a mathematical expression of the divine process of emanation of the cosmos from the primordial One. And he also regarded it as an acceptable expression of the mathematical properties of the line in his commentary on the *Elements*.

Thus, if Foix-Candale avoided using genetic definitions which attributed a translational or rotational motion to geometrical objects, it was for other reasons than merely its conjectural character. It was, first of all, because such definitions raised philosophical issues that would impede the properly rational and human understanding and investigation of geometrical notions, notably with regard to the composition of continuous quantity. It was also because it set forth an accidental and quasi-mechanical mode of generation of geometrical objects, rather than one that would be conform to its essential condition and properties. In effect, as we have seen through Fine's case, the genetic definition of the line through the flow or motion of a point could be interpreted as an abstract representation of an instrumental process, such as the dragging of the pointed edge of a stylus on a wax tablet. This definition of the line would therefore have corresponded, for Foix-Candale, to a *descriptio*, which would teach the rule or *modus operandi* followed by the geometer in

---

[168] Foix-Candale (1566, sig. e2v). See *supra*, n. 158, p. 132 and Foix-Candale (1579, p. 186): "Et par ce que tant que l'homme sera en corps materiel, il ne peut venir a pleine & parfaicte liberté de ces vertus divines a cause que, comme dict sainct Pol, corruption ne peut posseder incorrution, nous userons du semblable remede a cognoistre ce que nous pourrons de la nature de ceste divine unité, qui à cause de son integrité, nous est confuse, a celuy que nous usons en la Geometrie pour acquerir la cognoiscence de l'unité, qui nous represente la quantité confuse, a cause de son integrité. Ce remede est de luy aproprier discretion de nombres, c'est-à-dire combien qu'elle ne porte en soy aucune fracture ou division, toutes fois pour en avoir intelligence, nous luy apliquons des nombres, par lesquels nous departons la quantité confuse & entiere, signifiée par cette unité, en diverses & plusieurs unitez, combien qu'elle ne souffre aucune division, offensant son integrité, qui font un nombre representant la substance & principalle nature de la quantité proposée par l'unité, tellement que par la consideration particuliere des autres unitez composants le nombre, que nous avons apliqué à ceste unité, nous retirons intelligence plus familiere par la distribution, discretion, ou despartement de cette unité confuse: ce que nous n'avons peu faire pendant, que nous l'avons considerée une seule entiere, indivise & sans aucune partie."

order to produce a line concretely or imaginarily. Hence, the notion of flow of the point would be closer, in the way it represented the generation of magnitude, to the rotation of the semicircle producing the sphere in Df. XI.14 than to the expansion of the sphere's magnitude from its centre to its circumference. As said, the dismissal of *descriptiones* as proper definitions on account of their instrumental character would be confirmed by Foix-Candale's rejection of mechanical processes in geometry.[169] His rejection of superposition on account of its interpretation as an instrumentally-performed procedure would corroborate this.[170]

Thus, if, for both Foix-Candale and Peletier, the definitions through which geometers express the essential properties of the circle and of the sphere may only be regarded as an attempt to capture, in spatial terms, the proper essences and geneses of these figures, which are in their true state complicated in God's mind and which exceed the limits of human understanding, it remains that, while Foix-Candale dismissed the Euclidean definition of the sphere by rotation of a semicircle in favour of the Theodosian definition, for Peletier the conjectural character of mathematical definitions precisely legitimated the use in geometry of a diversity of genetic definitions. As Peletier wrote in his commentary on Euclid's *Elements*, since all geometrical forms, in their divine state, have appeared simultaneously, the geometer would be perfectly free to admit the straight line as the cause of the circle or, on the contrary, to assume the circle as the cause of the straight line according to the needs of his research and practice.

4.11   Motion in Foix-Candale's Commentary on Book II

On a very different note, I now turn to Foix-Candale's commentary on Euclid's book II, and more specifically on Df. II.1, where he found useful, as most of the commentators considered here, to appeal to motion. Now, despite the fact that he defended the use of numbers to clarify the principles of geometry in the paratext of his commentary on the

---

[169] Foix-Candale (1566, sig. e3r): "Propositiones demùm à principiis genitae, hoc praescripto disponantur, ut subsequentes à prioribus, non autem à posterioribus demonstrandae sint, ac earum demonstrationes solis disciplinae legibus construendae tanta religione decorentur, *ne ullum in eis demonstrandis intercedat mecanici instrumenti iuvamen.*" (My emphasis.)

[170] Foix-Candale (1566, fol. 5v), Prop. I.4: "Alteram demonstrationem huic quartae exhibere cogimur, ne praebeatur aditus, quo ulla mechanicorum usuum instrumenta in demonstrationes incidant." *Cf.* Foix-Candale (1566, fol. 6v), Prop. I.8 and Foix-Candale (1566, fol. 27r), Prop. III.24: "Quoniam Theon & Campanus hanc demonstrare conati sunt, aut hi à quibus demonstrationes sumpserunt, instrumento ferè mechanico, nempe coaptata figura supra figuram, quod indignum traditione mathematica supramodum existimatur." See *supra*, n. 24, p. 101. See also Axworthy (2018, pp. 13–14 and 32–33) and Mancosu (1996, p. 30).

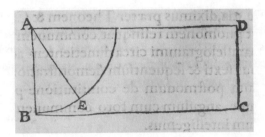

Fig. 4.1 François de Foix-Candale, *Euclidis Megarensis mathematici clarissimi Elementa, Libris XV*, 1578, p. 33, Df. II.1. Diagram illustrating the extent of the variation of the angle DAE. It represents the angle BAD in the cases when AD does not stand at right angles with AB and shows the extent of the variability of the quadrilateral generated thereby by the motion of the line AD along AE. Courtesy Max Planck Institute for the History of Science, Berlin

Elements,[171] he did not however appeal to an arithmetical interpretation of this definition, nor of any other part of Book II for that matter.[172]

In his commentary on Df. II.1, he appealed to motion in order to explain why the parallelograms considered by Euclid in the context of Book II are necessarily rectangular.

> The right angle, as it consists in a unique inclination of lines, expresses the same breadth on account of its length, admitting no other measure. This surely is not the case of the acute or obtuse angle, since there could be, for each, infinite differences in the inclination of the lines. Euclid says therefore that the rectangular parallelogram is formed under the two [lines] containing the right angle, so that one of them, that is, its length or its breadth, expresses the other without any error. [...] If we wanted however to understand this through motion, we would say that, as far as the point A will move along the line AB until [it reaches] the point B, as much each part of the straight line AD will simultaneously be produced until [it reaches] the straight line BC; and this motion will describe the rectangular parallelogram ABCD, since the motion of the length takes place at right angles. If however the point A is moved along AE according to an oblique angle, the motion of the line AD will similarly describe a rhombus or a rhomboid parallelogram (Fig. 4.1).[173]

---

[171] See *supra*, n. 167, p. 135.

[172] He only proposed an arithmetical treatment of magnitudes when dealing with the theory of ratios and proportions, as in his commentary on the definitions of Book V.

[173] Foix-Candale (1566, fol. 15r), Df. II.1: "Nam is rectus angulus unica inclinatione linearum compositus, eandem pro suae lineae longitudine exprimit latitudinis quantitatem, nihil sibi de alia dimensione sumens. Quòd quidem non efficit acutus aut obtusus, eò quòd utriusque eorum infinitae sint diversitates inclinationis linearum. Parallelogrammum igitur dixit Euclides rectangulum sub duabus rectum angulum continentibus exprimi, ut altera earum longitudinem reliqua verò eius latitudinem nulla fallacia exprimat. [...] Si verò per motum intelligere velimus, dicemus quantum efficiet motum signum *a* per lineam *ab* usque ad *b*, tantùm efficere simul singulas rectae *ad* partes, usque ad rectam *bc*, & is motus describet parallelogrammum *abcd*, & rectangulum, cùm motus fiat ad rectum angulum longitudinis. Si verò per obliquum moveatur angulum ut per *ae* signum *a*, is motus lineae, *ad* rhombum aut rhomboidem similiter describet."

The motion introduced here is not one that would properly generate the surface of a parallelogram, but rather one that would measure a parallelogram ABCD that is already given, since the given point A is said to move along the given line AB while the given line AD is said to move along the given line AB to reach the given line BC. This use of the motion of geometrical figures may be interpreted according to the Greek terms διεξέρχομαι, διέρχομαι and δίειμι, such as used by Autolycus of Pitane (c. 360—c. 290 BC) in his treatises *On the moving sphere* and *On risings and settings* and by Pappus, in the *Collections*, to designate the motion of a point on a curve, since then the point or the line is said to move along an already given line or surface.[174] The intention of the geometer when appealing to this notion may of course differ, as it relates in Autolycus to the motion of a celestial body along a circle, which represents its path in the firmament.

Also, if what is conceived as properly moved, in Foix-Candale's commentary on Df. II.1, is only one of the extremities (the point A) of the moving line-segment (AD), this point is also understood to carry with it the whole line-segment AD along the length of the line-segment AB, which is situated at right angles from AD. In this way, the motion of the point A could be imagined to draw out the whole area of the parallelogram ABCD. This procedure would then be understood as a reconstitution of the generative process through which the parallelogram would have been initially drawn out.

Hence, the motion through which the parallelogram is said to be measured in this context, given that it also takes place through the transversal motion of a line-segment across a given area or surface, would be quite similar, as for its ontological status, to the motion by which the semicircle is said to generate the sphere in Euclid's Df. XI.14 and which Foix-Candale considered as accidental. If this process were formulated as a genetic definition, it would be equivalent to the definition of the rectangular parallelogram as generated by the transversal flow or motion of a line, then corresponding to a *descriptio*.[175] Thus, for Foix-Candale, this motion would be different in its nature and ontological status from the *progressus* or expansion referred to in the first definitions of Book I, and which was presented as intrinsic to the generated magnitude, even if it only corresponds to a humanly apprehensible image of a metaphysical process. And that would be the case whether it is the point A or the whole line-segment AD that is conceived as the mobile element and whether this motion is conceived as a properly generative process or rather as a recreation of the figure in the aim of measuring its area.

As was noted above, even if Foix-Candale explicitly presented the two sides of the parallelogram as measuring its area, as well as mutually measuring each other (as would be

---

[174]Mugler (1958–1959, I, p. 139–140), under διεξέρχεσθαι (Autolycus, *On the moving sphere*, Df. 1–2 and Pappus, *Collections* VI.22), διέρχεσθαι (see, for instance, Autolycus, *On the moving sphere*, § 10 and Pappus, IV.31), and διέναι (Autolycus, *On risings and settings*, II, 7 and 11 and Pappus, VI.22 and 68).

[175]The assimilation of the generation of the parallelogram to the *descriptio* of the sphere according to Euclid's definition is also marked by Foix-Candale's use of the verb *describo* (*describet*) in this context.

expected in the context of Book II), he did not go as far as to compare, as did Fine and Peletier, the generation of the rectangular parallelogram to the multiplication of two numbers. Indeed, if he stated the usefulness of arithmetic to understand the principles of geometry in the paratext of his commentary, for which he justified Euclid's inclusion of arithmetical books in the *Elements*[176]—he considered indeed the *Elements* first and foremost as a geometrical treatise[177] –, he did not however consider it appropriate to directly express the propositions of Book II in arithmetical terms, as some of commentators did.[178] For that matter, according to Foix-Candale, the arithmetical content of the *Elements* only aimed to help the reader gain a full understanding of the properties and mutual relations of continuous quantities, which are by themselves source of confusion for the human intellect.[179] Therefore, if, at a philosophical and theological level, numbers and magnitudes are thought to have sprung from the same principle and are conjoined in God's mind, it remains that, for the human mind and in a mathematical context, discrete and continuous quantities themselves are to be clearly distinguished. This, by itself, confirms that one of the reasons why Foix-Candale avoided the notion of flow of the point, beside its quasi-instrumental and non-essential character, is that it would overthrow the distinction between discrete and continuous quantities, as for their proper modes of composition.

* * *

As we have seen, Foix-Candale provided, in his commentary on Euclid, genetic definitions of the line, of the surface and of figures in general, but not as generated by the flow or motion of a point or a line, despite his assertion of the causal role of the point in relation to continuous quantity. Instead, he defined geometrical objects as originating from the *progressus* or expansion of magnitude in one, two or three dimensions and by the

---

[176] (Foix-Candale 1566, sig. e3v–e4r): "Reperimus Euclide discreta Arithmetices exequia in auxilium continuorum exponendorum implorante, aliquos eius interpretes, Arithmetices cum Geometriae rudimentis principia confudisse, ac utraque eadem interpretatione elucidanda forè existimasse, discreti à continuo latam illam non venerantes discrepantiam, tum & natura, & methodo ipsis ab aeterno insitam magnitudinis scilicet naturam pro actione quae ipsi elucidandae confertur sumentes, in id ferè delusi sunt, ut pro substantia accidens exponendum esse arbitrati sint, nempe multiplicem magnitudinem pro ipsis collata multiplicatione recipiendam esse coniecerint, ac alia plura quae latius posthac Altissimi nutu ostensuri sumus, ab impropria rudimentorum expositione genita. Caeterùm tres numerorum libri sextum è regione sequentes, aequè ut reliqui ab Euclide in rudimentorum Geometricorum gratiam elucidandorum, non in Arithmeticem edocendam prolati sunt: is nanque numeros ad arcanam continuarum & ideo confusarum quantitatum intelligentiam exponendam sua discretione sumpsit, tum etiam ut illam exprimat inter inter quantitates naturae distantiam, qua quantitates habitudinem certam ac perceptibilem inter se habent, desciscunt ab iis quae incertum, indeterminatum, ac penitus ob omni tuta intelligentia detrusum inter se habent respectum."

[177] See *supra*, n. 167, p. 135.

[178] Examples of this approach were provided by Billingsley and Clavius. See *infra*, pp. 161–164 and 221–225.

[179] See *supra*, p. 132.

interruption of this process by the corresponding extremity or boundary (point, line or surface).

If Foix-Candale did present the point as the efficient principle of magnitude in his commentary on Df. I.1, it would be linked to his philosophical conception of the point-unit as a common principle of geometry and arithmetic, to which he attributed a divine status. This was expressed in more explicit terms in his commentary on the first treatise of the *Corpus hermeticum*, the *Poimandres*. In this context, the unit, presented as the principle of quantity in general, is what allows us to understand, to the best of our abilities, the nature of God, which exceeds the limits of the human mind. Just as all things are indistinctly present in God in an intelligible manner, the unit virtually contains all magnitudes, as well as all numbers, within it.

Although this discourse is to be placed on a metaphysical rather than on a mathematical level, it resonates with the notion of expansion of the figure that was presented in the commentary on Euclid. As Foix-Candale expressed it when commenting on Euclid's definition of the sphere, Theodosius' definition, which suggests the expansion of the sphere's magnitude from its centre, would offer an image of the eternal and ineffable essence of this geometrical figure, which, as made clear in the *Pimandre*, surpassed all geometrical figures in perfection. According to this Hermetic text, the universe would have been modelled by God in the shape of a sphere in view of its ability to express omnipotence and unity. Within the universe, the process of expansion of the sphere from a point would be represented by the expansion of light from the Sun, which represented the physical expression of the divine light that causes and makes all things known.

Foix-Candale's representation of the sphere echoed in various ways Peletier's representation of the circle, which stemmed from their common Platonic and Neoplatonic influences, as related to Renaissance Christian Hermeticism and Cusanian theology and epistemology. Hence, in line with Proclus' characterisation of the circle, Peletier and Foix-Candale ultimately asserted the coexistence, within the intelligible circle or sphere, of the centre, the figure and the boundary, as well as the procession of the figure from itself according to a non-temporal and non-spatial mode. In this framework, the power of the point to unravel into the circular or spherical figure and to enclose its own magnitude into a uniform whole would be identified with the power of the divine One from which all things would have proceeded and in which all things coexist indistinctly according to the Neoplatonic doctrine.

While most of the medieval and Renaissance scholars who compared the two definitions of the sphere by Euclid and Theodosius distinguished these in view of the former's appeal to motion, considering that the latter did not imply any motion, Foix-Candale would have considered both as involving motion. And while the motion implied by the Theodosian definition of the sphere would only provide a conjectural representation of this ontologically superior state and mode of causation, it would still be situated at a greater degree of truth compared to Euclid's definition of the sphere. Foix-Candale actually did not regard this definition as a proper definition given that it would only present the accidental and externally-determined properties and mode of generation of the sphere, for which it was designated as a *descriptio*. The same would have been held by Foix-Candale of the more traditional genetic definitions of the line, surface and solid as generated by the flow of a point, line or surface, respectively, as these could all be taken to evoke their mechanical generations.

Thus, while it could be conceded that a *descriptio* may be useful or even necessary to account for the spatial properties of the line, as was asserted by Fine and Peletier, Foix-Candale did not generally appeal to these notions in this aim, save in Book II, where he presented the area of the rectangular parallelogram as generated by the transversal motion of one of its sides. Yet, in this context, the fact of presenting the rectangular parallelogram as produced by the motion of a line-segment, and not as resulting from the expansion of magnitude in two dimensions, would be acceptable insofar as Foix-Candale did not intend then to provide a universal definition of this figure, but rather to express the operation performed by the geometer when assessing the quantitative relation of its sides to its area.

Yet, contrary to Fine and Peletier, and to the other commentators considered here, Foix-Candale did not go as far as to directly compare the generation of the rectangular parallelogram to the multiplication of two numbers in his commentary on Df. II.1. Although he considered that the understanding of geometrical properties required the study of arithmetical principles, he mostly avoided appealing to arithmetical notions in the context of Euclid's geometrical propositions.

Therefore, like Peletier, Foix-Candale acknowledged the ontological and epistemological boundary between the mathematical treatment of geometrical objects and their philosophical consideration while attempting as much as possible to make them mutually compatible. In doing so, he acknowledged the difference between properly essential and accidental, or intrinsic and extrinsic, kinematic processes in order to avoid attributing a concrete and somewhat mechanical type of motion to intelligible beings. He nevertheless diverged from Peletier insofar as the latter considered it perfectly legitimate to define the circle according to the mode of a *descriptio* in the context of geometry on account of the necessarily conjectural character of the geometer's knowledge concerning the true essence and geneses of his objects of study.

Henry Billingsley

5

5.1   The Life and Work of Henry Billingsley and the Significance of His Commentary on the *Elements*

Henry Billingsley (London,[1] c. 1538—London, 1606) was the son of a haberdasher from London. He studied at St. John's College, Cambridge in 1550–1551 and at Oxford, where he would have developed his taste for mathematics. He initially followed his father's path, having been made in 1560 a freeman of the Company of Haberdashers.[2] He later became sheriff of London in 1584, alderman of Tower Ward from 1585 to 1592 and, among other functions, one of Queen Elizabeth's customs collectors in 1589. He later acceded to the function of Lord Mayor of London in 1596 and received knighthood in 1597. He died in 1606.

Apart from a translation of the commentary on the *Epistle to the Romans* of St. Paul by Pietro Martire Vermigli (1499–1562), published in 1568,[3] Billingsley produced an English

---

[1] This is assumed by his biographers on the basis that his father's activity as haberdasher was settled in London.

[2] On Billingsley's life and career, see Lee (1886) and McConnell (2008).

[3] Vermigli (transl. Billingsley 1568). Billingsley mentioned, in his commentary on the *Elements*, that he had completed an English translation of the *Spherics* of Theodosius, which he intended to publish, but I have not found any traces of this translation: Billingsley (1570, fol. 315v), Df. XI.12: "Theodosius in his booke *De Sphericis* (a booke very necessary for all those which will see the groundes and principles of Geometrie and Astronomie, *which also I have translated into our vulgare tounge, ready to the presse*)". (My emphasis.)

© The Author(s), under exclusive license to Springer Nature Switzerland AG 2021
A. Axworthy, *Motion and Genetic Definitions in the Sixteenth-Century Euclidean Tradition*, Frontiers in the History of Science,
https://doi.org/10.1007/978-3-030-95817-6_5

translation and commentary of the fifteen books of Euclid's *Elements*, published in 1570.[4] This work was prefaced, partly corrected and augmented by John Dee. Billingsley would have used, for his translation and commentary, a copy of a 1558 edition of Zamberti's Latin translation (from the Greek), which he heavily annotated.[5] This copy was bound with the 1533 *editio princeps* of Euclid's text edited by Simon Grynaeus (1493–1541), which also contained a few annotations by the hand of Billingsley.[6] For this reason, it was believed that Billingsley produced his English translation directly from the Greek.[7]

Billingsley referred in his commentary on Euclid to many ancient and modern commentators of Euclid's *Elements*, and, among these, to many French authors, i.e. Fine, Peletier, Foix-Candale, as well as Pierre de Montdoré, who published a commentary on the tenth book of the *Elements* in 1551.[8] He also took up Foix-Candale's additional sixteenth book on regular polyhedra (*The sixteenth booke of the Elementes of Geometrie added by Flußas*[9]). He indicated, in his epistle to the reader, that he included in his work "manifolde additions, Scholies, Annotations, and Inventions (. . .) gathered out of many of the most famous & chiefe Mathematiciens, both of old time, and in our age".[10] Furthermore, he introduced each book with a preface on the topic of the considered book, the ontological value of its objects and the usefulness of its propositions. He also presented the methods then used by Euclid, the history of that particular book, the knowledge and tools that are necessary to understand the propositions and the mode of exposition he himself adopted henceforth to facilitate its comprehension by the reader.

This commentary was of significant importance for the diffusion of Euclid's *Elements* in the vernacular, as it was the first English translation of Euclid's text. With respect to its style and content, this commentary displays both a will to reconstitute the true meaning of

---

[4] On this English translation and commentary on Euclid's *Elements*, see Halsted (1879), Archibald (1950), Simpkins (1966, pp. 230–235), Mandosio (2003, pp. 480–481), Harkness (2007, pp. 109–115).

[5] Zamberti (1558). The fact that this volume belonged to and was annotated by Billingsley is indicated by the presence of the manuscript inscription *Henricus Billingsley* on the title page. This title page is reproduced in Archibald (1950, p. 447). This was previously noted in Halsted (1879), but R. Archibald was able to verify that the style of the inscription corresponds to Billingsley's handwriting.

[6] Archibald (1950, p. 448). See also Halsted (1879).

[7] De Morgan (1837, pp. 38–39), Halsted (1879) and Simpkins (1966). The fact that Billingsley made his translation from the Greek text has been rather convincingly refuted by R. Archibald (1950, p. 449).

[8] See for example, for Fine, Billingsley (1570, fol. 95v), for Peletier (fol. 22v), for Foix-Candale (fol. 31r) and for Montdoré (fol. 137v). For this latter reference, see Montdoré (1551). For a more complete list of the sources mentioned by Billingsley, see Archibald (1950, p. 449).

[9] See *supra*, n. 14, p. 100.

[10] Billingsley (1570, sig. 2v).

Euclid's discourse[11] and to adapt it to the interests of a new readership among the English population, in particular among people who were versed in practical and applied mathematics.[12] The result is a unique and voluminous work of nearly 1000 pages. Among its most remarkable features are paper solids to cut out, fold and paste to help the reader better understand the stereometrical concepts taught by Euclid in Book XI. It also includes a 50 page-long preface by John Dee. In this preface, Dee offered a full reassessment of the scope and objects of mathematics according to a philosophical and esoteric perspective, as well as an extensive promotion of the technical uses of mathematics in a variety of domains, which he displayed through a tree diagram in the *Groundplat* annexed to his *Mathematicall praeface*.

After looking at the status and designation of motion and genetic definitions in Billingsley's commentary on Euclid, I will consider his discourse on the two definitions of the sphere, comparing it with that of Foix-Candale. I will also present the different meanings Billingsley attributed to the notion of "description", as well as his kinematic treatment of magnitudes in his commentary on Df. II.1.

This analysis will reveal, in particular, that Billingsley, in his consideration of genetic definitions, did not so much point to a metaphysical interpretation of the origin of geometrical objects, but emphasised rather the relationship between the abstract and concrete modes of production of geometrical objects. We will also see that, although Billingsley admitted a similar distinction as Foix-Candale between genetic definitions and definitions by property, he endowed genetic definitions with a stronger epistemic value than his French predecessor did. Additionally, this chapter will show how Billingsley saw in the motions of figures not only an object of geometry in its own right, but also a key to understanding some of the principles of arithmetic and algebra.

5.2   Motion in Euclidean Geometry According to Billingsley

In a manner quite similar to Proclus,[13] Billingsley wrote, in his preface on Book XI (i.e. the first of Euclid's books on solid geometry), that Euclid's teaching on plane and solid geometry aimed to display not only the properties of geometrical objects, but also their generations.

---

[11] Billingsley notably took care to indicate the different parts of each proposition in the margin (e.g. construction, demonstration, things given, things required. . .).

[12] On practical mathematical culture in early modern England, see Harkness (2007, chap. 3).

[13] Proclus (Friedlein 1873, p. 57) and (Morrow 1992, p. 46): "Let us next speak of the science itself that investigates these forms. Magnitudes, figures and their boundaries, and the ratios that are found in them, as well as their properties, their various positions and motions—these are what geometry studies". See *supra*, n. 55, p. 12.

Hitherto hath Euclide in these former bookes [I–X] with a wonderfull Methode and order entreated of such kindes of figures superficial, which are or may be described in a superficies or plaine. And hath taught and set forth their properties, natures, *generations*, and *productions* even from the first roote, ground, and beginning of them: namely, from a point, which although it be indivisible, yet is it the beginning of all quantitie, and of it and of the *motion* and *flowing* therof is produced a line, and consequently all quantitie continuall, as all figures playne and solide what so ever. [. . .]. Now in these bookes [XI–XIII] following he entreateth of figures of an other kinde, namely, of bodely figures: as of Cubes, Piramids, Cones, Columnes, Cilinders, Parallelipipedons, Spheres and such others: and sheweth the diversitie of them, the *generation*, and *production* of them, and demonstrateth with great and wonderfull art, their proprieties and passions, with all their natures and conditions.[14]

Hence, according to this passage, the generations of lines and figures constitute one of the objects of Euclidean geometry. The flow of the point, which is explicitly presented as a motion in the proper sense, is here defined as the origin and foundation of all magnitudes, from lines to the variety of solids presented in the last books of the *Elements*.[15]

In his commentary on Df. I.2, Billingsley presented the genetic definition of the line as one definition of the line among others, as had done Proclus.[16]

A lyne is the movyng of a poynte, as the motion or draught of a pinne or a penne to your sense maketh a lyne.[17]

In this context, the generative process originating in the point is not designated as a flow, but only as a motion (a "movyng").

5.3   The Ontological Status of Geometrical Motion

When Billingsley provided the above-quoted genetic definition of the line in his commentary on Df. I.2, he made no explicit or implicit reference to any metaphysical understanding of the generative role of the point. Rather, this process was interpreted in a quasi-instrumental sense, since it was compared to the motion of a pin, a stylus or a pointed tool drawing out a concrete line as it is dragged on a material plane surface. This comparison evokes the concept of *vestigium* left by the point in its motion to generate the line which was presented by Fine, as by Peletier for the definition of the circle, only in a much more concrete manner.

---

[14] Billingsley (1570, fol. 311v).

[15] Although he wrote that the flow of the point is the principle of quantity in general at first ("from a point, which although it be indivisible, yet is it the beginning of *all quantitie*"), he afterwards limited it to continuous quantity ("of the motion and flowing therof is produced a line, and consequently *all quantitie continuall*").

[16] Proclus (Friedlein 1873, p. 97). See *supra*, n. 107, p. 25.

[17] Billingsley (1570, fol. 1v), Df. I.2.

Billingsley also implicitly made the connection between the genetic definition of the line and its concrete production in his commentary on Df. I.16, where he set forth the kinematic definition of the circle, and related it to the genetic definitions of the line, the surface and the body.

> For the more easy declaration, that all the lines drawn from the centre of the circle to the circumference, are equall, ye must note, that although a line be not made of pointes: yet a point, by his motion or draught, describeth a line. Likewise a line drawen, or moved, describeth a superficies: also a superficies being moved maketh a solide or bodie. Now then imagine the line AB, (the point A being fixed) to be moved about in a plaine superficies, drawing the point B continually about the point A, till it returne to the place where it began first to move: so shall the point B, by this motion, describe the circumference of the circle, and the point A being fixed, is the centre of the circle (Fig. 5.1). Which in all the time of the motion of the line, had like distance: from the circumference, namely, the length of the line AB. And for that al the lines drawn from the centre to the circumference are described of that line, they are also equal unto it, & betwene themselves.[18]

In this context, the line and the surface are said to be "drawn or moved", causing in their motion a surface and a solid, respectively, in the same way as the point was said, in Df. I.2, to describe a line by its "motion or draught". Hence, Billingsley then clearly equates the action of drawing a line from a point to another (as in the context of a construction) to the motion attributed to the point in the context of the genetic definition of the line. It may be additionally noted that, as Billingsley outlined here the process through which the circle is produced by the motion of a line-segment, he made a reference to time ("in all the time of the motion of the line"), which would tend to assimilate geometrical motion to a mechanical or physical process.

Thus, instead of relating the generative motion of the point to a metaphysical process, the former representing a spatial and imaginary representation of the latter as in Proclus, Billingsley pointed to the correspondence or the relation between the spatial structure of geometrical lines and that of physical lines. He related them, in particular, to lines that may be concretely drawn out by the student when studying the properties of geometrical figures. This is not only suggested by the comparison made in Df. I.2 between the geometrical line and the instrumentally-produced line as for their respective modes of production, but also by Billingsley's more general emphasis on the hands-on approach to geometrical learning in his commentary on Euclid.[19] It is then not surprising that Billingsley referred to the genetic definition of the line in his commentary on Euclid's second postulate, directly identifying the geometrical line to a "draught from one point to another",[20] and that he

---

[18] Billingsley (1570, fol. 3r), Df. I.16.

[19] Harkness (2007, pp. 132–136), Barany (2010), Taylor (2011) and Friedman (2018, pp. 73–76).

[20] Billingsley (1570, fol. 5v–6r), Post. 2: "And a line is a draught from one point to an other, therfore from the point b, which is the ende of the line ab, may be drawn a line to some other point, as to the point c, and from that to an other, and so infinitely."

Fig. 5.1 Henry Billingsley, *The Elements of Geometrie of the most auncient Philosopher Euclide of Megara*, 1570, 3r, Df. I.16. Diagram illustrating the generation of the circle by the rotation of the line-segment AB around its fixed extremity A. Courtesy History of Science Collections, University of Oklahoma Libraries

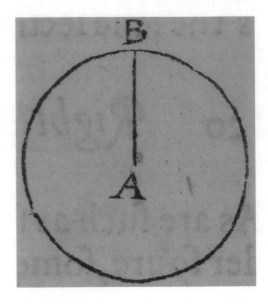

referred to the instrumental construction of the circle by the means of the compass in his commentary on the third postulate.

> A playne superficies may in compasse be extended infinitely: as from any pointe to any pointe may be drawen a right line, by reason wherof it commeth to passe that a circle may be described upon any centre and at any space or distance.[21]

The explicit reference which Billingsley made here to the use of the compass contrasts however with the assertion of the potentially infinite extension of the drawn circle, and therefore of the potentially infinite opening of the legs of the compass. But this is due to Billingsley's will to explain the meaning of Euclid's postulate, which states that it is possible to draw a circle with any center and *with any radius*. This manifests the theoretical scope of his discourse, which he nevertheless intended to support through a hands-on approach to Euclid's geometrical notions to allow students and unlearned readers to understand these more easily.

Indeed, in spite of his reference to the instrumental procedures as a means of illustrating genetic definitions, Billingsley admitted the ontological distance between geometrical and physical objects, since, as he stated in his commentary on Df. I.7, a partless point, a line without breadth, or a surface without depth cannot be found in nature.

> Here must you consider when there is in Geometry mention made of pointes, lines, circles, triangles, or of any other figures, ye may not conceyve of them as they be in matter, as in woode, in mettall, in paper, or in any such lyke, for so is there no lyne, but hath some breadth,

---

[21] Billingsley (1570, fol. 6r), Post. 3.

and may be devided, not points, but that shal have some partes, and may also be devided, and so of others. But you must conceive them in mynde, plucking them by imagination from all matter, so shall ye understande them truly and perfectly, in their owne nature as they are defined. As a lyne to be long, and not broade, and a poynte to be so little, that I shall have no part at all.[22]

The geometrical point, line or surface cannot be found in nature, but may be apprehended by and within the imagination, which separates quantity from matter, conforming to Aristotle's ontology of mathematics.[23] This assertion suggests that Billingsley did not, in this context, attribute an ontological precedence to mathematical objects over physical beings, as in the Pythagorean, Platonic and Neoplatonic doctrines, and that he did not conceive the generative motion of the point as the spatial image of a non-spatial and suprasensible process. It would furthermore corroborate the fact that he did not, contrary to Peletier, make any difference between the flow and the motion of geometrical objects, nor between the mode of generation of the line and those of other geometrical objects as for their philosophical implications.

However, if the point, for Billingsley, may be conceived as moving and as causing the line through its imaginary motion, it is because, unlike the arithmetical unit, it would possess a certain materiality, which is why it may be described as a unit with a position, according to the definition of the point that was attributed to the Pythagoreans by Aristotle.[24]

A signe or point is of Pithagoras Scholers after this manner defined. A poynt is an unitie which hath position. Numbers are conceaved in mynde without any forme & figure, and therefore without matter wheron to receave figure, & consequently without place and position. Where-fore unitie beyng a part of number, hath no position, or determinate place. Wherby it is manifest, that number is more simple and pure then is magnitude, and also immateriall: and so unity which is the beginning of number, is lesse materiall then a signe or poynt, which is the beginnyng of magnitude. For a poynt is materiall, and requireth position and place, and therby differeth from unitie.[25]

The interpretation of the geometrical point as material, which constrasts with the essential simplicity and purity Billingsley here attributes to the arithmetical unit and to number, evokes again Proclus' notion of imagination as intelligible matter. As we have seen earlier,

---

[22] Billingsley (1570, fol. 2r–v), Df. I.7.

[23] See *supra*, p. 10.

[24] This Pythagorean definition of the point as a unit which is endowed with a position is mentioned by Aristotle in *De anima* I.4, 409a6 (transl. Barnes 1995, I, p. 652): "for a point is a unit having position", and, in *Metaphysics* M.8, 1084b26, by defining the unit as a point without position (Barnes 1995, vol. II, p. 1714): "for the unit is a point without position." See also Proclus (Friedlein 1873, p. 95) and (Morrow 1992, p. 78): "Since the Pythagoreans, however, define the point as a unit that has position, we ought to inquire what they mean by saying this."

[25] Billingsley (1570, fol. 1r), Df. I.1.

it is indeed, for Proclus, the intelligible matter of the imagination that would allow the concepts of geometry to admit spatiality and divisibility, without which they would remain unknowable to the geometer given their non-spatial and indivisible state as intelligible forms.[26] More specifically, this passage appears to indirectly quote a passage of Proclus' commentary on Euclid's Df. I.1, where it is said that the unit and number exist only in thought, while the point and magnitudes are projected onto the matter of the imagination.[27] In fact, for Proclus, as here for Billingsley, numbers and the unit cannot be spatially represented in the imagination, which is why they should be regarded as devoid of matter and place or position. Hence, if the geometrical point ressembles the arithmetical unit inasmuch as it is indivisible and is the origin of its proper type of quantity, the fact that it may be attributed motion and that it may be understood to generate magnitude would be precisely due to its quasi-material nature and to its possessing position and place.

---

[26] See, for instance, Proclus (Friedlein 1873, pp. 54–55) and (Morrow 1992, pp. 43–44): "When, therefore, geometry says something about the circle or its diameter, or about its accidental characteristics, such as tangents to it or segments of it and the like, let us not say that it is instructing us either about the circles in the sense world, for it attempts to abstract from them, or about the form in the understanding. For the circle [in the understanding] is one, yet geometry speaks of many circles, setting them forth individually and studying the identical features in all of them; and that circle [in the understanding] is indivisible, yet the circle in geometry is divisible. Nevertheless we must grant the geometer that he is investigating the universal, only this universal is obviously the universal present in the imagined circles. Thus while he sees one circle [the circle in imagination], he is studying another, the circle in the understanding, yet he makes hid demonstrations about the former. For the under-standing contains the ideas but, being unable to see them when they are wrapped up, unfolds and exposes them and presents them to the imagination sitting in the vestibule; and in imagination, or with its aid, it explicates its knowledge of them, happy in their separation from sensible things and finding in *the matter of imagination* a medium apt for receiving its forms." (My emphasis.)

[27] Proclus (Friedlein 1873, pp. 95–96) and (Morrow 1992, p. 78): "Since the Pythagoreans, however, define the point as a unit that has position, we ought to inquire what they mean by saying this. That numbers are purer and more immaterial than magnitudes and that the starting-point of numbers is simpler than that of magnitudes are clear to everyone. But when they speak of the unit as not having position, I think they are indicating that unity and number—that is, abstract number—have their existence in thought; and that is why each number, such as five or seven, appears to every mind as one and not many, and as free of any extraneous figure or form. By contrast the point is projected in imagination and comes to be, as it were, in a place and embodied in intelligible matter. Hence the unit is without position, since it is immaterial and outside all extension and place; but the point has position because it occurs in the bosom of imagination and is therefore enmattered."

5.4   The Two Definitions of the Sphere: Essential Versus Causal Definition

The status of the generative motion of the point is also specified in Billingsley's commentary on Euclid's definition of the sphere (Df. XI.14). In this context, the mode of generation of the line is directly related to the rotative motion of the semicircle generating the sphere, as it was previously related to the generation of the circle in the commentary on Df. I.16.[28]

> A *Sphere is a figure which is made, when the diameter of a semicircle abiding fixed, the semicircle is turned round about, untill it returne unto the selfe same place from whence it began to be moved.* To the end we may fully and perfectly understand this definition, how a Sphere is produced of the motion of a semicircle, is shall be expedient to consider how quantities Mathematically are by imagination conceaved to be produced, by flowing and motion, as was somewhat touched in the beginning of the first booke. Ever the lesse quantitie by his motion bringeth forth the quantitie next above it. As a point moving, flowing, or gliding, bringeth forth a line, which is the first quantitie, and next to a point. A line moving produceth a superficies, which is the second quantitie, and next unto a line. And last of all, a superficies moving bringeth forth a solide or body, which is the third & last quantitie (Fig. 5.2).[29]

By comparing the Euclidean mode of definition of the sphere to the conception of the line as resulting from the motion, gliding or flow of a point,[30] Billingsley established here a direct relation, as for their ontological and epistemological status, between the generation of the line by the flow of the point and the generation of the sphere, the cone or the cylinder, as defined by Euclid. This is all the more important in this context as Billingsley compared the Euclidean and the Theodosian definitions of the sphere in the same way as Foix-Candale, who, as previously shown, had dismissed translational generations of lines, surfaces and solids as proper definitions.

> This definition of Theodosius is more essentiall and naturall, then is the other geven by Euclide. The other did not so much declare the incard nature and substance of a Sphere, as it shewed the industry and knowledge of a producing of a Sphere, and therefore is a causall definition geven by the cause efficient, or rather a description then a definition. But this definition is very essentiall, declaring the nature and substance of a Sphere.[31]

As Foix-Candale, Billingsley stated that the definition commonly ascribed to Theodosius sets forth the essence of the sphere, while Euclid's definition would rather present its mode

---

[28] See *supra*, p. 147.

[29] Billingsley (1570, fol. 315r), Df. XI.12.

[30] The combined use of the verbs *move, flow* and *glide* to describe the generation of the line from the point tends to suggest that the flowing of the point is here understood in a quasi-physical manner, at least as relatable to concrete types of local motion.

[31] Billingsley (1570, fol. 315v), Df. XI.12.

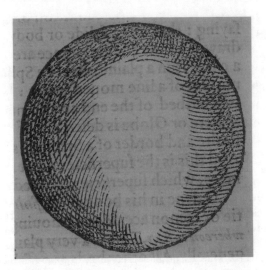

Fig. 5.2 Henry Billingsley, *The Elements of Geometrie*, 1570, 315r, Df. XI.12. Diagram representing the geometrical sphere. It may be noted how the three-dimensionality of the figure is suggested by the use of shadows, which gives the drawn figure a quasi-material character and is coherent with Billingsley's will to compare imaginary and concrete productions of geometrical figures in his commentary on the *Elements*. Courtesy History of Science Collections, University of Oklahoma Libraries

of production, which he called a description, as had done Fine for the circle and Foix-Candale for the sphere.

5.5   The Description of Geometrical Figures According to Billingsley

The notion of description appears in different contexts in Billingsley's commentary and with different meanings. While it clearly referred, in his commentary on Df. XI.14, to the generation of the figure within the imagination, and to the genetic definition itself thereby, the term *description* was also used in other parts of his commentary on Book XI to designate the concrete diagram that may be drawn out by the student on paper with the straightedge and compass.[32] It also designated the printed diagram itself, and ultimately the concrete image engraved on a woodblock and imprinted onto the book's paper.

---

[32] See, for instance, Billingsley (1570, fol. 57r), Prop. I.46: "[*In marg.* To *describe* a square mechanically] This is to be noted that if you will mechanically and redily, not regarding demonstration *describe* a square upon a line geven, as upon the line *ab*, after that you have erected the perpendiculer *line ca* upon the line *ab*, and put the line *ae* equall to the line *ab*: then open your *compasse* to the wydth of the line *ab* or *ae*, & set one foote thereof in the point *e*, and *describe* a peece of the circumference of a circle: and againe make the centre the point *b*, and *describe* also a piece of the circumference of a circle, namely, in such sort that the peece of the circuference of the one may cut

For that matter, when dealing with the definitions of the regular polyhedra, Billingsley first compared the two types of printed diagrams to which he resorted in this context. One is the type that was traditionally found in Latin and Greek editions of the *Elements*, which abstractly displayed the edges and angles of the figure as straight lines only, without indication of depth through shadows or of hidden edges through dotted lines (Fig. 5.3). The other is the type that offered a more concrete representation of the hollow polyhedron (Fig. 5.4), making use of linear perspective and representing the edges of the figure as endowed with thickness, shadows and even texture. This is the style in which Leonardo da Vinci (1452–1519) drew the polyhedra depicted in the *De divina proportione* of Luca Pacioli (c. 1445–1517), in which these figures were represented as if modelled on man-made material figures.[33] In Billingsley's commentary on the definition of the dodeca-hedron, both diagrams (the more abstract one and the one drawn in a more concrete and realistic manner) are designated as "descriptions".

> As in these two figures here set you may perceave. Of which the first (which thinge also was before noted of a Cube, a Tetrahedron, and an Octohedron) is the common description of it in a plaine, the other is *the description of it by arte upon a plaine* to make it to appeare somwhat bodilike. *The first description* in deede is very obscure to conceave, but yet of necessitie it must so, neyther can it otherwise, be *in a plane described* to understand those Propositions of Euclide in these five books following which concerne the same.[34]

After his commentary on the definitions of the polyhedra, Billingsley placed the nets of the corresponding figures, which he intended for the reader to redraw on paper, to cut out and to fold in order to build their three-dimensional models. In this context, a "description" is what allows the reader to produce a given solid object in three dimensions.

> Because these five regular bodies here defined are not by these figures here set, so fully and lively expressed, that the studious beholder can throughly according to their definitions conceyve them. I have here geven of them other *descriptions* drawn in a playne, by which ye may easily attayne to the knowledge of them. For if ye draw the like formes in matter that wil bow and geve place, as mist aptly ye may do in fine pasted paper, such as pastwives make womens pastes of, and then with a knife cut every line finely, not through, but halfe way only, if then ye bow and bende them accordingly, ye shall most plainly and manifestly see the formes and shapes of these bodies, even as their definitions shew. And it shall be very necessary for you to have store of that pasted paper by you, for so shal you upon it describe the formes of other bodies, as Prismes and Parallelipopedons, and such like set forth in these five bookes following, and see the very formes of these bodies there mencioned: which will make these

---

the peece of the circumference of the other, as in the point *d*: and from the point of the intersection, draw unto the points *e* & *b* right lines: & so shalbe *described* a square." (My emphasis.)

[33] Pacioli (1509b).

[34] Billingsley (1570, fol. 319r), Df. XI.24. (My emphasis.)

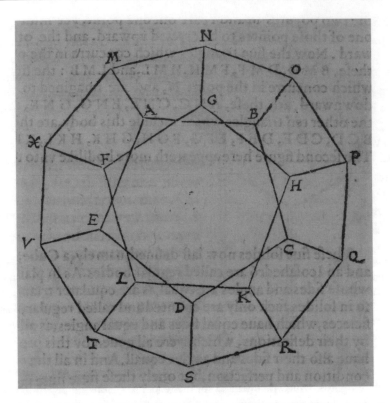

Fig. 5.3 Henry Billingsley, *The Elements of Geometrie*, 1570, 319r, Df. XI.24. A two-dimensional diagram of the dodecahedron in the style of the Latin and Greek editions of the *Elements*. Courtesy History of Science Collections, University of Oklahoma Libraries

> bokes concerning bodies, as easy unto you as were the other bookes, whose figures you might plainly scc upon a playne superficies.[35]

These models, once drawn on "pasted paper",[36] cut out and folded[37] (Figs. 5.5 and 5.6), would enable the reader to gain a more adequate apprehension of the solid figures presented in the definitions, as these would be more difficult for beginners to apprehend through their representation in two dimensions. This would furthermore allow the reader to accustom himself to interpret two-dimensional figures (also provided by Billingsley), as these were

---

[35] Billingsley (1570, fol. 320r), Df. XI.25.

[36] According to H. Smith (Smith 2017), "pasted paper" ("such as paste-wives make womens pastes of"), would correspond to moulded paper that was used to support fashionable headgear.

[37] These were then not only to be made out of nets that were reproduced by the reader on pasted paper, as those shown in appendix of Df. XI.25 (fol. 320v–322v), but also out of printed paper slips provided in the book, which were to be cut out, folded and glued directly onto the printed two-dimensional diagram, as in the case of the diagrams of Df. XI.2–5, 9–11 (fol. 312r–313r and 314r–v).

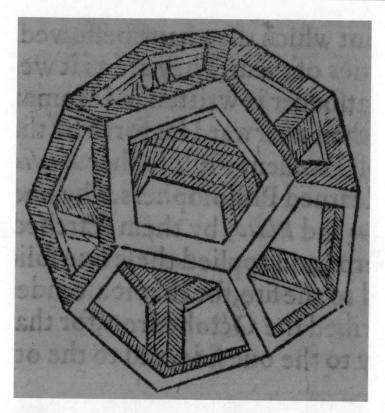

Fig. 5.4 Henry Billingsley, *The Elements of Geometrie*, 1570, 319r, Df. XI.24. A three-dimensional diagram of the dodecahedron, representing the edges and angles of the solid as if made out of wood or metal. This representation is conform to the style of the representations of the polyhedra drawn by Leonardo da Vinci for Pacioli's *De divina proportione* (Pacioli 1509b, Plate XXVIII). Courtesy History of Science Collections, University of Oklahoma Libraries

commonly used in geometrical treatises and in editions of Euclid's *Elements* in particular, because of their greater exactness. This is notably why Billingsley only added paper models in the eleventh book and not in the following books on solid geometry. In this context, the "description" properly corresponds to the two-dimensional diagram of the net, which is to be drawn on paper and then cut out and folded.

If the description of a figure chiefly designated here its concrete diagram, and not its generation in the imagination or its genetic definition, it remains that these various acceptations of the term *description* were connected in Billingsley's commentary. And what connected them was the fact that they all imply a notion of causation of a figure (mostly a particular figure) through a kinematic process, whether it is mechanical, imaginary or intellectually assumed, conceived as intrinsic or extrinsic to geometrical objects themselves. Hence, while the concrete figure presupposes the process of drawing out lines with the tip of a drafting tool on paper (or with a carving tool when engraved on a

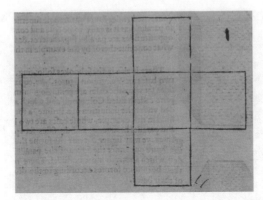

Fig. 5.5 Henry Billingsley, *The Elements of Geometrie*, 1570, 320v, Df. XI.25. Net of a cube or hexahedron provided by Billingsley in appendix of his commentary on Df. XI.25 and which is to be redrawn, cut out and folded into a three-dimensional figure by the reader, as instructed in the margin. Courtesy History of Science Collections, University of Oklahoma Libraries

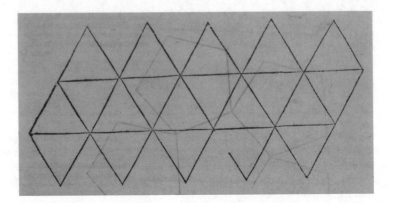

Fig. 5.6 Henry Billingsley, *The Elements of Geometrie*, 1570, 321v, Df. XI.25. Net of an icosahedron provided by Billingsley in appendix of his commentary on Df. XI.25 and which is to be redrawn, cut out and folded into a tridimensional body by the reader, as instructed in the margin. Courtesy History of Science Collections, University of Oklahoma Libraries

woodblock), and may be also generated by fashioning solid matter (as paper, clay, wood or metal) in the case of three-dimensional figures, the abstract figure merely presupposes the motion of an imaginary point, line or surface onto or within the intelligible matter of the imagination.

Both meanings of the term *description* (concrete and abstract) seemed to have been intended in the penultimate quotation, where both the abstract and "body-like" printed diagrams are compared as to their ability to convey Euclid's geometrical discourse in the clearest and most direct manner. The knowledge to attain here, according to Billingsley, does not only pertain to the spatial properties of the figure, but also to its mode of generation, which is itself important to gain knowledge of the properties of the figure. And its mode of generation is to be taken both as the intellectually-reconstituted process by which the defined magnitude is assumed to have come about and as the process by which the figure may be drawn out and represented concretely, even if the former is conceived as more essential than the latter. In a didactic context, the concrete figure would help jog the proper mode of generation and features of the intelligible figure in the student's imagination. But this concrete diagram or model (especially in the case of the paper diagrams) is always meant to lead the intellect to the immaterial figure situated in the imagination and which, as Billingsley clearly stated, is itself devoid of physical matter.

Thus, when stating that Euclid's definition of the sphere is "rather a description than a definition",[38] Billingsley may have understood the term description as referring both to an imaginary process and to the concrete and instrumental production of a sphere. This would be corroborated by the fact that he accompanied his commentary on Euclid's definition of the sphere with the representation of a lathe (Fig. 5.7), which is an instrument used by artisans to shape matter through a process of rotation and, in this case, specifically adapted to produce a sphere. This illustration intended to show how the shape of a sphere may be concretely produced by the rotation of a semicircle. It was used beforehand in sixteenth-century commentaries on Sacrobosco's *Sphaera* to visually explain Euclid's definition of the sphere, which (as said) was compared in this context with the definition of Theodosius. This image was first introduced by Jacques Lefèvre d'Étaples in his 1495 commentary on the *Sphaera*[39] (Fig. 5.8).

5.6   Definition and Description in Foix-Candale and Billingsley

Contrary to Foix-Candale, but like most commentators of the *Elements* or of Sacrobosco's *Sphaera*,[40] Billingsley did not consider Theodosius' definition as implying any motion. Yet, he did not compare these two traditional definitions of the sphere by pointing to

---

[38] See *supra*, p. 151.

[39] Oosterhoff (2020). See also Lefèvre's words on this analogy in his commentary on Sacrobosco (Lefèvre 1495, sig. a4r): "Et haec profecto mire efficaciae descriptio est, quae aperte docet (quantum sensibilis materia recipere valet) artificialem constituere sphaeram, cuius utilem commodamque intelligentiam nostrae tempestatis artifices multis auri pondo comparare deberent: qui metallo, ligno, aut alia materia figuras torno exprimere volunt. Si itaque in levi calybe aut ferro, sumpto circino supra quancunque lineam semicirculus educatur qui ab arcu ad diametrum usque excavetur, quin immo et medium diametri interstitium, et mox ad arcum circumferentiamque excavatur ut ea ex parte ad scindendum secandumque fiat aptus, exurget instrumentum tornandis sphaeris (haud secus quam circinus circulis) aptissimum. Hanc utilitatem sua descriptione nobis attulit Euclides, illamque intendebat cum dicere sphaeram esse transitum dimidii circuli, quae (fixa diametro) quousque ad locum suum redeat circumducitur, abditam, occultamque tamen, ut solis studiosis pateret."

[40] Billingsley did not himself write a commentary on Sacrobosco's *Sphaera*, but he did refer to it when dealing with Euclid's definition of the sphere. Billingsley (1570, fol. 315v): "And it is fully

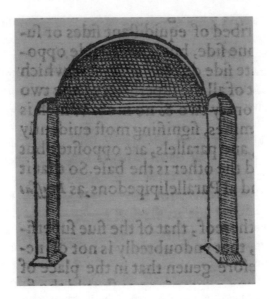

Fig. 5.7 Henry Billingsley, *The Elements of Geometrie*, 1570, 315v, Df. XI.12. Billingsley's use of the image of the lathe, common to many commentaries on Sacrobosco's *Sphaera* that were written after the publication of Lefèvre's commentary in 1495, to illustrate Euclid's definition of the sphere in his commentary on the *Elements*. Note the concreteness of Billingsley's representation in comparison with the illustration found in Lefèvre's treatise (Fig. 5.8). Courtesy History of Science Collections, University of Oklahoma Libraries

Euclid's appeal to motion, as was often done in commentaries on Sacrobosco, but rather by hinting at their different functions in the geometer's discourse. While Theodosius' definition would set forth the essence or nature of the defined figure, Euclid's definition would exhibit its mode of generation and, more fundamentally, its efficient cause.[41] In

---

round and solide, for that it is described of a semicircle which is perfectly round, *as our countrey man Iohannes de Sacro Busco in his book of the Sphere*, of this definition which he taketh out of Euclid, doth well collecte." (My emphasis.) In this context, Billingsley attacked Sacrobosco's rendering of Euclid's definition of the sphere, which states that only the circumference of the rotating semicircle, and not its whole surface, generates the sphere: "But it is to be noted and taken heede of, that none be deceived by the definition of a Sphere even by Iohannes de Sacro Busco: A Sphere (sayth he) is the passage or moving of the circumference of a semicircle, till it returne onto the place where it beganne, which agreeth not with Euclide. Euclide plainly sayth, that a Sphere is the passage or motion of a semicircle, and not the passage or motion of the circumference of a semicircle: neither can it be true that the circumference of a semicircle, which is a line should describe a body. It was before noted that every quantitie moved, describeth and produceth the quantities next unto it. Wherefore a line moved can not bring forth a body, but a superficies onely, which is the superficies and limite of the Sphere, and should not produce the body and solidity of the Sphere."

[41] Billingsley (1570, fol. 315v), Df. XI.12: "This definition of Theodosius is more essentiall and naturall, then is the other geven by Euclide. The other did not so much declare the incard nature and substance of a Sphere, as it shewed the industry and knowledge of a producing of a Sphere, and

Fig. 5.8 Jacques Lefèvre
d'Étaples, *In astronomicum
introductorium Ioannis de
Sacrobosco commentarius*, in
Sacrobosco 1531 et al., 127v,
Book I. The representation of a
lathe and its use to carve out a
sphere out of matter to illustrate
the Euclidean definition of a
sphere in Lefèvre's commentary
on Sacrobosco's *Sphaera*. The
present diagram drawn here
from the 1531 Venetian edition
is conform to that present in
Lefèvre's 1495 Parisian edition.
Courtesy Max Planck Institute
for the History of Science, Berlin

Aristotelian terms, this efficient cause of the sphere would correspond to the agent of its production or, in other words, that which causes the sphere to be produced or generated.[42] Whether it is the rotating semicircle or the geometer who represents the efficient cause of the sphere is not said here.

Although this discourse is sensibly close to that presented by Foix-Candale in the same context, Euclid's *description* of the sphere did not, for Billingsley, attribute an inadequate mode of generation to this figure, which would be distinct from a more truthful mode of generation, as was implied by his French predecessor. It would simply present the cause by which it may be conceived to be generated, which accounts for its essential properties, as opposed to merely stating the latter. For this reason, Billingsley designated the Euclidean definition of the sphere, and, with it, all genetic definitions of geometrical figures (since he placed all of these on the same ontological and epistemological level) as "causal definitions".

---

therefore is a causall definition geven by the cause efficient, or rather a description then a definition. But this definition is very essentiall, declaring the nature and substance of a Sphere." See also *supra*, p. 151.

[42] Aristotle, *Physics* II.3, 194b30–32 (Barnes 1995, I, fol. 333): "The primary source of the change or rest; e.g. the man who deliberated is a cause, the father is cause of the child, and generally what makes of what is made and what changes of what is changed."

To a certain extent, this assertion allows us to determine Billingsley's position in the sixteenth-century debate over the certainty and the scientificity of mathematical demonstrations,[43] as it indicates that mathematical knowledge does entail a causal form of knowledge, against what was argued by Alessandro Piccolomini (1508–1579) and Benito Pereira (1535–1610), among others.[44] It also anticipated later assertions of the intrinsic connection between motion and causality in geometry held notably by Barrow, Hobbes and Spinoza.[45] Thus, for Billingsley, genetic definitions, which are then associated to a properly spatial and even quasi-physical mode of generation of figures, are not only relevant to geometry, but are essential to understand the cause through which each geometrical object is properly constituted.

On a different (but related) level, the notion of efficient cause also points to the development of the mechanistic approach to nature in the seventeenth century, which limited the scientific investigation of physical phenomena to the discovery of material and efficient causes (i.e. matter and motion), to the exclusion of formal and final causes.[46]

---

[43] On this debate, in which was challenged the notion that mathematical demonstrations represented the highest form of demonstrations (*demonstrationes potissimae*), i.e. demonstrations that display simultaneously the fact and its cause, see De Pace (1993, pp. 21–120), Mancosu (1996, pp. 10–33) and Higashi (2018). See also Feldhay (1998).

[44] Higashi (2018, pp. 123–276 and 364–393).

[45] See, for instance, Barrow (1734, V, p. 83): "it seems plain that *Mathematical Demonstrations are eminently Causal,* from whence, because they only fetch their Conclusions from Axioms which exhibit the principal and most universal Affections of all Quantities, and *from Definitions which declare the constitutive Generations and essential Passions of particular Magnitudes.* From whence the Propositions that arise from such Principles supposed, must needs flow from the intimate Essences and Causes of the Things." (My emphasis.) On these authors and their attitude toward genetic definitions in geometry, see Mancosu (1996, pp. 94–100).

[46] This attitude is canonically expressed by Francis Bacon (1561–1626) in the *New Organon* II, Aphorism 9, in Bacon (transl. Silverthorne 2000, p. 109): "A true division of philosophy and the sciences arises from the two kinds of axioms which have been given above [...]. The inquiry after *forms,* which are (at least by reason and their law) eternal and unmoving, would constitute *metaphysics*; the inquiry after the *efficient* and *material causes, the latent process* and *latent structure* (all of which are concerned with the common and ordinary course of nature, not the fundamental, eternal laws) would constitute *physics*; subordinate to these in the same manner are two practical arts: *mechanics* to *physics*; and *magic* to *metaphysics* (in its reformed sense), because of its broad ways and superior command over nature." (Emphasis is proper to the quoted edition.) or by Descartes, in his *Traité du monde* (see *infra*, p. 266).

5.7 The Modes of Generation of Numbers and Magnitudes According to Billingsley's Commentary on Book II

As many of the commentators considered here, Billingsley appealed to a kinematic understanding of the generation of geometrical objects in his commentary on the definitions of Book II.

More precisely, Billingsley depicted the generation of the rectangular parallelogram in his commentary on Df. II.1 according to the same formulation and terminology as he described the genetic definition of the line, but, unlike Fine, he did not use the term *flow* as a synonym of motion or draught in this context.[47] He also compared then the generation of the rectangular parallelogram to the multiplication of two numbers,[48] as had also done Fine and Peletier in their own commentaries on Df. II.1.

> The parallelogramme is imagined to be made by the draught or motion of one of the lines into the length of the other. As if two numbers should be multiplied the one into the other.[49]

The aim of the present comparison is to explain what Euclid meant, in Df. II.1, when he wrote that the parallelogram is "contained" by the two sides adjacent to the right angle, which could appear to contradict the Euclidean axiom[50] according to which two straight lines do not enclose a surface.[51] It also aims thereby to demonstrate, as in Peletier's commentary, the equality of rectangular parallelograms which have equal adjacent sides and whose equality may be easily expressed by the commutative property of multiplication in arithmetic.

> And if we imagine the line AC to be *drawen or moved* directly according to the length of the line AB, or contrary wise the line AB to be moved directly according to the length of the line AC, you shall produce the whole rectangle parallelogramme ABCD which is sayde to be contayned of them: even as one number multiplied by an other produceth a plaine and righte angled superficiall number, as ye see in the figure here set, where the number of sixe or sixe unities: of which multiplication are produced 30, which number being set downe and described by his unities representeth a playne and a right angled number. Wherefore even as equall numbers multipled by equal numbers produce numbers equall the one to the other: so rectangle parallelogrames which are comprehended under equal lines are equal the one to the other.[52]

---

[47] *Cf.* Billingsley (1570, fol. 1v), Df. I.2: "A lyne is the movyng of a poynte, as the motion or draught of a pinne or a penne to your sense maketh a lyne".

[48] On the relation between genetic definitions and the arithmetical understanding of magnitudes, notably within Billingsley's commentary on Euclid, see also Malet (2006, pp. 73–74).

[49] Billingsley (1570, fol. 60r), Df. II.1.

[50] This axiom was a later addition to Euclid's text (De Risi 2016b, p. 15).

[51] Vitrac (1990, I, p. 325).

[52] Billingsley (1570, fol. 60r–v), Df. II.1.

Fig. 5.9 Henry Billingsley, *The Elements of Geometrie*, 1570, 60r, Df. II.1. Diagram of a rectangular number expressing the correspondence between the generation and containing of rectangular numbers by the multiplication of two numbers and the generation and containing of a rectangular plane surface by two line-segments that are mutually touching at right-angles. Courtesy History of Science Collections, University of Oklahoma Libraries

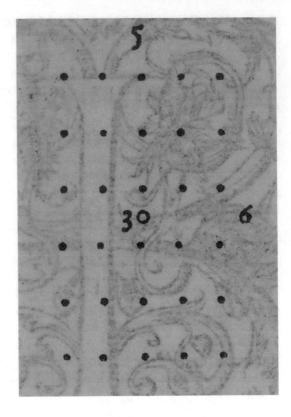

As shown by this passage, Billingsley explicitly founded the legitimacy of the comparison between the geometrical parallelogram and the number resulting from the multiplication of two different numbers on the Pythagorean notion of plane geometrical or figurate number. This notion is then visually depicted as a set of discrete points disposed in juxtaposed rows to form a rectangular number. As in Fine's commentary on Df. II.1,[53] the diagram is captioned with Arabic numerals to indicate the amount of point-units on each of the two adjacent sides and the total number of point-units forming the area of the rectangular number, obtained by multiplying one of the sides by the other (Fig. 5.9).

Billingsley did not take care to mention here that the analogy stops with the comparison between the modes of composition of numbers and magnitudes, given that the point is not a part of the line (or the line, a part of the surface) in the way the unit is a part of a number, as he had already mentioned in his commentary on Df. I.3.[54] In this context, he rather

---

[53] See *supra*, p. 67.

[54] Billingsley (1570, fol. 1v), Df. I.3: "For a line hath his beginning from a point, and likewise endeth in a point: so that by this also it is manifest, that pointes, for their simplicitie and lacke of composition, are neither quantitie, nor partes of quantitie, but only the termes and endes of quantitie. [...] And herein differeth a poynte in quantitie, from unitie in number for that although unitie be the beginning

classically stated that although both the point and the unit can be considered as the indivisible principles of their respective types of quantity, the point, as it is deprived of composition, cannot be considered as a component of the line. The line, as opposed to number, is indeed not a collection of indivisible points, but a continuous quantity and may always be divided further. In other words, the point can be considered as the beginning and end of a line, but not as one of its parts, just as the instant can be considered as the beginning and end of time, but not as a component of time. Hence, although he did not express it as clearly as Peletier did, Billingsley seems to have considered motion as a mode of generation only proper to points and magnitudes among mathematical objects, guaranteeing that the line is understood as caused by the point without having to be considered as composed of points.

For this reason, in the passage quoted above, the correspondence between the generation of the parallelogram and the multiplication of two numbers only has a didactic function. The situation is similar to that of Peletier in his commentary on Df. II.1, where this comparison mainly serves to illustrate the quantitative properties and relations of parallelograms with equal adjacent sides. This is confirmed by the fact that the parallelogram is then only said to be produced by the motion of one of its sides (along and according to the length of its other side) "as if" one multiplied a number by another in order to obtain a plane number. Plane numbers would as such not be inherently homogeneous to plane figures, that is, they would not have the same mereological properties, given that they do not belong to the same type of quantity.

In remains that, through the notion of figurate number, the motion of the side of the parallelogram on the other allowed Billingsley to attribute arithmetical properties to magnitudes on an operative level, such as the aforementioned commutative property of the multiplication of numbers. This certainly enabled Billingsley to justify the arithmetical and algebraic interpretation of Book II, in which he actively engaged by adding to his commentary the arithmetical reduction of the first ten propositions of Book II by Barlaam of Seminara (c. 1290–1348).[55]

---

of nombers, and no number (as a point is the beginning of quantitie, and no quantitie) yet is unitie a part of number. For number is nothyng els but a collection of unities, and therfore may be devided into them, as into his partes. But a point is no part of quantitie, or of a lyne neither is a lyne composed of pointes, as number is of unities. For things indivisible being never so many added together, can never make a thing divisible, as an instant in time, is neither tyme, nor part of tyme, but only the beginning and end of time, and coupleth & joyneth partes of tyme together." In this discourse, Billingsley is quite close to Foix-Candale in the passage quoted above, n. 32, p. 105.

[55] Billingsley (1570, fol. 62r), Prop. II.1: "Because that all the Propositions of this second booke for the most part are true both in lines and in numbers, and may be declared by both: therefore have I have added to every Proposition convenient numbers for the manifestation of the same. And to the end the studious and diligent reader may the more fully perceave and understand the agrement of this art of Geometry with the science of Arithmetique, and how nere & deare sisters they are together, so that the one cannot without great blemish be without the other, I have here also ioyned a little booke of Arithmetique written by one Barlaam, a Greeke authour a man of greate knowledge. In whiche booke

This also allowed him, in return, to legitimate the use of geometry to learn certain properties of numbers, as well as certain arithmetical and algebraic rules, since commutativity could be geometrically interpreted as the equality of the parallelograms contained by equal sides.

> In this second booke Euclide sheweth, what is a Gnomon, and a right angled parallelogramme. Also in this booke are set forth the powers of lines, devided evenly and unevenly, and of lines added one to an other. The power of a line, is the square of the same line: that is, a square, every side of which is equall to the line. So that here are set forth the qualities and proprieties of the squares and right lined figures, which are made of lines & of their parts. The Arithmetician also out of this booke gathereth many compendious rules of reckoning, and many rules also of Algebra, with the equations therein used. The groundes also of those rules are for the most part by this second booke demonstrated.[56]

Thus, although magnitudes are essentially distinct from numbers as for their modes of composition, they are nevertheless admitted here as a source of information on numbers and on certain procedures proper to arithmetic and algebra. The generative motion of magnitudes is considered as a guide to understand the operation of multiplication, and, through the spatial understanding of numbers associated with the Pythagorean theory of numbers, as one of its foundational notions.[57] Hence, while the arithmetical operation of multiplication would enable us to understand the quantitative relations between quadrilateral figures, geometry, and notably the generations of figures, would also condition our comprehension of certain key arithmetical and algebraic concepts.

The generative motion of the point, of the line or of the surface was therefore not only a useful metaphor, for Billingsley, to apprehend the mutual relations of magnitudes, but also a key to understanding the kinship between arithmetical and geometrical operations through an arithmetical interpretation of the *Elements*, Book II. For this reason, in his commentary on Prop. II.1, Billingsley designated arithmetic and geometry are "sisters".[58]

* * *

---

are by the authour demonstrated many of the selfe same proprieties and passions in number, which Euclide in this his second boke hath demonstrated in magnitude, namely, the first ten propositions as they follow in order. Which is undoubtedly great pleasure to consider, also great increase & furniture of knowledge. Whose Propositions are set orderly after the propositions of Euclide, every one of Barlaam correspondent to the same of Euclide." On Barlaam of Seminara's version of Euclid's Book II, see Corry (2013).

[56] Billingsley (1570, fol. 60r), Book II, preface.

[57] However, there is no direct use here of the geometrical term *ductus* (or of an equivalent English term) to directly designate the arithmetical operation of multiplication, in spite of the fact that it was commonly used to speak of the generation of numbers by multiplication, especially in the case of square or cubic numbers. See *supra*, p. 68–69.

[58] See *supra*, n. 55, p. 163.

In his commentary on Euclid's *Elements*, Billingsley presented the generations of geometrical magnitudes as a genuine part of the objects considered by the geometer, legitimating thereby the introduction of motion within geometrical definitions. In this context, the generative motion of the point, whether designated as a flow or as a motion, was always understood as a properly spatial process, the point being attributed a position and a quasi-material character. In other words, rather than referring to a metaphysical type of emanation of the multiple from a divine unity, the generation of the line from the flow or motion of the point would first relate, for Billingsley, to a concrete and instrumental process. This understanding was confirmed by his frequent appeal to instrumental comparisons and references to concrete devices in his commentary on the *Elements* to help the reader grasp the abstract concepts taught by Euclid.

It remains that Billingsley clearly admitted an ontological distinction between geometrical and material objects, asserting the abstract and imaginary status of the former, though not their ontological transcendence over the latter. As shown by his commentary on Euclid's definition of the sphere, Billingsley placed the generative flow of the point (ῥύσις) on the same ontological and epistemic level as the translations or rotations of lines and surfaces to generate plane or solid figures (φορά). On this aspect, he joined Fine, but differed from Peletier.

In his commentary on the definition of the sphere, Billingsley compared the two definitions of the sphere by Euclid and by Theodosius, as had done Foix-Candale in his own commentary on Euclid, presenting the definition of Theodosius as a proper definition and that of Euclid as a description. For Billingsley, a description meant altogether the concrete drawing of a geometrical diagram, its unravelling in the imagination, as well as the genetic definition itself. However, all of these would aim to provide the student of geometry with the knowledge or the understanding of the essential properties of figures by allowing the reader to imaginarily, as well as concretely, reconstitute the processes through which the figures necessarily come about with the properties stated in the definitions. Thus, while the definition of Theodosius would state the essential properties of the figure, the definition of Euclid would display the efficient cause of the figure, wherefore Billingsley designated it as a causal definition.

Billingsley did not, as Foix-Candale, consider Theodosius' definition of the sphere as implying motion and did not therefore distinguish the two traditional definitions of the sphere according to the types of motion or modes of generation they implied. He mainly distinguished them by their logical function, Theodosius' definition stating its essential properties and Euclid's definition displaying the mode of production or of causation of the sphere, which (as said) he regarded as part of the objects investigated by the geometer.

Within his commentary on the definitions of Book II, Billingsley described the mode of generation of the rectangular parallelogram in the same manner as he described the mode of generation of the line from the flow of a point, relating them on an ontological and epistemological level. Although he explicitly distinguished the mode of composition of the line from the point and that of number from the unit respectively, he compared the production of the rectangular parallelogram's surface to the multiplication of two different numbers in order to explain the quantitative relation between the sides of the rectangular parallelogram to the area they contain. The comparison was presented as a didactic tool to explain the quantitative properties and relations of rectangular quadrilaterals, but it also represented a means of deriving knowledge of certain arithmetical and algebraic operations. Motion, although proper to geometrical objects, would thus help understand certain properties of quantity in general.

John Dee

6

6.1   The Life and Work of John Dee and the Significance of His Contributions to the Euclidean Tradition

John Dee (London, 1527—Mortlake, 1608/09) was a Welsh-English mathematician, astronomer, geographer, astrologer, occult philosopher and advisor to Queen Elizabeth I.[1] He was the son of a London mercer and studied at St John's College in Cambridge in 1542, after attending Chelmsford grammar school. He was one of the founding members (as a fellow and as an under-reader in Greek) of Trinity College, Cambridge, where he obtained a Master's degree in 1548. He is also said to have received a doctorate of medicine in Prague around 1585. In his autobiographical writings of 1592, he claimed to have lectured on Euclid's *Elements* in Paris (at the Collège de Rheims) in front of a crowded assembly while he was travelling around Europe in the summer of 1550.[2] He wrote that he pleased his audience so much that he was proposed a Royal lectureship in mathematics for

---

[1] On Dee's life and career, see Roberts (2006), Easton (2008), Debus and Dee (1975, pp. 2–33), Dee (Crossley 1851), Dee (Halliwell-Phillipps 1842) and French (1987).

[2] Dee (Crossley 1851, p. 7): "From Lovayne I tooke my journey towardes Paris A. 1550, the 15 day of July, and came to Paris the 20 day of that moneth. Where, within a few daies after (at the request of some English gentlemen, made unto me to doe somewhat there for the honour of my country) I did undertake to read freely and publiquely Euclide's *Elements* Geometricall, *Mathematicè, Physicè, et Pythagoricè*; a thing never done publiquely in any University of Christendome. My auditory in Rhemes Colledge was so great, and the most part elder then my selfe, that the mathematicall schooles could not hold them; for many were faine, without the schooles at the windowes, to be auditors and spectators, as they best could helpe themselves thereto."

A. Axworthy, *Motion and Genetic Definitions in the Sixteenth-Century Euclidean Tradition*, Frontiers in the History of Science,
https://doi.org/10.1007/978-3-030-95817-6_6

200 French crowns, a proposition he would have turned down.[3] He also claimed that a readership in mathematics was offered to him at Oxford University in 1554.[4]

Dee built one of the largest collections of scientific books and instruments in Europe.[5] He would have in vain urged Queen Mary to create a royal library to salvage all the ancient manuscripts contained in monastic libraries, which were threatened to be scattered or destroyed at the occasion of the Reformation.[6] He also exchanged, throughout his career and various travels, with many important mathematicians and cartographers, among whom Oronce Fine, but also Girolamo Cardano (1501–1576), Thomas Digges (1546–1595), Gemma Frisius (1508–1555), Gherardus Mercator (1512–1594), Conrad Gessner (1516–1565) and Abraham Ortelius (1527–1598).[7] He produced works (printed and unpublished)[8] on navigation,[9] on natural and occult philosophy (which combine altogether elements of Hermeticism, alchemy, magic and kabbalistic exegesis)[10] and wrote the famous *Mathematicall praeface* to the previously considered 1570 English translation and commentary of Euclid's *Elements* by Henry Billingsley,[11] in which he presented the

---

[3]Dee (Crossley 1851, p. 8): "In that University of Paris, were at that tyme above forty thousand accounted studientes; some out of every quarter of Christendome being there. Among these very many of all estates and professions were desirous of my acquaintance and conference, as Orontius, Mizaldus, Petrus Montaureus, Ranconetus, Danesius, Jacobus Sylvius, Jacobus Goupylus, Turnebus, Straselius, Vicomercatus, Paschasius Hamelius, Petrus Ramus, Gulielmus Postellus, Fernelius, Jo. Magnionus, Johannes a Pena, &c. as by letters lying on the table may partly appeare. There I refused to be one of the French kinge's mathematicall readers, with 200 French crownes yearely stipend offred me, if I would stay for it."

[4]Dee (Crossley 1851, p. 10): "Of the University of Oxford, some of the chiefe studientes (Doctors of Divinity and Masters of Art) caused a yearely good stipend to be offered unto me to read the mathematicall sciences there."

[5]Dee (Halliwell-Phillipps 1842), James (1921) and French (1987, pp. 40–61). See also Sherman (1995, pp. 44, 83 and 95).

[6]John Dee, *A supplication to Q. Mary, By John Dee, for the recovery and preservation of ancient writers and monuments* and *Articles concerning the recovery and preservation of the ancient monuments and old excellent Writers*, in Dee (Crossley 1851, pp. 46–49). On this episode, see also Sherman (1995, pp. 36–37).

[7]Dee (Crossley 1851, pp. 5–8).

[8]See Dee's list of published and unpublished works in Dee (Crossley 1851, pp. 73–78).

[9]Dee (1577). On Dee and navigation, see Baldwin (2006).

[10]John Dee, *Propaedeumata Aphoristica Ioannis Dee, Londinensis, de Praestantioribus quibusdam naturae virtutibus*, London: H. Sutton, 1558 (repr. R. Wolf, 1568); *Monas Hieroglyphica*, Antwerp, Gulielmus Silvius, 1564. On these works and their contribution to occult philosophy, see Josten (1964), Clulee (1977), French (1987, pp. 62–125), Clulee (1988, parts I–II, pp. 19–135), Clulee (2005), Dunn (2006), De Léon-Jones (2006) and Norrgrén (2005).

[11]John Dee, *Mathematicall Praeface*, in Billingsley (1570, n.p., sig A4v). On Dee's *Mathematicall praeface*, see Debus and Dee (1975, Introduction); Clulee (1988, chap. VI, pp. 143–176), Rambaldi (1989), Mandosio (2003), Rampling (2011) and Johnston (2012).

objects, virtues and division of mathematics.[12] He also made occasional additions to Billingsley's commentary on Books XI–XIII.[13] He collaborated moreover with Commandino on an edition of the *De superficierum divisionibus* attributed to Machometus Bagdedinus (Muḥammad al-Baghdādī) (1050–1141).[14]

The *Mathematicall praeface* was written by Dee after Billingsley had completed his translation and commentary and therefore complemented, as well as introduced, this work. To a certain extent, it provided this commentary with a philosophical foundation, even if Billingsley made punctual ontological and epistemological remarks on the objects and methods of Euclid's geometry, as was previously shown. It is notably why, although this text precedes Billingsley's commentary within the book, I have chosen to discuss Dee's case afterwards.

Even if Dee did not himself provide a commentary on Euclid, he was nevertheless an active contributor to the sixteenth-century Euclidean tradition throughout his career. This is marked in particular by his substantial contributions to Billingsley's translation and commentary on the *Elements* and by the teaching he allegedly gave on Euclid's geometry. He also, less directly, contributed to the promotion of ancient mathematics, and of Euclid's *Elements* in particular, through his large collection of scientific manuscripts and printed works, which contained many editions of the *Elements* and, among these, certain editions he thoroughly annotated.[15] Given his extensive network within European mathematical circles, Dee also represents a key figure of sixteenth-century mathematical culture.

As mentioned earlier, I will consider, in this chapter, not only Dee's *Mathematicall praeface*, but also his *Monas hieroglyphica*,[16] composed in 1564, as it provides useful

---

[12] This text was reedited independently from Billingsley's English commentary on Euclid three times over the seventeenth century (in 1651, 1661 and 1699). See Mandosio (2003, p. 486 and n. 53).

[13] In Dee (Crossley 1851, 73), *My labors and paines bestowed at divers times, to pleasure my native Country: by writing of sundry Bookes, and Treatises: some in Latine, some in English, and some of them, written, at her Majesties commandement*: "Of which Bookes, and Treatises, some are pinted, and some unprinted. The printed Bookes, and Treatises are these following: [. . .] My divers and many Annotations, and Inventions Mathematicall, added in sundry places of the foresaid English Euclide, after the tenth Booke of the same. Anno 1570." On Dee's annotations to Billingsley's Euclid, and more generally on his contribution to the Euclidean tradition, see Mandosio (2003) and Johnston (2012).

[14] Commandino and Dee (1570). Dee would have notably provided Commandino with a Latin copy of this work in 1563. On this collaboration between Dee and Commandino, see Rosen (1970) and Rose (1972). Rosen and Rose disagreed on the date when Dee visited Urbino and provided Commandino with his copy of the manuscript. On this treatise attributed to Muḥammad al-Baghdādī, see Moyon (2011).

[15] Sherman (1995, p. 44). See also Mandosio (2003, pp. 478–479).

[16] Dee (1564). On this work, see Josten (1964), Clulee (1988, part II, pp. 75–142) and Clulee (2005). On its reception, see Forshaw (2005) and Campbell (2012).

information on Dee's understanding of genetic definitions in geometry, as well as on their uses and interpretation outside of geometry.

Beside Dee's uses and formulations of genetic definitions, I will be examining here the role of motion in his comparison of number and magnitude within his *Mathematicall praeface*. I will also look at the connections between genetic definitions of geometrical objects and Dee's metaphysical, physical and esoteric understanding of nature.

This analysis will reveal, in particular, that Dee had a quite original interpretation of the difference between the point and the unit in their causal relation to their respective types of quantity. It will also show that Dee attributed a key role to the generative deployment of magnitudes in the constitution of the cosmos. Hence, the status of genetic definitions offers a perfect illustration of the continuity between the mathematical, metaphysical and esoteric aspects of Dee's thought.

6.2   The Place and Function of Genetic Definitions in Dee's Teaching of Euclidean Geometry

Dee is important to consider here not only because of his discussion of the status of genetic definitions in geometry, but also because of the approach he claimed to have adopted to Euclid's geometrical teaching. As he explained in his autobiography, he would have attempted, in his teaching on Euclid, to explain the *Elements* altogether "mathematically, physically and pythagorically" (*Mathematicè, Physicè, et Pythagoricè*), going beyond the first exposition or reading of Euclid's propositions.[17] It may be assumed that Dee would have done so by explaining to his audience the alleged underlying meaning and scope of Euclidean geometry in a mathematical, physical and metaphysical (or theological) perspective, given that the Pythagoreans regarded mathematical objects, and mainly numbers, as divine entities. He claimed in particular to have provided his public with a visible testimony of the first four definitions of the *Elements*, which, as he said, "by imagination only are to be conceived".[18] In this framework, the genetic definitions of geometrical objects would have played an important role in the intuition of these primary geometrical notions of point, line and straight line. As the discourse he held on geometry within his mathematical and philosophical writings will confirm it, Dee saw in the generative motion of the point a process that has altogether a mathematical, physical and metaphysical significance. It

---

[17] Dee (Crossley 1851, p. 7): "I did also dictate upon every proposition, beside the first exposition." J.-M. Mandosio (Mandosio 2003, 478) also interprets this assertion according to Proclus' classification of the parts of a proposition in Euclid's *Elements*, the exposition corresponding to the ἔκθεσις, which is the part that follows the enunciation or "proposition" (πρότασις).

[18] Dee (Crossley 1851, pp. 7–8): "And by the first foure principall definitions representing to the eyes (which by imagination onely are exactly to be conceived), a greater wonder arose among the beholders, than of my Aristophanes *Scarabeus* mounting up to the top of Trinityhall in Cambridge *ut supra*."

would, as such, reveal the relations and correspondences between these various types and levels of reality within the universe.[19]

In this regard, it is important to note that Dee, like Peletier and Foix-Candale, was strongly influenced by the Platonic, Neoplatonic and Pythagorean conceptions of mathematics.[20] As his French predecessors, but contrary to Billingsley, he thus admitted the distinction between the various levels of interpretation (mathematical, physical and metaphysical) of the generative motion of geometrical objects and was eager to show the connections and correspondences between them.

Along the same lines, although Dee clearly distinguished pure and applied mathematics,[21] he also aimed to display, in his annotations to Billingsley's commentary on Euclid, the connections between the mathematician and the "mechanical practiser". Indeed, while he asserted the value of theoretical propositions, he emphasized the value of applied mathematics and of empirical knowledge.[22] In this sense, while Billingsley would have mainly used mechanical images and means to help the reader understand the meaning of

---

[19] Mandosio (2003, pp. 488–490).

[20] The latter is suggested by the adverb *pythagoricè* used to describe his teaching on the *Elements* in his autobiography.

[21] In his *Mathematicall praeface*, Dee made sure to distinguish geometry, as a science, from the activity of the field-measurer, advising to call geometry by the name of *megethica* or *megethologia* rather than *geometria*, which literally means measure of the earth: Billingsley (1570, sig. a2r–v): "This Science of *Magnitude*, his properties, conditions, and appertenances: commonly, now is, and from the beginnyng, hath of all Philosophers, ben called *Geometrie*. But, veryly, with a name to base and scant, for a Science of such dignitie and amplenes. [...] The people then, by this art pleasured, and greatly relieved, in their landes just measuring: & other Philosophers, writing Rules for land measuring: betwene them both, thus, confirmed the name of *Geometria*, that is, (according to the very etimologie of the word) Land measuring. [...] An other name, therefore, must nedes be had, for our Mathematicall Science of Magnitudes: which regardeth neither clod, nor turff: niether hill, nor dale: neither earth nor heaven: but is absolute *Megethologia*: not creping on ground, and daddeling the eye, with pole perche, rod or lyne: but liftyng the hart above the heavens, by indivisible lines, and immortall beames meteth with the reflexions, of the light incomprehensible: and so procureth Joye, and perfection unspeakable. Of which true use of our *Megethica*, or *Megethologia*, *Divine Plato* seemed to have good taste, and judgement: and (by the name of *Geometrie*) so noted it: and warned his Scholers therof [...]."

[22] Dee (Billingsley 1570, fol. 386v), Prop. XII.18, *An advise*: "In noting or signifying of Spheres, sometimes we use by one and the same circle, in plaine designed, to represent a Sphere and also the greatest circle in the same contained: and likewise, by a segment of that circle, signifie a segment of the same Sphere, as by a straight line, we often signifie the circle, which is the base of a segment of a Sphere, Cone, or Cylinder: and so in such like. Wherin, consider our suppositions: and take heede when we shift from one signification to an other, in one and the same designation: and withall remember the principall intent of our drift: and such light thinges, can not either trouble or offend thee. Compendiousnes and artificiall custome, procureth such meanes: sufficient, to stirre up imagination Mathematicall: or to informe the *practiser Mechanicall*." (My emphasis.) On this aspect of Dee's annotations, see Mandosio (2003, pp. 483–484). On Dee's interest in practical mathematics and in the mechanical arts, see French (1987, pp. 160–187).

Euclid's theoretical discourse, Dee would have additionally pointed to the usefulness of Euclidean propositions to craftsmen.

6.3   The Motion of the Point and Its Ontological Implications

In Dee's *Mathematicall praeface*, the generation of the line from the flow of the point is, as in the commentaries of Fine and Billingsley, explicitly and directly described as a motion, corresponding to a spatially determined and properly mathematical process.

> A Point, is a thing Mathematicall, indivisible, which may have a certayne determined situation. If a Poynt move from a determined situation, the way wherein it moved, is also a Line: mathematically produced. Whereupon, of the auncient Mathematiciens, a Line is called the race or course of a Point.[23]

While describing the genetic definition of the line as resulting from the spatially-determined motion of a point, Dee also mentioned another more ancient terminology, which identified the line to the "race" or "course" of a point and which could aim to refer to the ancient notion of $\rho \upsilon \sigma \iota \varsigma$ or flow of the point, though these terms primarily evoke a local form of motion. If the notions of "race" or "course" of the point aimed here to convey the ancient notion of $\rho \upsilon \sigma \iota \varsigma$, it would then not be interpreted on a metaphysical level, but merely according to a mathematical understanding.

In a later passage, in which Dee compared the geometrical point and the arithmetical unit, he attributed to the point "a certain motion" and explained, as had done Billingsley, that a point may be attributed motion because it is endowed with a position or a situation,[24] contrary to the arithmetical unit,[25] which can be attributed neither.

> We defined an Unit, to be a thing Mathematicall Indivisible: A Point, likewise, we said to be a Mathematicall thing Indivisible. And farder, that a Point may have a certaine determined Situation: that is, that we may assigne, and prescribe a Point, to be here, there, yonder, etc. Herein, (behold) our Unit is free, and can abyde no bondage, or to be tied to any place, or seat: divisible or indivisible. Again, by reason, a Point may have a Situation limited to him: a certain motion, therfore (to a place, and from a place) is to a Point incident and appertaining. But an Unit, can not be imagined to have any motion.[26]

---

[23] Dee (Billingsley 1570, sig. *1r).

[24] As noted above, this definition of the point as endowed with a position was attributed to the Pythagoreans by Aristotle, in *De anima* I. 4, 409a4–7 (Barnes 1995, I, p. 652): "a point is a unit having position". See *supra*, n. 122, p. 28.

[25] It is interesting to note that Dee, in his *Mathematicall praeface*, used the term "unit" while Billingsley used the term "unitie", which appears to have been the common term for the arithmetical unit in the English language at the time. J.-M. Mandosio (2003, 488) considers that Dee introduced the term unit (which is still in use today to designate the elementary component of natural numbers) in order to distinguish the mathematical concept of arithmetical unit from its empirical notion.

[26] Dee (Billingsley 1570, sig. a2r).

In this passage, the motion of the point is considered independently from its generative function and from its relation to the line, which is unusual for the tradition considered here. The spatial character of the motion of the point is then clearly asserted. The point is indeed said to move from one specific place to another and its motion or mobility is attributed to it as an essential property ("incident and apperrtaining"). In a sense, the mobility of the point is not only presented as a consequence of its spatial situation or "situatedness", but also as what determines its situation in space (at least, in relation to the spatiality of a given magnitude or figure).

As Dee presented it immediately after, this motion of the point is what produces a geometrical line.

> A Point, by his motion, produceth, Mathematically, a line: (as we said before) which is the first kinde of Magnitudes, and most simple: An Unit, can not produce any number.[27]

6.4   The Mobility of the Point Versus the Immobility of the Unit

In the above-quoted passage, the mobility of the point is again what distinguished it from the arithmetical unit, since the unit was presented as unable to produce a number. This last assertion may seem surprising, given that the geometrical point was often compared to the arithmetical unit since Antiquity on account of the fact that they both held a generative function with regard to their respective kinds of quantity. Yet, for Dee, the point differs from the unit insofar as it produces the line, but does not compose it—this is another way of saying that the line does not consist of an aggregation of points –, while the unit composes number, but does not generate it.

> A Line, though it be produced of a Point moved, yet, it doth not consist of pointes: Number, though it be not produced of an Unit, yet doth it consist of units, as a materiall cause. [. . .] And so, every number, may have his least part, given: namely, an Unit: But not of a Magnitude, (no, not of a Lyne,) the least part can be given: by cause, infinity, division therof, may be conceived.[28]

Motion is here used as a means of distinguishing the modes of composition of number and magnitude, as it was for Peletier and Billingsley (although the latter expressed it in less explicit terms). But, unlike them, Dee did not present the unit as producing number at all, not even by multiplication or by successive addition. This would be because he would then be describing the number-idea or "Number Numbryng" present in the human soul, such as

---

[27] Dee (Billingsley 1570, sig. a2r).

[28] Dee (Billingsley 1570, sig. a2v).

described in the *Mathematicall praeface*.[29] This number, which would be ontologically superior to the "Number Numbred" present within the creatures, would however differ from the divine numerical principle through which all things were created. Indeed, while the number present in the human soul (as well as in the minds of spiritual beings and angels) would have the capacity to distinguish and discern, it would not however be able to produce anything, unlike the number that exists within God.[30] As presented here, addition and multiplication would not be considered as generative processes, unlike motion. Motion thus represents the sole creative principle in the domain of quantity. If the unit is here designated as the "material cause of number", it is because it is that in which consist numbers, as its material cause in the Aristotelian sense, and not as its efficient cause.[31]

Hence, in the context of geometry, the generative motion of the point was defined by Dee altogether as a spatial process, as a properly mathematical notion and as a property attributed to the point essentially, in virtue of its situatedness. It is also crucial to explain and establish the distinction between numbers and magnitudes, such as apprehended by the human mind. Being described here as the sole possible principle of generation or efficient cause in the domain of quantity, it actually deprives numbers of their ability to be produced or caused by the unit, although units make up numbers. Since the relation of the point to the line is, in this sense, radically different from that of the unit to number, the generative motion of the point would not be considered as raising any mereological aporia. Indeed, in this framework, the line cannot be held as composed of points in the way numbers are composed of units.

If Dee clearly distinguished the modes of composition of number and magnitude, it may however be noted that, in another passage of the *Mathematicall praeface*, he pointed to the fact that, in a practical context, and more precisely in the art of the reckoner (that is, in the context of logistics and not in the theory of numbers),[32] the arithmetical unit may be

---

[29] Dee (Billingsley, 1570, sig. *1v): "Of my former wordes, easy it is to be gathered, that *Number* hath a treble state: One, in the Creator: an other in every Creature (in respect of his complete constitution:) and the third, in Spirituall and Angelicall Myndes, and in the Soule of man. In the first and third state, *Number*, is termed *Number Numbryng*. But in all Creatures, otherwise, *Number*, is termed *Number Numbred*. And in our Soule, Number beareth such a swaye, and hath such an affinitie therwith: that some of the old *Philosophers* taught, *Mans Soule, to be a Number movyng it selfe*."

[30] Dee (Billingsley, 1570, sig. *1v): "And in deed, in us, though it be a very Accident: yet such an Accident it is, that before all Creatures it had perfect beyng, in the Creator, Sempiternally. *Number Numbryng* therfore, is the discretion discerning, and distincting of thinges. But in God the Creator, this discretion, in the beginnyng, produced orderly and distinctly all thinges." On this aspect of Dee's representation of number, see French (1987, pp. 105–106).

[31] As was shown in the previous chapter, Billingsley already hinted at the Aristotelian determination of the causes of substances as he mentioned the notion of efficient cause in his discussion of Euclid's kinematic definition of the sphere. See *supra*, p. 151.

[32] Dee (Billingsley 1570, sig. *2r): "And by this reason: the Consideration, doctrine, and working, in whole numbers only: where, of an Unit, is no lesse part to be allowed: is named (as it were) an Arithmetike by it selfe."

conceived as infinitely divisible, making it comparable to magnitude itself, rather than to its indivisible principle.[33] In this context, numbers would also be taking on the properties of magnitudes, such as incommensurability and spatiality, by becoming linear, plane and solid.[34] Yet, these geometrical properties are not those of the figurate numbers of Pythagorean arithmetic, but rather those conferred to numbers in the context of practical arithmetic and algebra.

6.5   Human Versus Divine Geometry in the *Mathematicall praeface* and in the *Monas hieroglyphica*

In a different part of the *Mathematicall praeface*, Dee referred more specifically to numbers as the divine principles of all things, to which he therefore attributed a generative power. As said, this conception of numbers as divine principles stemmed from the Pythagorean arithmetical theology, according to which the arithmetical unit may be regarded as the mathematical image of the primordial One from which the universe would have proceeded.[35] Assuming that Dee considered this divine procession to have taken place according to the mode of a suprasensible flow or emanation (as in ancient Pythagorico-Platonic cosmogonic narratives), he would nevertheless have made an ontological distinction between the generative power of the divine One and the generative motion of the point in geometry given that the latter would only represent a mathematical and spatial representation of this divine and non-spatial process, as was also held by Peletier and Foix-Candale.

---

[33] Dee (Billingsley 1570, sig. *2r): "But farder understand, that vulgar Practisers, have Numbers, otherwise, in sundry Considerations: and extend their name farder, then to Numbers, whose least part is an Unit. For the common Logist, Reckenmaster, or Arithmeticien, in hys using of Numbers: of an Unit, imagineth lesse partes: and calleth them *Fractions*. As of an Unit, he maketh an halfe, and thus noteth it, ½ and so of other, (infinitely diverse) partes of an Unit. Yea and farder, hath, Fractions of Fractions. &c."

[34] Dee (Billingsley 1570, sig. *2r): "Practise hath led Numbers farder, and hath framed them, to take upon them, the shew of Magnitudes propertie: Which is Incommensurabilitie and Irrationalitie. (For in pure Arithmetike, an Unit, is the common Measure of all Numbers.) And, here, Numbers are become, as Lynes, Playnes and Solides: some tymes Rationall, some tymes Irrationall. And have propre and peculier characters, (as $\sqrt{z}$, $\sqrt{\&}$ and so of other. Which is to signifie Rote Square, Rote Cubik: and so forth:) & propre and peculier fashions in the five principall partes: Wherfore the practiser, estemeth this, a diverse Arithmetike from the other."

[35] Dee (Billingsley 1570, sig. *1r–v): "All things (which from the very first originall being of thinges, have bene framed and made) do appeare to be Formed by the reason of Numbers. For this was the principall example or patterne in the minde of the Creator. [. . .] By Numbers propertie therefore, of us by all possible meanes (to the perfection of the Science) learned, we may both winde and draw our selves into the inward and deepe search and vew, of all creatures distinct vertues, natures, properties, and Formes. And also, farder, anise, clime, ascend and mount up (with Speculative winges) in spirit, to behold in the Glas of Creation, the Forme of Formes, the Exemplar of all thinges Numerable, both visible and invisible, mortall and immortall, Corporall and Spirituall."

This reading is supported by the ontological status Dee attributed to mathematical objects in his *Mathematicall praeface*, which was influenced by Proclus' commentary on the first book of Euclid's *Elements*.[36] This appears in particular in the passage where he asserted the intermediary status of mathematical objects between natural and divine beings at the beginning of this text.[37] He wrote there that mathematical objects are immaterial and indivisible by essence (as are divine beings), but are able to be represented in the imagination as composed and divisible (through which they may be related to physical beings). The motion of the point, which is unravelled in the imagination of the mathematician, would thus differ from the emanation of the multiple from the divine One (in which there is fundamentally no distinction between the point and the unit), but would nevertheless refer to this ontologically superior and non-spatial generative process in a spatial manner.

This reading is also confirmed by Dee's *Monas hieroglyphica*,[38] which J.-M. Mandosio holds as a form of esoteric reinterpretation of Euclid's *Elements*.[39] In this work, Dee aimed

---

[36] Dee knew Proclus' *Commentary on the first book of Euclid's Elements*, as shown by the general structure of his preface (see the comparative table established by Clulee between the prefaces of Proclus and Dee, in Clulee 1988, 158). He possessed copies of Proclus' commentary, both in the Greek edition published by Grynaeus in 1533 and in the Latin translation of Francesco Barozzi printed in 1560 (Rose 1972). More generally, on Proclus' influence on Dee's philosophy of mathematics, see Clulee (1988, chap. VI).

[37] Dee (Billingsley 1570, sig. *1v): "All thinges which are, & have beyng, are found under a triple diversitie generall. For, either, they are demed Supernaturall, Naturall, or, of a third being. Thinges Supernaturall, are immateriall, simple, indivisible, incorruptible, & unchangeable. Things Naturall, are materiall, compounded, divisible, corruptible, and chaungeable. Thinges Supernaturall, are, of the minde onely, comprehended: Things Naturall, of the sense exterior, ar hable to be perceived. In thinges Naturall, probabilitie and conjecture hath place: But in things Supernaturall, chief demonstration, & most sure Science is to be had. By which properties & comparasons of these two, more easily may be described, the state, condition, nature and property of those thinges, which, we before termed of a third being: which, by a peculier name also, are called Thynges Mathematicall. For, these, beyng (in a maner) middle, betwene thinges supernaturall and naturall: are not so absolute and excellent, as thinges supernatural: Nor yet so base and grosse, as things naturall: But are thinges immateriall: and neverthelesse, by materiall things hable somewhat to be signified. And though their particular Images, by Art, are aggregable and divisible: yet the generall Formes, notwithstandyng, are constant, unchaungeable, untransformable, and incorruptible. Neither of the sense, can they, at any tyme, be perceived or judged. Nor yet for all that, in the royall mynde of man, first conceived. But, surmountyng the imperfection of conjecture, weenyng and opinion: and commyng short of high intellectuall conception, are the Mercurial fruite of Dianoeticall discourse, in perfect imagination subsistyng. A mervaylous newtralitie have these thinges Mathematicall and also a straunge participation betwene thinges supernaturall, immortall, intellectual, simple and indivisible: and thynges naturall, mortall, sensible, compounded and divisible."

[38] Dee (1564). On this work, see Josten (1964), French (1987, pp. 76–81), Clulee (1988, part II, pp. 75–142) and Clulee (2005). On its reception, see Forshaw (2005) and Campbell (2012).

[39] This is notably marked by the division of the propositions in theorems. On this aspect, see Clulee (1988, pp. 77 and 120).

to explain "mathematically, magically, cabbalistically, and anagogically" (*mathematicè, magicè, cabalisticè, anagogicèque*)[40] a symbol he called the hieroglyphic monad (*monas hieroglyphica*), or, more precisely, "a hieroglyphic figure thereof, after the manner (called) Pythagorean".[41] This symbol, which resulted from the composition of several geometrical, astrological and esoteric symbols, would concentrate and deliver the knowledge of all the causes (mathematical, physical, occult and divine) involved in the Creation.[42]

As shown by Mandosio,[43] the geometrical notions of point, straight line and circle, and their interrelation, played a foundational role in the constitution of the symbol, expressing the mutual relation of all beings in the universe.

> *Theorem I.* The first and most simple manifestation and representation of things, non-existent as well as latent in the folds of Nature, happened by means of straight line and circle.[44]
>
> *Theorem II.* Yet the circle cannot be artificially produced without the straight line, or the straight line without the point. Hence, things first began to be by way of a point, and a monad. And things related to the periphery (however big they may be) can in no way exist without the aid of the central point.[45]

In these two first theorems, the causal dependence of the straight line on the point and also of the circle on the straight line clearly refers to the genetic definitions of the straight line as produced by the undeviating flow or motion of the point and of the circle as produced by the rotation of the line-segment.[46] In this context, the emergence of the straight line from the point and that of the circle from the straight line (and ultimately from the point), aims to identify the geometrical point with the principle of the hieroglyphic monad, which is constituted of points, straight lines and circles. More fundamentally, it also relates the

---

[40] Dee (1564, fol. 12r); Josten (1964, p. 155).

[41] Dee (1564, fol. 5r), Josten (1964, p. 119): "Hieroglyphicum Typum, ad Pythagoricam (dictam) appingemus literam."

[42] See Josten (1964, pp. 99–111), French (1987, pp. 78–80), Clulee (1988, pp. 86–96), Clulee (2005) and Mandosio (2003).

[43] Mandosio (2003).

[44] Dee (1564, fol. 12r), transl. C.H. Josten, in Josten (1964, pp. 154–155): "*Theorema I.* Per lineam rectam, Circulumque, Prima, Simplicissimaque fuit Rerum, tum, non existentium, tum in Naturae latentium Involveris, in Lucem Productio, representatioque."

[45] Dee (1564, fol. 12r), transl. C.H. Josten, in Josten (1964, pp. 154–155): "*Theorema II.* At nec sine Recta, Circulus; nec sine Puncto, Recta artificiosè fieri potest. Puncti proinde, Monadisque ratione, Res, & esse coeperunt primò: Et quae peripheria sunt affectae, (quantaecunque fuerint) Centralis Puncti nullo modo carere possunt Ministerio."

[46] These generations naturally point to the constructive postulates enunciated by Euclid (see *supra*, n. 60, p. 13).

geometrical point to the monad, which corresponds to the indivisible principle of the whole universe, all things having originally sprung from it.[47]

This conception echoes what Dee wrote in the preface of this treatise, namely, that the letters with which all the sacred texts were written (Latin, Greek, Hebrew) were designed by mathematicians through points, straight lines and circles.[48] The mathematical origin and structure of the elements of words is thereby founded on the mathematical origin and structure of the universe and its components.

In a later passage of the same work, Dee explained the natural place and motion of each of the four elements (earth, water, air and fire) in the universe through the composition of the hieroglyphic monad. He then explicitly appealed to the geometrical notion of flow of the point as the principle of the line.

> Theorem VII. As dislocated homogeneous parts of the elements will teach an experimenter, the elements, removed outside their natural habitations, return to them along straight lines. It will, therefore, not be absurd [to assume] that the mystery of the four elements (into which their several compounds can be ultimately resolved) is intimated by the four straight lines going forth from one indivisible point and into opposite directions. You will also carefully note that the geometricians teach that a line is produced by the flowing of a point. We assert that here [things] happen in a similar way, insofar as, in [the process of] our mechanical magic, our lines signifying the elements are produced by the continuous fall of [successive] drops becoming a flow (if we consider drops to be like mathematical points).[49]

---

[47] As noted by A. Campbell (Campbell 2012, p. 521), Dee returned to this idea in the very last theorem of his work (Dee 1564, fol. 27v): "nostrum huius libelli exordium, a puncto, recta, circuloque coepimus."

[48] Dee (1564, fol. 4v): "Cum ipse, qui omnium mysteriorum Author est *Solus*, ad Primam et Ultimam, *Seipsum* Comparavit Literam. (Quod non in Graeca solum esse intelligendum Lingua: sed in Hebraea, tum in Latina, variis, ex Arte ista, demonstrari potest viis.)"; *ibid.* fol. 5r: "(Quicquid humana iactare solet Arrogantia) Earumque omnium Figuras, ex Punctis, Rectis lineis et Circulorum peripheriis, (mirabili, Sapientissimoque dispositis artificio) prodiisse"; *ibid.* fol. 5v: "Sed dimissis, hoc modo, Literarum istis, et Linguae Philosophis; *Mathematicos* meos, Raritatis istius nostri Muneris, adducam sincerissimos testes." On the interpretation of this passage, see Clulee (1988, p. 92), Mandosio (2003, pp. 488–489) and Campbell (2012, p. 521). On the relation between language, mathematics and the order of the universe in Dee's *Mathematicall praeface* and his esoteric works, see Knoespel (1987).

[49] Dee (1564, Theorem VII, fol. 13r), transl. C.H. Josten, in Josten (1964, pp. 158–159): "Elementis, extra suas Sedes naturales, dimotis: Suos ad ad easdem Reditus, naturaliter, per Rectas facere lineas, Dislocatae eorundem homogenae Partes, experientem docebunt: Absurdum igitur non erit, per 4 rectas, ab unico Puncto, Individuoque in Contrarias excurrentes partes, *Quatuor elementorum*, (in quae Elementata, singula, tandem resolvi possunt) innere Mysterium. Hoc etiam notabis diligenter; Geometras docere, *Lineam, ex puncti fluxu*, produci: Nos hic simili ratione, fieri monemus: Dum Elementares nostrae lineae, es *Stillae*, (tanquam Puncti, Physici) continuo Casu, (quasi *fluxu*) in Mechanica nostra producantur Magia."

In this passage, the flow of the point generating the geometrical line is compared to the continuous falling or cascading of droplets, which echoes the physical understanding of the term ῥύσις in ancient Greek as a stream of water. It is however more likely interpreted here as the falling of the atoms composing the matter of natural bodies (since these are compared to physical points) in ancient atomistic doctrines of nature.[50] In the composition of the hieroglyphic monad, the geometrical flow of the point is literally materialised through the four straight lines generated from a central point, each being led toward one of the four cardinal points and forming all together a cross. The motion implied by the generation of these four straight lines from a common point would symbolise the mode of generation of the four elements from one indivisible principle, but also the motion through which each element aims to rejoin its natural place according to the Aristotelian theory of nature.[51]

This text therefore brings together a conception of the generative motion of geometrical objects which combines geometry with the principles of natural philosophy, theology, astrology and esoteric doctrines. This allows us to confirm that Dee, contrary to Billingsley, saw in the generative motion of the geometrical point the image of a metaphysical process, similarly to what Proclus implied in his commentary on Euclid's first postulates.[52] Stating, in the second theorem, that it is "by virtue of the point and the Monad that all things commence to emerge in principle" (12r), the generation of magnitude from the point would even be understood as an active principle of the creation of the universe and not merely as a mathematical representation of this primordial process.

Hence, the esoteric natural philosophy which Dee drew up in the *Monas hieroglyphica* not only displays the connection between mathematics, theology and natural philosophy (in a broad sense, as it includes the consideration of magic and occult causes), but also confirms that Dee, just as Peletier and Foix-Candale, saw the generative motion of the mathematical point as the image of a higher creative process originating from a transcendent One. In the context of geometry, Dee did not distinguish the flow of the point from a spatial process and from a motion in a straightforward sense. He presented it furthermore as a properly mathematical definition of the line.

---

[50] J. M. Mandosio (Mandosio 2003), who supports this interpretation, points to the fact that Lucretius, in his *De natura rerum*, assimilated atoms to falling rain drops: Lucretius (transl. Rouse, 1924, p. 113), II, 220–224: "For if they were not apt to incline, all would fall downwards like raindrops through the profound void, no collision would take place and no blow would be caused amongst the first-beginnings: thus nature would never have produced anything."

[51] Aristotle, *Meteorologica*, I, 2, 339a11–19 (Barnes 1995, I, p. 555): "We have already laid down that there is one principle which makes up the nature of the bodies that move in a circle, and besides this four bodies owing their existence to the four principles, the motion of these latter bodies being of two kinds: either from the centre or to the centre. These four bodies are fire, air, water, earth. Fire occupies the highest place among them all, earth the lowest, and two elements correspond to these in their relation to one another, air being nearest to fire, water to earth." Aristotle, *De caelo*, I.2, 269a 16–17 (Barnes 1995, I, p. 448): "if the natural motion is upward, it will be fire or air, and if downward, water or earth."

[52] See *supra*, p. 31.

Moreover, Dee did not, in either of the considered texts, assert the separation between the motion of mathematical objects and of physical substances, but rather established (at least in the *Monas hieroglyphica*) an ontological correspondence between the two. Indeed, as pointed out by Mandosio, Dee explained in the preface to his *Monas hieroglyphica* that mathematical objects such as numbers are not to be interpreted as they are by mathematicians (and notably by himself in the *Mathematicall praeface*), that is, as abstract from material and sensible beings,[53] but rather as concrete and corporeal. Along the same line, the kinematic processes through which the line and the circle are thought to come about in the *Monas hieroglyphica*, are more than mere mathematical symbols of physical or magical processes taking place in nature, but are also to be regarded as active principles of the constitution of the universe and of all its components.

* * *

Thus, although Dee did not himself produce a commentary on Euclid's *Elements*, he proposed an interpretation of the generative motion of geometrical objects in at least two important works, the *Monas hieroglyphica* and the *Mathematicall praeface*. In both texts, he concentrated on the generation of the line, leaving aside other types of generations, notably those presented within Euclid's own genetic definitions (in Book XI). Yet, he implicitly referred, in the first theorems of the *Monas hieroglyphica*, to the genetic definition of the circle as produced by the rotation of a line-segment, in addition to that of the line.

While, in the *Mathematicall praeface*, Dee focused on the mathematical understanding of the generation of the line, where the motion of the point is described as what determines the spatial properties of the line, in the *Monas hieroglyphica* on the other hand, he presented this notion in a larger context of the interpretation of an esoteric symbol synthetising all the relations at play in the cosmos on a physical, metaphysical, magical as well as mathematical level. In this context, the very notion of generative motion of the point was attributed an active role in the creation and composition of the physical world.

When comparing the modes of composition of numbers and of magnitudes in the *Mathematicall praeface*, Dee distinguished the point from the arithmetical unit insofar as it possesses a situation or position, in conformity with the Pythagorean definition of the geometrical point as a unit with a position. He thereby attributed to the point the ability to move from one place to the other within the imagination. The fact that the unit, contrary to the point, may not be attributed a position or a situation, and therefore may not be held to move from one place to the other, would also make it incapable of generating number, although numbers consist of units. The fact that only the point was regarded as a generative principle for its kind of quantity enabled Dee to remove any tension related to the notion of line as caused by a point. In this, Dee differed from Billingsley, who maintained the arithmetical unit as the generative principle of number.

---

[53] This is the case in the *Mathematicall praeface*, where both number and magnitude are said to be immaterial. Dee (Billingsley 1570, sig. *1r): "Neither Number, nor Magnitude, have any Materialitie".

Now, if in the context of theoretical mathematics, as dealt with in Euclid's *Elements*, the modes of generation and composition of numbers and magnitudes were clearly distinguished, Dee admitted that, in the practice of the reckoner, the unit and numbers are treated in the manner of geometrical magnitudes. Indeed, in this context, the unit is potentially subject to infinite division and numbers may be dealt with as sides of squares, quadrilaterals or cubes. Hence, through these geometrical treatments of magnitude, the arithmetical unit would be implicitly regarded, for Dee, as a generative principle of number and as endowed with the ability to generate quantity, as the point in geometry. However, this confirms that, for Dee, only continuous quantity may be properly regarded as generated by an indivisible principle, since it would be through its imitation of continuous quantity that number would be conceived as generated in a mathematical context.

Thus, while Dee was aware of the various interpretations (mathematical, physical and metaphysical) of the generative motion of mathematical objects, he did not express any reservations toward the introduction of motion in geometrical definitions. Genetic definitions played indeed a foundational role in geometry by displaying the mode of constitution of lines and figures from the point. These ultimately allowed the geometer to establish the truth of his knowledge on the fundamental connection between mathematical, metaphysical and physical (even magical) principles in the constitution of nature.

Federico Commandino

7

7.1 The Life and Work of Federico Commandino and the Significance of His Commentary on the *Elements*

Federico Commandino (Urbino, 1509—Urbino, 1575)[1] was an Italian mathematician and humanist, close to the Court of the Dukes of Urbino. He studied philosophy and medicine at Padua and Ferrara, before returning to Urbino, where he was appointed tutor and medical advisor to the Duke Guidobaldo II della Rovere (1514–1574, reigned 1538–1574). He also worked under the patronage of cardinal Ranuccio Farnese (1530–1565), among others,[2] and taught mathematics to Guidobaldo del Monte (1545–1607) and to Bernardino Baldi (1553–1617).

Commandino translated and commented a great number of ancient Greek mathematical treatises, beside Euclid's *Elements* (in Latin in 1572 and in Italian in 1575),[3] among which several works of Archimedes,[4] the *Planisphere* and *Analemma* of Claudius Ptolemy (c. 85–165),[5] Apollonius' *Conics* and Eutocius' commentary on Archimedes,[6] the treatise on the sizes and distances of the Sun and Moon of Aristarchus of Samos (c. 310–c.

---

[1] On Commandino's life, career, works and contribution to the advancement of science, see Rose (1973), Rose (1975, pp. 185–219), Rosen (2008) and Zerlenga (2016).

[2] On Commandino's various patrons, see Rose (1973) and (1975, pp. 185–219).

[3] Commandino (1572a and 1575a).

[4] Commandino (1565a and 1588a).

[5] Commandino (1558 and 1562).

[6] Commandino (1566).

230 BC),[7] Hero of Alexandria's *Pneumatics*,[8] as well as Pappus of Alexandria's *Mathematical collection*.[9] As mentioned in the previous chapter, Commandino also published an edition of the *De superficierum divisionibus* by Machometus Bagdedinus (Al-Baghdādī), on which he worked with John Dee, the latter having provided him in 1563 with his copy (dating from 1559).[10] In addition to these editions, translations and commentaries, he authored a treatise on the calibration of sundials (published together with his edition of Ptolemy's *Planisphere*)[11] and one on the centres of gravity of solid bodies.[12] His intense and philologically precise work as a translator and commentator of ancient mathematical works earned him the title of *Restaurator mathematicarum* in the historiography of Western mathematics.[13]

His Latin translation from the Greek of Euclid's *Elements* is the most philologically accurate of all pre- and early modern translations and was used as the standard Latin version of this canonical work until the nineteenth century.[14] The quality of his translation of Euclid was acknowledged by Clavius.[15] In his epistle dedicated to the duke Francesco Maria della Rovere of Urbino (1549–1631), Commandino mentioned the commentaries of Fine and Peletier, stating their inadequacy on account of their neglect of the Greek version of the *Elements* and, in particular, of Peletier's dependence on the Latin translation from the Arabic used by Campanus.[16] He also mentioned Foix-Candale for his dismissal of the Greek demonstrations in view of their lack of elegance.[17] Commandino's translation and

---

[7] Commandino (1572b).

[8] Commandino (1575b).

[9] Commandino (1588b).

[10] See *supra*, n. 14, p. 169.

[11] Commandino (1562).

[12] Commandino (1565b).

[13] Del Monte (1577, sig. **1v) and Renn and Damerow (2010, p. 52): "Solis instar Federicus Commandinus, qui multis doctissimis elucubrationibus amissum mathematicarum patrimonium non modò *restauravit*, verùm etiam auctiùs, & locupletiùs effecit" and Rose (1975, p. 185).

[14] De Risi (2016b, p. 598). Vitrac (in Vitrac 2021, pp. 42–43) shows that, in spite of its undeniable philological accuracy, it presented certain imperfections with regard to form and content, notably as he took up some of the mathematical mistakes contained in the Greek edition of the *Elements* by Grynaeus.

[15] Clavius (1611–1612 I, p. 10): "Federicum Commandinum Urbinatem, Geometram non vulgarem excipio, qui nuper Euclidem Latin redditum in pristinum nitorem restituit." On the significance and reception of Commandino's translation and commentary on Euclid, see Vitrac (2021, pp. 36–47).

[16] Commandino (1572a, sig. *2v): "Nam ut pauca de hac re loquar, Orontius quidem Phinaeus haud obscuri nominis auctor priores tantum sex libros nulla graeci codicis ratione habita edidit. Iacobi vero Peletarii in eadem re labor eo etiam minus probatur, quòd Campani editionem ex arabica conversam lingua, magis, quàm graecam sequi voverit. Alii autem peracuti sanè ingenii homines ἀναλύσεις geometricas in priores sex libros conscripserunt, cetera tamen non sunt prosecuti."

[17] Commandino (1572a, sig. *2v–*3r): "At Candalla vir & generis nobilitate, & rerum cognitione insignis, licet omnes Elementorum libros, qui postulari à latinis videbantur, latinos fecerit,

commentary was also accompanied by translations of many ancient scholia[18] and by a long preface[19] in which he distinguished, on the basis of Proclus' commentary on Euclid, the author of the *Elements* from Euclid of Megara, thereby dissipating a confusion which had been perpetuated by the prior Euclidean tradition.[20]

As in the previous chapters, I will first look at the place and function of genetic definitions in Commandino's commentary on the *Elements*, as well as the ontology that underlies his interpretation of their role in geometry. I will consider afterwards the use he made of the genetic definition of the rectangular parallelogram in his commentary on Df. II.1.

Commandino, among all the commentators of Euclid considered here, offered perhaps the least original treatment of genetic definitions. Yet, his case will show how the Proclus' treatment of genetic definitions could be taken up and applied to a different ontological model. Commandino's commentary on the second book of the *Elements* will notably show how he attempted to remain faithful to Euclid's distinction between discrete and continuous quantity while indirectly hinting at the possibility of connecting the modes of generation of numbers and magnitudes through a kinematic understanding of magnitudes.

7.2 Formulation and Distribution of Genetic Definitions in Commandino's Commentary on Euclid

Commandino first mentioned the genetic definition of the line in his commentary on Df. I.2, where he presented different modes of definitions of the line, as Proclus had done in his commentary.[21] He then directly quoted the Greek form of the expression "flow of the point".

---

locupletaveritque, parum tamen (ut audio) eo nomine commendatur, quod longius iter ab Euclide averterit; & demonstrationes, quae in graecis codicibus habentur, velut inelegantes, & mancas suis appositis reiecerit."

[18] Vitrac (2021). As shown there by Vitrac, the portions of Commandino's commentary on Euclid which are entitled "Scholium" contain either proper scholia translated from ancient manuscripts, additions or alternative proofs. These are distinguished from his own commentaries, which are entitled "F.C. Commentarius", as are the passages commented on here.

[19] On this preface, see Homann (1983), Rambaldi (1989), Rommevaux (2004) and Claessens (2015).

[20] See, for example, Zamberti (1505), *Euclidis megarensis philosophi platonici mathematicarum disciplinarum Ianitoris*; Fine (1536), *In sex priores libros Geometricorum elementorum Euclidis Megarensis demonstrationes*; Tartaglia (1543), *Euclide megarense philosopho: solo introduttore delle scientie mathematice*; Billingsley (1570), *The Elements of Geometrie of the most auncient Philosopher Euclide of Megara*. See Rose (1972, p. 206), Homann (1983, p. 243) and Billingsley (1993).

[21] See *supra*, n. 103, p. 25.

There are some who defined the line in a different manner: indeed, others have said that it is a σημείου ρύσιν, that is, a flow of the point.[22]

In his commentary on the postulates, Commandino used both Latin terms *fluxus* and *motus* to express the generation of the line from the point.

The fact of leading a straight line from any point to any point follows the definition which teaches that the line is the flow (*fluxus*) of the point and that the straight line is its equal and undeviating flow (*fluxus*). For if we think that the point is moved according to a motion (*motus*) that is equal and the briefest of all, we will arrive to the other point and the first postulate will have been fulfilled, surely without anything complicated for us to understand.[23]

One may clearly see in this passage a paraphrase of Proclus' commentary on the first three postulates of Euclid.[24] Now, this is the very passage in which Proclus established the relation between the intelligible flow of the line from the point (σημείου ρύσις) and the local motion of a point producing a line by leaving a trace as it moves from one place to another (for which was then used the term κίνησις). As was said earlier, the points, lines and figures of the geometer, and their imaginary motion, would aim, for Proclus, to imitate in a spatial manner what is by essence indivisible and only exists on a non-spatial mode, as would the emanation of the multiple from the primordial One according to the Pythagorean

---

[22] Commandino (1572a, fol. 1r), Df. I.2: "fuerunt qui lineam aliter diffinirent: alii enim σημείου ρύσιν, hoc est puncti fluxum dixerunt".

[23] Commandino (1572a, fol. 6r), Post. 3: "A quovis puncto ad quodvis punctum rectam lineam ducere, sequitur eam diffinitionem, quae tradit lineam esse puncti fluxum, & rectam lineam aequabilem, & non declivem fluxum. Si igitur intelligamus punctum aequabili, et brevissimo motu ferri in alterum punctum incidemus, & primum postulatum factum erit, nihil utique varium intelligentibus nobis."

[24] Proclus (Friedlein 1873, p. 185) and (Morrow 1992, p. 145): "The drawing of a line from any point to any point follows from the conception of the line as the flowing (ρύσις) of a point and of the straight line as its uniform and undeviating flowing. For if we think of the point as moving (κίνησις) uniformly over the shortest path, we shall come to the other point and so shall have got the first postulate without any complicated process of thought." See *supra*, p. 48. The rest of Commandino's commentary on Post. 3, which explains the two following postulates, is also drawn from Proclus. Commandino (1572a, fol. 6r), Post. 3: "Si vero recta linea puncto terminata, similiter intelligamus eius terminum brevissimo, & aequabili motu ferri, secundum postulatum facili, simplicique aggressione comparatum erit. Quòd si rursus terminatam rectam lineam manere quidem ex altera parte, ex altera autem moveri circa manens punctum intelligamus, tertium fiet postulatum; centrum namque erit punctum manens, intervallum vero recta linea; & quanta ea fuerit, tantum erit intervallum à centro ad omnes circumferentiae partes." *Cf.* Proclus (Friedlein 1873, p. 185) and (Morrow 1992, p. 145): "And if we take a straight line as limited by a point and similarly imagine its extremity as moving uniformly over the shortest route, the second postulate will have been established by a simple and facile reflection. And if we think of a finite line as having one extremity stationary and the other extremity moving about this stationary point, we shall have produced the third postulate; for the stationary point will be the center and the straight line the distance, and whatever length this line may have, such will be the distance that separates the center from all parts of the circumference."

doctrine. Commandino quite conventionally translated the term ῥύσις by *fluxus* and the term κίνησις by *motus*, as had done Francesco Barozzi in his translation of Proclus' commentary on Euclid.[25] As Commandino also had access to Proclus' ontological discourse on the status of geometrical objects, either through Barozzi's translation or through Grynaeus' Greek edition of Proclus' commentary (or even through Giorgio Valla's *De expetendis et fugiendis rebus opus*[26]), one may wonder whether Commandino also aimed here to take up Proclus' ontological distinction between the two states of the generation of the line, intelligible and imaginary.

7.3   The Ontological and Epistemological Status of Geometrical Motion

In the long preface to his commentary, Commandino actually also borrowed from Proclus certain assertions concerning the mode of existence and of apprehension of mathematical objects. For instance, he affirmed the necessity for mathematical objects to be considered as endowed with matter because of the weakness of the human mind, as it would not be able to directly seize the properties of the concepts enfolded in the intellect without projecting them in the (intelligible) matter of the imagination.[27]

It is however unlikely that Commandino intended thereby to follow Proclus' ontological doctrine and that he attributed a metaphysical dimension to the notion of flow of the point in the way intended by the Pythagorean and Neoplatonic philosophical doctrines. Indeed, even if Commandino took up parts of Proclus' discourse concerning the gnoseological status of mathematical objects, he also presented mathematical objects as abstracted from

---

[25] Proclus in (Barozzi 1560, p. 106): "Nam illa quidem ab omni Signo ad omne Signum rectam Lineam ducere, eam consequitur definitionem, quae Lineam Signi *fluxum* esse ait, & Rectam indeclivem, atque inflexibilem *fluxum*. Si igitur Signum indeclivi, brevissimoque *motu* moveri intellexerimus, in alterum Signum incidemus, & prima Petitio facta est, nilque varium intellexerimus." (My emphasis.) *Cf.* Commandino (1572a, 6r), see *supra*, n. 23, p. 186.

[26] This was very likely Fine's source on Proclus' philosophy of mathematics. See *supra*, n. 28, p. 45.

[27] Compare Proclus (Friedlein 1873, pp. 54–55) and (Morrow 1992, p. 44): "For the understanding contains the ideas but, *being unable to see them when they are wrapped up*, unfolds and exposes them and presents them to the imagination sitting in the vestibule; and in imagination, or with its aid, it explicates its knowledge of them, happy in their separation from sensible things and *finding in the matter of imagination a medium apt for receiving its forms*." (My emphasis.) and Commandino (1572a, n. p. 1): *Federici Commandini in elementa Euclidis prolegomena*: "while [mathematical things] are absolutely deprived of matter, if you carefully examine their true condition by an attentive study, they appear to be somewhat endowed with matter, since they cannot be known without a certain addition of matter due to the weakness of the intellect" ("tum quòd omni vacant materia, si accuratiori studio veram illarum conditionem inspexeris, tum quòd materia praeditae quodam modo videantur, quia sine aliqua eius adiunctione ob ingenij nostri imbecillitatem cognosci nequeunt"). On this correspondence, see Claessens (2015, pp. 501–504).

natural substances, stating that "the mathematician considers quantity and the forms separable from matter *by abstraction* and teaches their definitions without mentioning matter" ("Mathematicus ἐκ τῆς ἀφαιρέσεως circa quantitatem, formasque à materia separabiles versatur: et earum diffinitiones tradit, materiam non attingens").[28] In this passage, one may clearly recognise a reference to Aristotle's ontological doctrine, according to which mathematical objects would not have a separate existence outside of material beings (apart in the mathematician's mind, which separates them from matter) and are therefore not to be considered as divine principles of natural substances.[29] The connection with Aristotle's ontology is marked also by Commandino's assertion that, in doing so, the mathematician does not corrupt mathematics by bringing falsehood into it ("nec illarum imaginatione aliquo contaminantur mendacio mathematicae disciplinae"), which distinctly resonates with the Stagirite's discourse on the topic in the *Physics*.[30]

As G. Claessens has shown, Commandino was able to take up elements of Proclus' projectionism, notably his concept of imagination as a material medium for the study of geometrical concepts (which he described as a drawing or reckoning board (*abacus*) for the mathematician[31]) and to incorporate it into an abstractionist framework by following John Philoponus' (c. 490–c. 570) commentary on Aristotle's *De anima*.[32] In Philoponus' work,

---

[28] Commandino (1572a, n. p. 2): "explicanda nobis est ratio, & modus aperiendus, quo mathematicis quantitatem, & continuam, & discretam pro subiecto, eruditorum auctoritate substerni dicimus: neque enim de quoto, quod in sensilibus ipsis est, nec de quanto, quod circa corpora excogitatur, est absolute intelligendum; physici enim potius, quàm mathematici finibus continetur haec contemplatio. Eorum igitur quae naturali corpori insunt, nec ab eo separantur, alia quidem nec re, nec cogitatione removeri queunt, ut calor frigus, siccitas, quòd illa qua naturale est corpus, obtineat, alia vero etiam si re ipsa disiungi minime queant, animi tamen cogitatione fingimus abesse, eò quòd per accidens, non autem per se, nec quatenus natura praeditum est corpus, haec habeat, qualia sunt rectum, curvum, inflexum, ceteraque, id genus. *Mathematicus igitur hoc pacto ἐκ τῆς ἀφαιρέσεως circa quantitatem, formasque à materia separabiles versatur: & earum diffinitiones tradit, materiam non attingens.*" (My emphasis.)

[29] Aristotle, *Physics* II.2, 193b32–194a6 (see *supra*, n. 123, p. 28) and *De anima* III.7, 431b12–16 (Barnes 1995, I, p. 686): "The so-called abstract objects the mind thinks just as, in the case of the snub, one might think of it *qua* snub not separately, but if anyone actually thought of it *qua* hollow he would think of it without the flesh in which it is embodied: it is thus that the mind when it is thinking the objects of mathematics thinks of them as separate though they are not separate."

[30] Commandino (1572a, n. p. 2). *Cf.* Aristotle, *Physics* II.2, 193b31–35 (1995, I, p. 331): "and it makes no difference nor does any falsity result, if they are separated."

[31] Commandino used the term abacus, which primarily translates as reckoning board, but G. Claessens translates it (correctly, I believe) as drawing board in order to conform to the function of the imagination in geometry, which involves the drawing out of figures rather than the manipulation of numbers.

[32] Commandino (1572a, n. p. 2): "Quid est linea? μῆκος ἀπλατές, longitudo latitudinis expers. Quid est triangulum? Figura, quae tribus rectis lineis continetur. & circulus figura, quae ab una comprehenditur linea. Nulla hoc materiae mentio est, nullum eius vestigium ob allatam modo rationem. [. . .] nam imaginatione quidem Geometra tamquam abaco utitur, magnitudines dividendo, intervalla dimetiendo, & lineas describendo." *Cf.* Philoponus (Hayduck 1897, pp. 58 and 1–13) and (Van der Eijk 2006, p. 75), I.1: "For example, what is a triangle? A shape encompassed by three

the imagination is indeed presented as a tool to display geometrical concepts within an intelligible matter, but geometrical concepts are also admitted as derived from physical substances and sense perception.

It does not appear, therefore, that Commandino considered the flow of the point as a metaphysical process in the way Proclus intended it. As said, it is possible that he read Proclus through Barozzi's translation. Now, in Barozzi's translation of the above-mentioned commentary on the postulates, the ancient notion of ῥύσις σημείου is first and foremost presented as *a mode of definition* of the line among others and does not appear to say anything about the *mode of being* of the line.[33] Thus, it is possible that Commandino interpreted it then as a logical principle, as well as a foundational concept for the geometer's constructions and demonstrations, and not as a statement on the ontological status of the line. This would be corroborated by the fact that the Greek expression ῥύσις σημείου is textually set forth within Commandino's commentary on Euclid's *definition* of the line.[34] This enables us at least to confirm that, to him, genetic definitions corresponded to acceptable definitions in the context of geometry.

Hence, if, as it seems, the notion of line as the flow of a point (*puncti fluxus*) was understood here by Commandino as a purely logical concept, the motion of the point would correspond to the spatial interpretation of this logical concept in the imagination. In other words, the definition of the line as resulting from the flow of the point would intend to state the mode of generation of the geometrical line in a universal manner, without relating it to its apprehension by the geometer nor even to its mode of being. On the other hand, the postulate would appeal to the interpretation of this definition as a local process in the imagination (as *puncti motus*) to convey the operation by which the geometer should actually generate a line in the context of geometrical constructions, which corresponds to the procedure of *leading* (*duco* or *ductus*) a line from one point to another.[35] Accordingly, the definition of the line as the flow of the point, once it has been spatially translated as the actual motion of the point in the imagination, would serve as a foundation for the first postulates, establishing a relation between the constructions performed by the geometer and the definitions.

---

straight lines. What is a circle? A shape encompassed by one line. In these definitions he does not mention matter. [...] Now the geometrician discusses the extended forms in the imagination; for he uses the imagination as a kind of reckoning board, exercising activity in individual parts and measuring and dividing distances." It may be noted that Commandino mentioned Philoponus in his *prolegomena* (Commandino 1572a, sig. **r).

[33] Proclus in (Barozzi 1560, p. 106): "consequitur definitionem, quae Lineam Signi fluxum esse ait". See *supra*, n. 25, p. 187.

[34] See *supra*, p. 186.

[35] Commandino (1572a, fol. 6r), Post. 1: "Postuletur à quovis ad quodvis punctum rectam lineam *ducere*." (My emphasis.)

7.4   Commandino's Use of Kinematic Notions in His Commentary on Book II

As the other commentators considered here, Commandino also referred to a kinematic understanding of geometrical figures in his commentary on the first definition of Book II. There, he characterised the generation of the rectangular parallelogram by the transversal motion of a line-segment along another at right angles neither as a flow nor as a motion (*fluxus*, or *motus*), but through the above-mentioned operative term *ductus*. Now, this term is related to the verb *duco* that was used in Commandino's translation of Euclid to designate the production of a line from any point to any point in the first postulate.[36] As was shown earlier, this term also had an arithmetical meaning at the time, whereby it was used to designate the multiplication of a number by another.[37]

> What a rectangular parallelogram is has been said above. Moreover, it is said to be contained by the two straight lines which surround the right angle, since the area of the rectangle originates from the leading (*ductus*) of one of the two lines on the other, which does not occur in other parallelograms, that is, in those that are not rectangular. Indeed, let there be the rectangular parallelogram ABCD, and let, for example, the side AB be three feet long and the side BC, four feet long. Therefore, the total area of the rectangle will be equal to twelve square feet.[38]

The operation of leading one of the sides of the parallelogram along and according to the length of its adjacent side to produce and/or hypothetically measure the area of the parallelogram is not directly compared to a multiplication of numbers, as it was in the commentaries of Fine, Peletier and Billingsley. This may be due to Commandino's wish to closely follow and restitute the meaning of Euclid's text, in which a clear separation was made between arithmetic and geometry, although he did also, like Billingsley, include the arithmetical interpretation of the ten first propositions of Book II by Barlaam of Seminara in his commentary on Book IX.[39] Also, in the diagram that accompanies Euclid's text (Fig. 7.1), the measured parallelogram is only represented as divided into a number of

---

[36] See *supra*, n. 23, p. 186.

[37] See *supra*, pp. 68–69.

[38] Commandino (1572a, fol. 28v), Df. II.1: "Quid sit parallelogrammum rectangulum dictum est superius. Dicitur autem contineri duabus rectis lineis, quae sunt circa rectum angulum, quoniam ex ductu alterius in alteram provenit eius rectanguli area, quod non contingit in alijs parallelogrammis, quae rectangula non sunt. Sit enim parallelogrammum rectangulum *abcd*: & sit, exempli gratia, latus quidem *ab* pedum trium, latus vero *bc* quattuor erit totius rectanguli area pedum duodecim quadratorum."

[39] Commandino (1572a, fol. 114v–117r), IX.15: "Barlaam Monachi arithmetica demonstratio eorum, quae Euclides libro secundo in lineis demonstravit." The whole text of these propositions is to be found in Commandino (1572a, fol. 114v–117r). On these insertions, see Vitrac (2021, pp. 39–40 and 142–146).

square-units and not as a geometrical number (that is, as juxtaposed rows of point-units forming a rectangle). In this manner, Commandino could indeed maintain here the homogeneity of the surface of the parallelogram with its parts. He also showed thereby that the process of *ductus*, which then implied the multiplication of the number of length-units on one side with the number of length-units on the other side, should be primarily taken as a geometrical operation, even when used as related to the measurement of areas, for instance in the context of practical geometry.[40] Thus, if this passage hinted at the possibility of interpreting the propositions of Book II arithmetically, this would only be done in an indirect manner, as he did by discreetly referring in the same place to Johannes Regiomontanus' *De triangulis omnimodis*.[41] In this work, Regiomontanus explicitly set forth the operative correspondence between numbers and magnitudes in order to teach propositions relating to the measure of right-angled quadrilaterals and triangles.[42] This reference confirms that Commandino himself considered this interpretation as admissible, even if he postponed it to his commentary on Euclid's arithmetical books, so as to avoid openly subverting Euclid's demarcation between numbers and magnitudes in the *Elements*.

It remains nevertheless that the use Commandino made here of the term *ductus* (instead of *fluxus* or *motus*) to designate the generation of the rectangular parallelogram would not intend, in the first place, to point to the possibility of understanding Euclid's Book II arithmetically (even if it implicitly made this interpretation possible), but to mark the operative nature of this process. Indeed, *ductus* would specifically correspond here to a process carried out by the geometer imaginarily either to produce or to measure a rectangular parallelogram with a given area. It would, as such, tend to differ in its epistemological and methodological implications from the notions of *fluxus* or *motus*, and even from the verb *ducere* such as used to express the first postulate, which would rather relate this process to the logical concept of rectangular parallelogram or to its metrically undefined spatial construction within the imagination.

* * *

---

[40] This approach was also demonstrated in the commentary on Props. II.12 and II.13, where he taught how to measure the areas of triangles with specific dimensions.

[41] Commandino (1572a, fol. 28v), Df. II.1: "Verum parallelogrammi rectanguli aream provenire ex ductu laterum, quae circa rectum angulum sunt, in praesentia ponatur, quo ad ita esse manifesto apparebit. Demonstratur autem hoc à Ioanne Regiomontano in principio primi libri de triangulis".

[42] This is done for most propositions of the first book, within a separate and auxiliary part of the propositions designated as *Opus* or *Operatio*, to show their practical scope and to distinguish them from the first parts, which only properly deal with magnitudes and depend on Euclidean propositions. See, for instance, Regiomontanus (1533, p. 8), Prop. I.1: "*Omnis datae lineae quadratum erit cognitum.* [...] *Opus brevissimus.* Numerus secundum quem nota est linea, in se multiplicetur, & productus erit numerus secundum quem quadratum eius notum habebitur. Ut si *ae* sive *d* mensura fuerit in *ab* secundum numerum 5 multiplicatis 5 in se, producuntur 25 quadratellum igitur *af* in quadrato *ah* secundum numerum 25 reperitur & similiter in reliquis." See also *infra*, pp. 222–223. On the relation between numbers and magnitudes in Regiomontanus' *De triangulis omnimodis*, see Malet (2006) and Bos (2001, pp. 136–138).

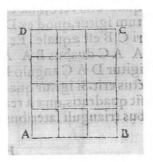

Fig. 7.1 Federico Commandino, *Euclidis elementorum: libri XV*, 1572, 28v, Df. II.1. Diagram illustrating the spatial relation of the sides of the rectangular parallelogram to its area. Contrary to nearly all the other commentators considered here, Commandino did not explain this relation by resorting to the notion of rectangular number, but through the subdivision of the rectangular area into square-units. Courtesy Max Planck Institute for the History of Science, Berlin

Commandino's commentary on Euclid included very few genetic definitions, in comparison with the other commentaries considered here. He appealed to the genetic definition of the line twice, each time following Proclus' commentary on the *Elements*. The first time was in the commentary on Df. I.2, where he quoted the Greek term ῥύσις σημείου as one of the ancient definitions of the line. The second time was in the commentary on the postulates, where this definition of the line as the flow of a point (*puncti fluxus*) was presented as what the geometer would aim to reconstitute in his imagination as he visualises a point moving from one place to another to produce thereby a geometrical line (*puncti motus*). As in Proclus' text, a distinction was therefore made between a non-spatial and properly essential concept of line and its spatial understanding as seized by the imagination. In other cases, he referred to the generation of magnitudes through the more operative notion of *ductus*, as in the commentary on Df. II.1, where this process applied to the generation and measurement of rectangular parallelograms.

Although Commandino conformed to Proclus for the way he described the generation of the line, he did not however share his views on the ontological status of mathematical objects. On this question, he rather followed the Aristotelian doctrine, such as interpreted by Philoponus, which allowed him to combine the Proclean notion of imagination as intelligible matter with an abstractionist perspective. It is thus highly unlikely that Commandino himself would have interpreted the notion of flow of the point as having a metaphysical dimension, although he would have been aware of Proclus' conceptions in this regard. The notion of *flow* of the point would only differ from that of *motion* of the point, according to Commandino, as for its logical function in geometry, the former representing a logical concept or definition of the line (and for that matter a genetic definition) and the latter displaying the spatial expression of this definition in a constructive aim. As a logical concept, the genetic definition of the line, which is for Commandino one type of definition among others of this geometrical object, would only be apprehended by the intellect, within which it would be deprived of spatiality. Only the "drawing board" of the imagination would enable to provide it with the spatiality that is necessary to deploy its proper qualities as a kinematic process.

In his commentary on Df. II.1, Commandino did not straightforwardly compare the production of a rectangular parallelogram to the operation of multiplying a number by another, although he indirectly pointed to this comparison, notably through his reference to Regiomontanus' *De triangulis omnimodis*. Generally speaking, Commandino seems to have taken care to avoid making a direct comparison between geometrical objects and numbers in the context of Euclid's geometrical books, so as not to subvert the boundary set by Euclid between geometry and arithmetic. On this account, he left aside the notion of figurate number (divisible in discrete point-units) to which Fine and Billingsley appealed in their commentaries on Df. II.1. He resorted instead to the representation of a parallelogram divided into square-units, which would enable him to observe the specific mode of composition and division of continuous quantities.

Hence, Commandino's treatment of genetic definitions does not enable us to clearly assess his attitude toward motion and genetic definitions in geometry, but it hints at a few key points concerning the epistemological and ontological status of such definitions. In this regard, he clearly considered generative motion as an acceptable means of definition of geometrical objects, but he did not, as some of his predecessors, attribute to it a metaphysical dimension, nor an explicit role in connecting numbers and magnitudes.

Christoph Clavius

8

8.1 The Life and Work of Christoph Clavius and the Significance of his Commentary on the *Elements*

Christoph Clavius (Bamberg, 1538—Rome, 1612) was a Jesuit mathematician and professor of mathematics.[1] After studying philosophy and theology in the Jesuit colleges of Coimbra and Rome, he succeeded Balthasar Torrès as a professor of mathematics in 1565 at the Collegio Romano, a position he maintained until his death in 1612. He was instrumental to the transformation of the scientific program of Jesuit colleges through his contribution to the *ratio studiorum*, as well as to the development of mathematical teaching at the Collegio Romano and within Jesuit institutions.[2] He also played an important role in the reform of the Julian calendar on which he worked in the early 1580s at the request of Pope Gregory XIII.

Clavius wrote and published a large number of mathematical works, most of which were reprinted several times.[3] Among these works were editions and commentaries of ancient and medieval scientific texts, such as Sacrobosco's *Tractatus de sphaera*,[4] Euclid's *Elements*[5] and Theodosius' *Sphaerics*[6]; manuals on various aspects of practical

---

[1] Knobloch (1988, 1990, 1995a) and Rommevaux (2012).

[2] On Clavius' role in the development of mathematical teaching in Jesuit colleges, see Romano (1999, chap. 2–3), Smolarski (2002) and Gatto (2006). On Clavius' importance for the constitution of Jesuit science, see Dear (1995a, chap. 2).

[3] Knobloch (1988) and Knobloch (1995b).

[4] Clavius (1570).

[5] First edition, Clavius (1574).

[6] Clavius (1586).

A. Axworthy, *Motion and Genetic Definitions in the Sixteenth-Century Euclidean Tradition*, Frontiers in the History of Science, https://doi.org/10.1007/978-3-030-95817-6_8

mathematics, that is, on gnomonic,[7] practical arithmetic,[8] practical geometry,[9] trigonometry[10] and algebra,[11] as well as works on the building and use of mechanical and astronomical instruments.[12] He also wrote texts on controversial subjects, in particular concerning the
reform of the calendar,[13] to which may be added his writings on the pedagogical program
of Jesuit colleges, in which he defended the importance of mathematical education.[14] He
extensively corresponded with important members of the scientific community such as
Francesco Barozzi, Guidobaldo del Monte, Galileo Galilei (1564–1642) and François
Viète (1540–1603). According to Bernardino Baldi, he was a close friend of
Commandino.[15]

In conformity with his strong pedagogical commitment, Clavius' commentary on the
Elements seems to have had a primarily didactic aim, having been written to introduce the
students of Jesuit colleges to Euclid's mathematical work. But it also features many parts
which could be regarded as superfluous to an elementary course on arithmetic and geometry, pertaining rather to a scientific and historical exegesis of Euclid's work and of its
ancient and modern tradition. Indeed, with regard to its content, this commentary is rich in
considerations on the scientific, methodological, epistemological, as well as pedagogical
value of Euclid's *Elements*, beside many scholia and alternative propositions. It also
contains Foix-Candale's sixteenth book on the regular polyhedra, as well as a comprehensive and critical recension of the various opinions of Clavius' predecessors on Euclid's
propositions,[16] including Fine, Peletier, Foix-Candale and Commandino.[17] The various
editions of this commentary were also the occasion for Clavius to enter into debates with

---

[7] Clavius (1581).

[8] Clavius (1583).

[9] Clavius (1604).

[10] Clavius (1607).

[11] Clavius (1608).

[12] Clavius (1593).

[13] Clavius (1588) and Clavius (1603).

[14] Clavius, *Ordo servandus in addiscendis disciplinis mathematicis*; *Modus quo disciplinae
mathematicae in scholis Societatis possent promoveri*; *De re mathematica instructio* and *Oratio de
modo promovendi in Societate studia linguarum, politioresque litteras ac mathematicas*, written
between 1581 and 1594 and published in Lukacs (1965–1992, VII). On these texts and their
significance for the constitution of the *ratio studiorum*, which established the rules for the
organisation of teaching in Jesuit schools and colleges, see Cosentino (1970), Cosentino (1971)
and Cosentino (1973), Gatto (1994), Moscheo (1998) and Romano (1999, chap. 1).

[15] Baldi (Nenci 1998, p. 517). P. Rose (Rose 1975, p. 207) considers however that this friendship was
most likely merely epistolary.

[16] On the structure of Clavius' commentary on Euclid's *Elements*, see Rommevaux (2005,
pp. 31–58).

[17] See, for example, Clavius (1611–1612, p. 107) for Fine and Peletier, (*ibid.* p. 610) for Foix-
Candale, and (*ibid.* p. 84) for Commandino. No mention was made, to my knowledge, to Billingsley's
commentary. On Clavius' relation to Commandino specifically, see Homann (1983, p. 236).

other commentators over Euclid's methods and concepts, as with Peletier, concerning the angle of contact and geometrical superposition.[18] As in Billingsley and Commandino's commentaries, Clavius' exposition of the *Elements* was introduced by a long preface presenting, among other topics, the nature of mathematical objects, the division and utility of mathematics and the structure of Euclid's work.[19] This commentary was used in Jesuit colleges long after Clavius' death[20] and remained a work of reference in the Euclidean tradition, and among Jesuit scientific works in general, at least until the nineteenth century.

In addition to Clavius' uses and terminology of genetic definitions, this chapter will examine his conceptions on the ontological status and mode of apprehension of the generative motion of geometrical objects and its relation to instrumental procedures. It will also analyse the epistemological status and function of genetic definitions in geometry, as well as Clavius' treatment of the quadratrix and his approach to the kinematic interpretation of superposition. The last section will consider his commentary on Df. II.1 and the way he used geometrical motion as a means to connect numbers and magnitudes.

As we will see, Clavius' use of genetic definitions in his commentary on Euclid displayed an inherently constructive approach to geometry. His use of genetic definitions was as extensive as that of Fine, but it was perhaps more pragmatic and more creative. Generally speaking, Clavius' approach to geometrical motion was, among all the commentaries presented here, the least affected by philosophical or philological constraints. He notably did not hesitate to compare the motion of geometrical objects to instrumental processes.

8.2   Formulation and Distribution of Genetic Definitions in Clavius' Commentary on Euclid

Like Billingsley and Commandino, Clavius presented, in his commentary on the *Elements*, the genetic definition of the line among various other definitions of the line, following the structure of Proclus' commentary on Df. I.2. And like Fine, Dee and Billingsley, he presented this definition of the line as properly relevant to the mathematician, its function being to allow the geometer to reach a proper understanding of the essential properties of the line.

> Mathematicians, in order to teach us the true understanding of the line, imagine the point (already defined by the previous definition) to be moved from one place to another. Indeed,

---

[18] On these debates between Peletier and Clavius, see *supra*, n. 8 and 9, p. 76.

[19] On this preface, also in comparison with those of Dee and Commandino, see Homann (1983), Rommevaux (2004), Axworthy (2004) and Higashi (2018, pp. 354–360).

[20] Sasaki (2003, pp. 2–3).

since the point is absolutely indivisible, what is left behind, through this imaginary motion, is a somewhat long trace that is entirely deprived of breadth.[21]

The generation of the line, as it is said here to result from a motion of the point (*puncti motus*) from one place to another (*è loco in locum*), is clearly presented as a spatial process. This process is described as taking place in the mathematician's imagination (*mathematici imaginantur*; *motus imaginarius*), within which and through which would be deployed and "materialised" the geometrical line, that is, a breadthless length.

In his commentary on the definition of the straight line (Df. I.4), Clavius used the expression *puncti fluxus* instead of *puncti motus*, but described this process in the same manner as in Df. I.2, presenting it as a spatial and imaginary process.

> As mathematicians conceive the line to be described by the imaginary flow of the point, they apprehend the quality of the described line through the quality of the flow of the point. For, indeed, if the point is conceived as flowing in a straight line through the shortest space, so that it does not deviate toward one part or toward another, but maintains an equal motion and progression, the described line will be called straight.[22]

In this passage, the spatial character of the flow of the point is marked in particular by the fact that it appears to be deployed in a space that is (hypothetically) metrically-determined ("the shortest space") and by the fact that it determines the formal structure of the line (i.e. its straightness), the flowing point being said here to maintain "an equal and unceasing motion" ("aequabilem quendam motum, atque incessum"). The flow of the point (*fluxus*) is also terminologically identified to a motion in the straightforward sense (*motus*).

---

[21] Clavius (1611–1612, I, p. 13), Df. I.2: "Mathematici quoque, ut nobis inculcent veram lineae intelligentiam, imaginantur punctum iam descriptum superiore definitione, e loco in locum moveri. Cum enim punctum sit prorsus individuum, relinquetur ex isto motu imaginario vestigium quoddam longum omnis expers latitudinis."

[22] Clavius (1611–1612, I, p. 14), Df. I.4: "Quemadmodum autem Mathematici per fluxum puncti imaginarium concipiunt describi lineam, ita per qualitatem fluxus puncti qualitatem lineae descriptae intelligunt. Si namque punctum recta fluere concipiatur per brevissimum spatium, ita ut neque in hanc partem, neque in illam deflectat, sed aequabilem quendam motum, atque incessum teneat, dicetur linea illa descripta, Recta".

8.3 The Mechanical and Empirical Character of Genetic Definitions

The notion of trace (*vestigium*) used here by Clavius,[23] which was also used by Fine in a similar context, as well as by Peletier in connection with the definition of the circle,[24] tends to relate the imaginary generation of geometrical figures to the concrete production of material figures by instrumental means (through drawing, engraving or carving). This also evokes Billingsley's own formulation of the genetic definition of the line, where the motion of the point is compared to the concrete motion caused by a drafting device.[25]

In the following section of the commentary on Df. I.4, Clavius indicates that the flow of the point may take place according to a great variety of paths, bringing the line to take on an infinite diversity of spatial structures, beyond those constructed and analysed in Euclid's *Elements*.

> If however the flowing point is thought to vacillate in its motion and to stagger from here to there, the described line will be called mixed. If finally the flowing point does not vacillate in its motion, but is carried about in a circle by a certain uniform motion and at a given distance from a certain determined point, the described line will be called circular. That is why if two points are moved according to absolutely similar motions, so that they are always equally apart from each other, two similar lines will be described by these, that is, if one of them is straight, the other one will also be straight; if however one of them is curved, the other one will be curved in the same manner, etc.[26]

---

[23] Clavius (1611–1612, I, p. 13), Df. I.2: "relinquetur ex isto motu imaginario *vestigium*"; Df. I.5: "*vestigium* relictum ex isto motu" and Df. XI.1: "*vestigium* quoddam longum, latum, atque profundum." See n. 21, p. 198, n. 28, p. 200 and n. 29, p. 201.

[24] See *supra*, n. 20, p. 42 and n. 21, p. 79.

[25] Billingsley (1570, fol. 1v), Df. I.2: "A lyne is the movyng of a poynte, as the motion or draught of a pinne or a penne to your sense maketh a lyne". It may also be interesting to note that Johannes Valentinus Andreae (1586–1654) referred to Clavius in the title of the geometrical part of his *Collectaneorum mathematicorum* (Andreae 1614, sig. A3r: *Geometriae tantum est inter scientias Imperium, ut Plato eius ignarum, iure Academiae suae haut intromiserit. Omnia metitur, omnia examinat. & tractanti quotidie magis ampla, infinitaque fit. Eius vera sedes Euclides; felix interpres Christophorus Clavius est: nos tamen hic Adrianus Metius Cl. vir, plurimum iuvit.*) and interpreted the line defined as resulting from the flow of the point as the instrumentally produced line, by the means of the straightedge or the compass (*ibid.*): "Ex huius [puncti] fluxu linea: quae vel recta est, & regula fit, vel circularis, fitque circino, vel mixta quae variè fit." I would like to thank Carsten Nahrendorf for drawing my attention to this source.

[26] Clavius (1611–1612, I, p. 14), Df. I.4: "Si vero punctum fluens cogitetur in motu vacillare, atque hinc inde titubare, appellabitur linea descripta, mixta: Si denique punctum fluens in suo motu, non vacillet, sed in orbem feratur uniformi quodam motu, atque distantia à certo aliquo puncto, circa quod fertur, vocabitur descripta illa linea, circularis. Itaque si duo puncta moveantur similibus prorsus motibus, ita ut semper aequaliter inter se distent; describentur ab ipsis duae lineae similes, hoc est, si una earum fuerit recta, erit & altera recta: si vero una fuerit curva, erit & altera eodem omnino modo curva, etc."

The motion that would be attributed to the point by the geometer is presented here as the direct cause of the structure or spatial quality of the resulting line, be it straight, curved, mixed or even parallel. This passage also seems to suggest that Clavius' conception of the line as generated by the flow or motion of a point involves a certain temporal factor, at least hypothetically. Indeed, in order to generate a specific type of mixed or broken line by the flow or motion of a point, it would be necessary to assume that the direction of the moving point changes at a given moment during this generative process. The length of the straight line would also, in principle, be determined by the duration of its generation.

Hence, in various ways, Clavius' description of the generation of lines in his commentary on Df. I.4, whether it is called a flow or a motion, indirectly evokes the mode of production of concrete lines. The generative processes described here appear in fact analogous to mechanical processes, the structure of the line being caused by the structure of the flow or motion of the point as if produced by a drafting tool. This conception is here confirmed by the illustration of a straightedge which Clavius appended to his commentary on Df. I.4 in order to explain the process through which a straight line is produced (Fig. 8.1). Through this illustration, Clavius established indeed a correlation between the generative motion of the point and the instrumental practice of the geometer,[27] since he directly related the imaginary undeviating flow of the point to its concrete production by the means of a straightedge.

Moreover, such as Clavius expressed it in his commentary on the first definitions of Book I, and notably in his commentary on Df. I.5 (definition of the surface), the mathematician's apprehension of the geometrical line or surface through the motion of a point or of a line would occur through a kind of sense-perception. Thanks to this imaginary process, the mathematician would be able to set the geometrical object so conceived "before our eyes".

> Mathematicians, in order to set the surface before our eyes, advise us to imagine a line that is transversally moved. Indeed, the trace left behind through its motion will be long, due to the length of the line, but also large due to the motion which has been transversally performed.[28]

---

[27] Clavius (1611–1612, I, p. 14), Df. I.4: "Sed quoniam lineas rectas regula ducere solemus, doceamus, qua ratione regulam propositam examinare possimus, num linea per illam descripta recta sit, nec ne. Sit ergo regula *ab*, secundum cuius latus *cd*, recta *cd*, describatur ex puncto *c*, in punctum *d*. Deinde convertatur regula, ut manente eadem parte superiore, punctum *c*, statuatur in *d*, & punctum *d*, in *c*, & secundum idem latus regulae *cd*, recta ducatur ex eodem puncto *c*, in punctum *d*. Nam si posterior haec linea priori omni ex parte congruet, dubitari non debet, quin regula *ab*, in lineis rectis ducendis fidere possimus: Si vero non congruet omni ex parte, latus illud *cd*, perfecte rectum non erit, sed corrigendum erit diligentius."

[28] Clavius (1611–1612, I, p. 15), Df. I.5: "Mathematici vero, ut nobis eam ob oculos ponant, monent, ut intelligamus lineam aliquam in transversum moveri: Vestigium enim relictum ex isto motu erit quidem longum, propter longitudinem lineae, latum quoque propter motum, qui in transversum est factus".

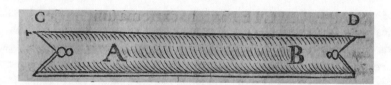

Fig. 8.1 Christoph Clavius, *Euclidis elementorum libri XV: Accessit XVI*, 1607, p. 5, Df. I.4. Representation of a straightedge to illustrate the mode of generation of a straight line. Courtesy Max Planck Institute for the History of Science, Berlin

As presented here, it would be the imaginary trace (*vestigium*) caused by the motion of the line that would allow us to "see" the geometrical surface.

A similar assertion is put forward in Clavius' introduction to the definitions of Book XI, where he compared the modes of generation of the line, of the surface and of the solid. In this context, he wrote that the surface, when it is imaginarily conceived as transversally elevated or moved from one place to the other and as leaving a trace of itself in this process, produces depth and allows us to set the solid figure *before our eyes*.

> *The solid is that which has length, breadth and depth.* [...] In order to allow us to correctly understand the line, mathematicians enjoin us to imagine a given point as moved from one place to another. The point describes a trace which is only long, that is, the line, because the point is entirely deprived of magnitude. But in order to allow us to perceive the surface, they advise us to imagine a line as transversally moved. Indeed, the line will describe a trace, which is only long and large. It is certainly long due to the length of the line, but large due to the motion which has been transversally performed. It however lacks depth, since the line is deprived of it. In this way also, in order to place before our eyes the body or solid, that is, the quantity endowed with three dimensions, they recommend us to conceive a surface uniformly and transversally raised or moved. According to this mode, a trace that is long, large and deep will be described. It is long and large because the surface is long and large, but it is deep or thick because of the elevation or the motion of the surface.[29]

---

[29]Clavius (1611–1612, I, p. 476), Df. XI.1: "*Solidum est, quod longitudinem, latitudinem, & crassitudinem habet.* [...] Porro quemadmodum Mathematici, ut recte intelligamus lineam, praecipiunt, ut imaginemur punctum aliquod è loco in locum moveri; haec enim describit vestigium quoddam longum tantum, hoc est, lineam, propterea quod punctum omnis est magnitudinis expers; ut autem percipiamus superficiem, monent, ut intelligamus lineam aliquam in transversum moveri; haec enim describet vestigium longum et latum dumtaxat; longum quidem propter longitudinem lineae, latum vero propter motum illum, qui in transversum est factus; carens autem profunditate, quod & linea illius sit expers: Ita quoque, ut nobis ob oculos ponant corpus, seu solidum, hoc est, quantitatem trina dimensione praeditam, consulunt, ut concipiamus superficiem aliquam aequaliter elevari, sive in transversum moveri; hac enim ratione describetur vestigium quoddam longum, latum, atque profundum; longum quidem & latum, ob superficiem, quae longa et lata existit; profundum vero seu crassum, propter elevationem illam, seu motum superficiei". *Cf.* (*ibid.*), Df. XI.2: "*Solidi autem extremum, est superficies.* Quemadmodum linea finita in extremitatibus puncta, superficies vero lineas recipit, ut in 1. lib. Docuit, ita nunc ait Solidi finiti extremum esse superficiem. Cum enim solidum, sive corpus efficiatur ex illo motu imaginario superficiei; perspicuum est extremas partes illius esse superficies."

The trace of the moving point, line or surface in the imagination, as it is said to set geometrical objects before our eyes, would establish a connection between genetic definitions of geometrical figures and the instrumental processes appealed to by geometers in the context of their teaching or research practice. Similarly to the instrumental procedures through which geometers materialise and manipulate their objects in order to study or teach their properties and relations, the generation of geometrical objects according to this quasi-material mode in the imagination would enable one to directly "experience" the proper mode of causation of magnitudes and gain a first-hand knowledge of their essential attributes and of their condition of possibility.[30] This mechanical interpretation of the generation of figures, as marked in particular by the illustration of the straightedge in the commentary on Df. I.4, would furthermore display the relation between theoretical and practical geometry.[31] Their connection was made by Clavius within his commentary on the *Elements* through the practical parts he added to most of Euclid's problems in Book I,[32] as well as through his 1604 *Geometria practica*,[33] which also drew a part of its content from his commentary on Euclid.[34]

8.4   The Mode of Apprehension of Geometrical Motion

Now, despite the quasi-physical or mechanical character Clavius seemed to confer to the generation of lines in his commentary on Euclid, it remains that, to him, when the geometer imagines the generation of a straight or broken line, he would not intend it as a process that occurs in space and time as the motion of physical realities, since it then takes place in the imagination only. Indeed, to Clavius, mathematicians consider objects that are related to physical realities as for their existence and properties, but which are nevertheless

---

[30] The quasi-empirical character of the imaginary causation of geometrical objects in Clavius' commentary on Euclid would corroborate P. Dear's representation of Jesuit science (in Dear 1995a, chap. 2), which he considered started with Clavius and which was based on a notion of evidence derived from universally accessible experience and gained through the guided reconstruction of specific objects of knowledge. In this regard, the discourse of geometers was presented as a model for physico-mathematical sciences such as optics or astronomy. This notion anticipates the discourse of Barrow, when he wrote that genetic definitions "not only explain the Nature of the Magnitude defined, but, at the same time, shew its possible Existence, and evidently discover the Method of its Construction: They not only describe what it is, but prove by Experiment, that it is capable of being such; and do put it beyond doubt how it becomes such." (Barrow 1734, XII, p. 223). See *supra*, n. 27, p. 5.

[31] On the relation between theoretical and practical geometry in Clavius' geometrical work, see Knobloch (2005) and Malet (2006).

[32] On these sections, see *infra*, pp. 214–216.

[33] Clavius (1604).

[34] Knobloch (1997).

considered independently from physical matter,[35] and thus from motion as determined by the qualities of physical bodies. Geometrical figures, and the processes through which they are thought to come about within the imagination, would not be determined spatially and temporally as are natural substances and physical motion. For this reason, in the consideration of the generation of mixed or straight lines, the time-factor would remain undetermined or uniform, not including any notion of velocity. And the spatiality in which this process would take place is not that of the physical world, but that of the imagination. This is notably confirmed by the fact that, in his commentary on the second postulate, Clavius wrote "we can imagine the point as being moved at an infinite distance",[36] which he would have deemed impossible to both experience and imagine in the case of physical motion. In his commentary on Sacrobosco's *Sphaera*, Clavius admitted in fact the finiteness of the physical world, though he left open the possibility for an infinite space to exist in the divine realm beyond the last sphere of the universe.[37]

Hence, in the context of definitions, the quasi-empirical apprehension of the generation of geometrical objects primarily corresponds to an internal form of experience, since Clavius mostly refers to the internal vision provided by the imagination or by a mental faculty of apprehension. Indeed, in his commentary on the definitions, he defined this motion as imaginary (*imaginarius motus*), in the same way as Fine had done in his *Geometria* and in his commentary on Euclid. Moreover, Clavius described the apprehension of this motion through verbs such as *percipio*, *imaginor*, *intelligo*, *concipio* and *cogito*, which all loosely refer, in this context, to a mental (rather than corporeal) faculty of apprehension.

The fact that Clavius does not appear to have made a distinction, in this framework, between perception, imagination, intellection, conception and understanding is notably

---

[35] Clavius (1611–1612, I, p. 5), *Prolegomena*: "Quoniam disciplinae Mathematicae de rebus agunt, quae absque ulla materia sensibili considerantur, quamvis reipsa materiae sint immersae; perspicuum est, eas medium inter Metaphysicam, & naturalem scientiam obtinere locum, si subiectum earum consideremus, ut recte à Proclo probatur. Metaphysices etenim subiectum ab omni est materia seiunctum & re, & ratione: Physices vero subiectum & re, & ratione materiae sensibili est coniunctum: Unde cum subiectum Mathematicarum disciplinarum extra omnem materiam consideretur, quamvis re ipsa in ea reperiatur, liquido constat hoc medium esse inter alia duo". See also Claessens (2009).

[36] Clavius (1611–1612, I, p. 22), Post. 2: "Quod si punctum illud ferri adhuc cogitaverimus motu directo, & qui omnis inclinationis sit expers, producta erit ipsa recta terminata, & nunquam erit finis huius productionis, cum punctum illud intelligere possimus moveri ad infinitam distantiam."

[37] Clavius (1570, p. 97): "Extra hunc vero mundum, seu extra caelum Empyreum, nullum prorsus corpus existit, sed est spacium quoddam infinitum, (Si ita loqui fas sit) in quo etiam toto Deus existit sua essentia, & in quo infinitos alios mundos perfectiores etiam hoc fabricare posset, si vellet, ut Theologi asserunt". See also Lattis (2010, pp. 84–85).

revealed by the identity of the structure and of the meaning of the different passages where these verbs are used, as in Df. I.2, Df. I.4, Df. I.5, Df. XI.1.[38]

Df. I.2	Mathematicians, in order to teach us the true understanding (*intelligentiam*) of the line, imagine (*imaginantur*) the point to be moved from one place to another.
Df. I.4	As mathematicians conceive (*concipiunt*) the line to be described by the imaginary flow of the point, they apprehend (*intelligunt*) the quality of the described line through the quality of the flow of the point. For, indeed, if the point is conceived (*concipiatur*) as flowing in a straight line through the shortest space...
Df. I.5	Mathematicians, in order to set the surface before our eyes (*ob oculos ponant*), advise us to imagine (*intelligamus*) a line that is transversally moved.
Df. XI.1	In order to allow us to correctly understand (*intelligamus*) the line, mathematicians enjoin us to imagine (*imaginemur*) a given point as moved from one place to another. ...in order to allow us to perceive (*percipiamus*) the surface, they advise us to imagine (*intelligamus*) a line as transversally moved. ...in order to place before our eyes (*ob oculos ponant*) the body or solid... they recommend us to conceive (*concipiamus*) a surface uniformly and transversally raised or moved.

Thus, when Clavius juxtaposed *imaginari* and *intelligere*, in the commentary on the definition of the straight line (Df. I.7), he aimed to suggest the semantic equivalence of these verbs.

> The [plane] surface will be the only one that we can *imagine* and *conceive* as described by the transversal motion of a straight line, which is made on two other straight lines.[39]

The structural similarity among the passages (quoted above) in which Clavius presented the generations of the line, the surface and the solid in Book I and XI shows furthermore that, just as Fine and Billingsley, he did not distinguish the mode of generation of the line

---

[38] Clavius (1611–1612, I, p. 13), Df. I.2: "Mathematici quoque, ut nobis inculcent veram lineae intelligentiam, *imaginantur* punctum iam descriptum superiore definitione, e loco in locum moveri"; (*ibid.*, p. 14), Df. I.4: "Quemadmodum autem Mathematici per fluxum puncti imaginarium *concipiunt* describi lineam, ita per qualitatem fluxus puncti qualitatem lineae descriptae *intelligunt*. Si namque punctum recta fluere *concipiatur* per brevissimum spatium..."; (*ibid.*, p. 15), Df. I.5: "Mathematici vero, ut nobis eam ob oculos ponant, monent, ut *intelligamus* lineam aliquamus in transversum moveri"; (*ibid.*, p. 476), Df. XI.1: "Mathematici, ut recte *intelligamus* lineam, praecipiunt, ut *imaginemur* punctum aliquod è loco in locum moveri... ut autem *percipiamus* superficiem, monent, ut *intelligamus* lineam aliquam in transversum moveri;...ut nobis ob oculos ponant corpus... consulunt, ut concipiamus superficiem aliquam aequaliter elevari, sive in transversum moveri".

[39] Clavius (1611–1612, I, p. 15), Df. I.7: "Haec autem superficies sola erit ea, quam *imaginari*, & *intelligere* possumus describi ex motu lineae rectae in transversum, qui super duas alias lineas rectas conficitur." (My emphasis.)

through the flow of the point from those proper to other types of magnitudes from an ontological, epistemological and gnoseological point of view, contrary to Peletier.[40] As the flow of the point was straightforwardly interpreted as a local motion of a point, it did not refer, for Clavius, to any metaphysical process, even indirectly. It rather referred, as shown above, to the instrumental procedures performed by the geometer to draw out concrete lines.

Regarding this issue, it is relevant to note that G. Claessens, in his account of Clavius' ontology of mathematics,[41] defends the view that the Jesuit professor considered geometrical objects as directly apprehended within sensible objects, although they are immaterial by definition. According to this thesis, this would occur through a form of "snapshot idealization" that consists in the consideration of the physical body *qua*-quantity only, that is, by mentally discarding the matter and material qualities of the physical body, in the manner of Aristotle's account of abstraction in *Metaphysics* M.[42] The mind would thereby find in these filtered quantitative properties of physical bodies a perfect instantiation of geometrical objects, whereby no proper ontological gap would be admitted by Clavius between mathematical and physical quantity when considered *in abstracto*.

This interpretation would be confirmed by Clavius' approach to motion and to genetic definitions in geometry, in the sense that it allowed him to establish a connection between the structure of geometrical and mechanical processes and that it presented the apprehension of the generation of figures as occurring according to the mode of sense-perception (*ob oculos*). This would enable us to regard the concrete experience of mechanical processes as the first instantiation of the generation of geometrical figures for Clavius. Claessens' interpretation, which is based on Clavius' assertion that mathematical objects "are immersed and found in matter, although they are considered without matter"[43] and on the fact that Clavius made no mention of any concept of imagination (*phantasia*) or of

---

[40] See *supra*, p. 95.

[41] Claessens (2009): "Clavius, Proclus, and the limits of interpretation: snapshot-idealization versus projectionism".

[42] Aristotle, *Metaphysics* M3, 1077b24–30 (Barnes 1995, II, p. 1704): "There are many formulae about things merely considered as in motion, apart from the essence of such thing and from their accidents, and as it is not therefore necessary that there should be either something in motion separate from sensibles, or a separate substance in the sensibles, so too in the case of moving things there will be formulae and sciences which treat them not qua moving but only qua bodies, or again only qua planes, or only qua lines, or qua divisibles, or qua indivisibles having position, or only qua indivisibles." This passage is quoted by Claessens in the article referred to here (Claessens 2009, p. 328) in order to explain Clavius' understanding of the mode of apprehension of mathematical objects.

[43] Clavius (1611–1612, I, p. 5), *Prolegomena*: "disciplinae Mathematicae de rebus agunt, quae absque ulla materia sensibili considerantur, quamvis re ipsa materiae sint immersae". See *supra*, n. 35, p. 203. See also (*ibid.*): "subiectum Mathematicarum disciplinarum extra omnem materiam consideretur, quamvis re ipsa in ea reperiatur".

imagined or intelligible matter in his *Prolegomena*,[44] certainly has the merit of underlining Clavius' admission of the close connection and the homology between mechanical and geometrical processes. Yet, the statement that Clavius' philosophy of mathematics lacks a concept of imagination, which seems mainly based on a reading of the *Prolegomena*, should be qualified, since it dismisses Clavius' repeated assertion, in the commentary, that the motion through which the mathematician apprehends the generation of magnitudes is imaginary (*imaginarius motus*).

Admittedly, in these passages, Clavius did not directly speak of imagination or of imaginary or intelligible matter as the medium within which the generations of figures are deployed (through projection or abstraction).[45] But he manifested a will to indicate the ontological distance between the generation of concrete magnitudes and those of geometrical magnitudes, and therefore between concrete and geometrical objects themselves, by asserting that the latter are imagined or caused by an imaginary motion. Moreover, the imagination of the moving point producing a line by leaving a trace, which allows us to set geometrical objects *before our eyes* and apprehend it in a quasi-empirical manner, would in a way work as the projection of the abstract concept of breadthless length onto the screen of the imagination.

In any case, for Clavius, this imaginary motion was connected, by its structure, to the concrete motion by which the line may be instrumentally drawn, the former corresponding to an idealised version of the latter.[46] On a mathematical level, the difference between the two would mainly be the greater degree of abstractness of the former.[47] This confirms that Clavius did not see in the flow or motion of the point the spatial image of a metaphysical process, and that he related it rather to a concrete form of generation, leaving aside any distinction between a motion that would be self-induced, or that would be regarded as proper to geometrical objects themselves, and one that would be externally-induced and originating in the operations performed (imaginarily or concretely) by the geometer.

8.5   Clavius' Constructivist Interpretation of Euclidean Geometry

What Clavius' approach to genetic definitions, in his commentary on Euclid's definitions, reveals is first and foremost his constructivist conception of geometry[48] insofar as, for him, the geometer would gain knowledge of his objects by constructing them. This is confirmed

---

[44] Claessens (2009, p. 329).

[45] Such a concept could however be suggested, to a certain extent, by the notion of a point leaving a trace as it is imagined to flow from one place to another.

[46] Bos (2001, pp. 165–166).

[47] On this issue, see Bos (2001, p. 166) and *infra*, p. 24.

[48] On this aspect of Clavius' approach to geometry, see also Dear (1995b).

by his treatment of Euclid's postulates,[49] which coincides with the first and main conception presented by Proclus on this topic. Indeed, to Proclus, the principal function of postulates is to found geometrical constructions.[50] For this reason, only Euclid's first three postulates should be considered as postulates in the proper sense.[51] Following this conception, Clavius only kept among the postulates the first three given by Euclid.[52] He also added a fourth postulate ("Given any magnitude, it is possible to take a magnitude greater or lesser than it"[53]), which equally pertained to the constructibility of geometrical objects.[54] As P. Dear expressed it, Clavius saw geometrical postulates as assertions of existence, or at least of the possibility of existence.[55] In this sense, postulates would relate to Aristotle's hypotheses, such as defined in the *Posterior analytics*.[56]

It is therefore not surprising that the notion of line as generated by the flow of the point also appears in his commentary on the first postulates, as it did in the commentaries of Proclus, Fine and Commandino.

> *It is requested that to draw a straight line from any point to any point be granted*. This first postulate is fully clear, if what we have written earlier about the line is correctly considered. For, since the line is the imaginary flow of a point and since the straight line is indeed the flow progressing along an absolutely straight path, it turns out that, as we will have conceived a

---

[49] On Clavius' approach to Euclid's postulates, see again Dear (1995b, pp. 49–51).

[50] Proclus (Friedlein 1873, p. 179) and (Morrow 1992, p. 140): "in a postulate we ask leave to assume something that can easily be brought about or devised, nor requiring any labor of thought for its acceptance nor any complex construction." and Friedlein (1873, p. 181) and (Morrow 1992, p. 142): "a postulate prescribes that we construct or provide some simple or easily grasped object for the exhibition of a character".

[51] Proclus (Friedlein 1873, p. 182) and (Morrow 1992, p. 143): "These, then are the three ways in which postulate and axiom are distinguished. According to the first—that which bases the distinction on the fact that the postulate produces and the axiom knows—clearly it is not a postulate that all right angles are equal. Nor is the fifth, that when two straight lines are intersected by a third making the two interior angles on one side of it less than two right angles, then the straight lines when extended will meet on that side on which the two angles are less than two right angles. For these statements are not assumed for the sake of any construction, nor do they demand that we produce anything".

[52] Clavius placed Euclid's fourth and fifth postulates (Heath 1956, I, pp. 154–155: "That all right angles are equal to one another." and "That, if a straight line falling on two straight lines make the interior angles on the same side less than two right angles, the two straight lines, if produced indefinitely, meet on that side on which are the angles less than the two right angles.") among the axioms (Clavius, 1611–1612, I, p. 25, CN 12 and 13). On Clavius' axiomatic, see De Risi (2016b).

[53] Clavius (1611–1612, I, p. 23): "quacunque magnitudine data, sumi posse aliam magnitudinem vel mariorem vel minorem." The English translation is taken from De Risi (2016b, p. 619).

[54] Clavius (1611–1612, I, p. 23): "Omnis enim quantitas continua per additionem augeri, per divisionem vero diminui potest infinite".

[55] Dear (1995b, pp. 50–51).

[56] On Aristotle's hypotheses, see *supra*, n. 53, p. 11.

given point as moved toward another rectilinearly, the straight line will have been correctly drawn from one point to another (Post. 1).[57]

And to produce a finite straight line continuously in a straight line. For if we think that a point is moved according to a direct motion, which is entirely deprived of deviation, a finite straight line will be produced and there will never be an end to this production, since we can understand this point to be moved at an infinite distance (Post. 2).[58]

In this context, the notion of line as generated by the flow of a point, and in particular the notion of straight line as generated by the undeviating flow of the point (as defined in Clavius' commentary on Df. I.4), is presented as directly relevant to the teaching of the first two postulates. As in Commandino's commentary, this passage appears to take after Proclus' commentary on the first three postulates.[59] Like Proclus, Clavius wrote here that the definition of the straight line as the undeviating flow of the point guides the geometer's production of the straight line in the framework of his constructions, explaining and legitimating thereby the first postulate. The genetic definition of the straight line would therefore play a foundational role with regard to the postulates, which, for Clavius, as expressed again by P. Dear, "assert the possibility of *doing* something — that is, performing constructions".[60] Thereby, genetic definitions would ultimately play a foundational role with respect to the constructions or constructive procedures required by Euclidean propositions. Indeed, in this context, as in the commentary on the definitions, the very fact of generating the line in the imagination is presented as the cause of its production ("describetur vestigium"; "producta erit [recta linea]").[61]

The main difference between the texts of Clavius and Proclus would then lie in the interpretation of the ontological status of the flow of the point, and this more explicitly than in Commandino's commentary. While Proclus' text suggests that the flow of the point corresponds to an ontologically superior process, which the imagination aims to imitate as

---

[57] Clavius (1611–1612, I, p. 22), Post. 1: "*Postuletur, ut à quovis in quodvis punctum, rectam lineam ducere concedatur.* Primum hoc postulatum planum admodum est, si recte considerentur ea, quae paulo ante de linea scripsimus. Nam cum linea sit fluxus quidam puncti imaginarius, atque adeo linea recta fluxus directo omnino itinere progrediens, fit ut si punctum quodpiam ad aliud directo moveri intellexerimus, ducta sane sit à puncto ad punctum recta linea."

[58] Clavius (1611–1612, I, p. 22), Post. 2: "*Et rectam lineam terminatam in continuum recta producere.* Quod si punctum illud ferri adhuc cogitaverimus motu directo, & qui omnis inclinationis sit expers, producta erit ipsa recta terminata, & nunquam erit finis huius productionis, cum punctum illud intelligere possimus moveri ad infinitam distantiam."

[59] Proclus (Friedlein 1873, p. 185) and (Morrow 1992, p. 145): "the drawing of a line from any point to any point follows from the conception of the line as the flowing of a point and of the straight line as its uniform and undeviating flowing." See *supra*, p. 48. See also Friedlein (1873, pp. 179–180) and (Morrow 1992, p. 141): "For example, drawing a straight line from a point to a point is something our thought grasps as obvious and easy, for by following the uniform flowing of the point and by proceeding without deviation more to one side than to another, it reaches the other point."

[60] Dear (1995a, 219). See also Dear (1995b, pp. 50–51).

[61] Clavius (1611–1612, I, p. 476), Df. XI.1 and (*ibid.*, p. 22), Post. 2.

it causes a point to move transversally from one place to the other, Clavius only conceived the flow of the point as a local motion of the point. Moreover, Clavius did not adhere to the ontology of mathematical objects defended by Plato or Proclus,[62] although many parts of his *Prolegomena* were inspired by Proclus' commentary on the first book of Euclid's *Elements*.[63] Accordingly, he would not have believed that geometrical objects are projections of suprasensible principles on the screen of the imagination, as they were for Proclus, but rather, as Commandino also conceived it, that the imagination plays the role of an intelligible screen for the concepts abstracted from the physical realm.

Thus, rather than interpreting Proclus' distinction between the flow and the motion of the point as a distinction between an intelligible or non-spatial process (ῥύσις), on one hand, and an imaginary and spatial process (κίνησις), on the other, Clavius seems to have interpreted it as Commandino did, in his commentary on the postulates. Commandino, as was shown above, would have considered the *flow* of the point as the intellectually conceived genetic *definition* of the straight line and the *motion* of the point as the *spatial representation* of this definition, which would then found the production of straight lines in the context of geometrical constructions.[64]

For Clavius, in fact, both processes would be imaginary and both would be equally represented by the motion of a point from one place to the other. But one—that which corresponds to a mode of definition—would be more universal and therefore prior to the other in the order of knowledge and in its logical function in the argumentative structure of the geometer's discourse. It would show as such the essential relation between the defined object, its properties and its mode of production. And it would guide the spatial representation and production of lines in the context of constructions.

In this context, the genetic definition of the straight line comes forth as what connects the actual operation of producing a straight line in the framework of a geometrical construction and the logical notion of straight line, playing a foundational role in the constitution of geometrical science. This function would also be imparted, for Clavius, to the genetic definition of the circle in relation to the constructive procedure authorised by the third postulate, which establishes that one can draw a circle of any centre and of any radius.[65] Indeed, in his commentary on the third postulate, which again evokes Proclus'

---

[62] See *supra*, n. 36, p. 48.

[63] It is notable however that Clavius drew many of the ideas presented in his *Prolegomena* from Proclus' commentary on the first book of Euclid's *Elements*. On this aspect, see Rommevaux (2004) and Axworthy (2004).

[64] See *supra*, p. 189.

[65] Clavius (1611–1612, I, p. 22), Post. 3: "*Item quovis centro, & intervallo circulum describere.* Iam vero, si terminatam rectam lineam cuiuscunque quantitatis mente conceperimus applicatam esse secundum alterum extremum ad quodvis punctum, ipsamque circa hoc punctum fixum circumduci, donec ad eum revertatur locum, à quo dimoveri coepit; descriptus erit circulus, effectumque, quod tertia petitio iubet. Exemplum habes in his quinque lineis *ab*, *ac*, *ad*, *ae*, *af*, quae singulae citra

discourse on this topic,[66] Clavius wrote that, from the mental representation of the motion of a line-segment of any length ("si terminatam rectam lineam cuiuscunque quantitatis mente conceperimus. . ."), a circle (of any size) will be described and made ("descriptus erit circulus, effectumque"). This discourse distinctly resonates with the genetic definition of the circle provided in his commentary on Df. I.15, where the circle is defined as a plane figure that is described by a finite straight line rotating around one of its fixed extremities until it returns to the place where it started to move.[67]

8.6   The Logical and Epistemological Status of Genetic Definitions

Clavius did not explicitly say whether he considered genetic definitions as proper definitions of geometrical objects, but it is clear that he regarded them as relevant and even as fundamental to geometry, and thus as essential to the teaching of geometrical principles. The fact that he very often complemented Euclid's definition by property with the corresponding genetic definition and, conversely, that he added the definition by property to Euclid's kinematic definitions of the sphere—he then presented Theodosius' definition after that of Euclid, as had done Billingsley[68]—indicates in any case that he regarded them as mutually complementary.

Looking at Clavius' commentary on Euclid, it is not clear whether he distinguished genetic definitions from definitions by property in the way that Foix-Candale distinguished them in his commentary on Df. XI.14.[69] As we have seen, in Df. I.2, Clavius presented the genetic definition of the line as one definition of the line among others, following the model set by Proclus' commentary on Euclid.[70] Yet, when dealing with Euclid's definition of the sphere, Clavius indirectly presented this definition as a *descriptio*[71] in the sense that he

---

centrum *a*, circumvolutae singulos circulos descripserunt iuxta quantitatem, seu intervallum ipsarum."

[66] Proclus (Friedlein 1873, p. 185) and (Morrow 1992, p. 145): "And if we think of a finite line as having one extremity stationary and the other extremity moving about this stationary point, we shall have produced the third postulate".

[67] Clavius (1611–1612, I, p. 18), Df. I.15: "Potest circulus etiam hac ratione describi. Circulus est figura plana, quae describitur à linea recta finita circa alterum punctum extremum quiescens circumducta, cum in eundem rursus locum restituta fuerit, unde moveri coeperat."

[68] Clavius (1611–1612, I, p. 480), Df. XI.14: "*Sphaera est, quando semicirculi manente diametro, circumductus semicirculus in se ipsum rursus revolvitur, unde moveri coeperat, circum assumpta figura.* [. . .] Quapropter ad similitudinem definitionis circuli, sphaera definiri poterit hoc etiam modo. *Sphaera est figura solida, una superficie comprehensa, ad quam ab uno puncto eorum, quae intra figuram sunt posita, cadentes rectae lineae inter se sunt aequales*".

[69] See *supra*, p. 109.

[70] See *supra*, n. 107, p. 25.

[71] Clavius (1611–1612, I, p. 480), Df. XI.14: "*Sphaera est, quando semicirculi manente diametro, circumductus semicirculus in se ipsum rursus revolvitur, unde moveri coeperat, circum assumpta*

wrote that the semicircle describes (*describit*) the sphere in its circumduction and that he compared Theodosius' definition of the sphere to Euclid's definition of the circle, which both were considered by Foix-Candale as proper definitions, as opposed to their kinematic pendants.[72]

In this context, however, the notion of *descriptio* (or the related verb *describo*) was then very likely interpreted by Clavius as expressing the production of the figure in general and not a specific type of characterisation of geometrical objects. The *descriptio* would represent any form of generation of geometrical objects, whether it is conceived as taking place in the imagination or on paper, as marked by his practical additions to Euclid's problems, in Book I.[73] In this regard, it is notable that the notion of *descriptio* was also applied to the generation of the quadratrix, which was dealt with also by Clavius in his *Geometria practica*.[74] This would also be conform to the way he presented other genetic definitions, such as the definition of the straight line in his commentary on Df. I.4, where he wrote that the line is "described by the imaginary flow of the point" ("per fluxum puncti imaginarium *describi* lineam") or that the straight line is the line described ("linea *descripta*") by an equal and unceasing motion of the point.[75] He also used this term in his commentary on Df. XI.1, where the imaginary motion of the point, the line or the surface are said to describe (*describit*) a trace in one, two or three dimensions.[76]

---

figura. Sicut linea recta circa alterum eius extremum quiescens revoluta *describit* circulum: ita & semicirculus circa alterum eius extremum, nempe circa diametrum, circumductus figuram *describit*, quam Geometrae sphaeram appellant." (My emphasis.)

[72] Clavius (1611–1612, I, p. 480), Df. XI.14: "Unde quemadmodum in circulo punctum assignatur, extremum videlicet illud quiescens, à quo omnes lineae rectae in peripheriam cadentes sunt aequales; propterea quod omnes aequales existunt illi lineae circumvolutae: Ita quoque in sphaera punctum reperitur, nempe medium diametri quiescentis, hoc est, centrum semicirculi circumducti, à quo omnes rectae cadentes in peripheriam sunt aequales; eo quod omnes sunt semidiametro dicti semicirculi aequales. Quapropter *ad similitudinem definitionis circuli*, sphaera definiri poterit hoc etiam modo." (My emphasis.)

[73] See, for example, Clavius (1611–1612, I, p. 36), Prop. I.10: "Praxis. [...] *descriptis* supra eam duobus arcubus sese intersecantibus in *c*, *describemus* ad easdem partes alios duos arcus sese intersecantes mutuo in *d*' or (p. 37) Prop. I.12: "Ex quovis puncto *a*, in linea data, & intervallo quolibet usque ad *c*, assumpto, arcus circuli *describatur*." (My emphasis.)

[74] This appears both in the general list of additional topics Clavius dealt with in his commentary on Euclid (Clavius (1611–1612, I, p. 11): "*Descriptio* inflexae cuiusdam lineae facillima, per quam & in circulo figura quotlibet laterum aequalium describitur: circulus quadratur..." and in the actual section presenting the construction of the quadratrix, where it is compared to the construction of conic sections (Clavius (1611–1612, I, p. 297): "Haec igitur est *descriptio* lineae Quadratricis, quae Geometrica quodammodo appellari potest, quemadmodum & conicarum sectionum *descriptiones*...". (My emphasis.) On Clavius' treatment of the quadratrix, see *infra*, p. 217–219.

[75] Clavius (1611–1612, I, p. 14), Df. I.4: "Si namque punctum rectà fluere concipiatur per brevissimum spatium, ita ut neque in hanc partem, neque in illam deflectat, sed aequabilem quendam motum, atque incessum teneat, dicetur linea illa *descripta*, recta." (My emphasis.)

[76] Clavius (1611–1612, I, p. 476), Df. XI.1: "Mathematici [...] praecipiunt, ut imaginemur punctum aliquod è loco in locum moveri; haec enim *describit* vestigium quoddam longum tantum [...] ut intelligamus lineam aliquam in transversum moveri; haec enim *describet* vestigium longum & latum

Looking at Clavius' commentary on Sacrobosco's *Sphaera*, which was first published in 1570, that is, four years before his commentary on Euclid, we find a more explicit characterisation of the respective status and functions of the two definitions of the sphere by Euclid and by Theodosius. In this context, Clavius stated that the Euclidean definition of the sphere is not a "formal predication" (that is, a mode of definition that would state the properties of the defined object), but a "causal predication" ("praedicatio causalis, minimè verò formalis").[77] The notion of *praedicatio causalis* distinctly evokes the notion of causal definition used by Billingsley in his commentary on the *Elements* to characterise Euclid's definition of the sphere.[78] If Clavius intended this notion of causal predication in the manner Billingsley understood his notion of causal definition, the function of Euclid's definition of the sphere would be, for him, to exhibit the efficient cause of the sphere, that is, the rotation of the semicircle around its fixed axis. In Euclid's geometry, these genetic definitions would have the role of displaying the chain of causality starting with the generative flow or motion of the point and ending with the generation of the most complex solid figures.

What the argumentative structure of Clavius' commentary on Df. XI.14 in the *Elements* tells us in this regard is that genetic definitions (or *praedicationes causales*) seem endowed, for the Jesuit professor, with a stronger epistemic value than definitions by property (or *praedicationes formales*). In this context, the Theodosian definition of the sphere, which Clavius added at the end of his commentary on Euclid's definition, is actually explained thanks to the genetic definition provided by Euclid.[79] The latter would indeed exhibit the necessity of the relation between the *definiendum* and its attributes by displaying

---

dumtaxat [. . .] ut concipiamus superficiem aliquam aequaliter elevari, sive in transversum moveri; hac enim ratione *describetur* vestigium quoddam longum, latum, atque profundum." (My emphasis.)

[77] Clavius (1570, p. 21): "Quibus rite intellectis, facile duae definitiones sphaerae percipientur. Ita namque habet prima definitio, quam auctor se desumpsisse testatur ab Euclide. [*Sphaera est transitus circunferentiae dimidij circuli, quae fixa diametro, eousque circunducitur, quousque ad locum suum redeat;*] Id est, ut auctor ipse declarat. [*sphaera est tale rotundum, seu solidum, quod describitur ab arcu semicirculi circunducto.*] Neque enim sphaera est transitus, seu revolutio ipsa, sed efficitur ex eiusmodi transitu, seu revolutione; Ita ut haec *praedicatio*, Sphaera est transitus, sit *causalis*, minime vero *formalis.*" (My emphasis.)

[78] It is unlikely that Clavius would then have taken up Billingsley, as these two commentaries were published the same year (1570). As I have found no reference to Billingsley in Clavius' own commentary on the *Elements*, even in later editions, it does not seem that he ever read Billingsley's own commentary on Euclid. See *supra*, p. 196.

[79] Clavius (1611–1612, I, p. 480), Df. XI.14: "Unde quemadmodum in circulo punctum assignatur, extremum videlicet illud quiescens, à quo omnes lineae rectae in peripheriam cadentes sunt aequales; propterea quod omnes aequales existant illi lineae circumvolutae: Ita quoque in sphaera punctum reperitur, nempe medium diametri quiescentis, hoc est, centrum semicirculi circumducti, à quo omnes rectae cadentes in peripheriam sunt aequales; eo quod omnes sunt semidiametro dicti semicirculi aequales. Quapropter ad similitudinem definitionis circuli, sphaera definiri poterit hoc etiam modo. *Sphaera est figura solida, una superficie comprehensa, ad quam ab uno puncto eorum, quae intra figuram sunt posita, cadentes rectae lineae inter se sunt aequales.*"

its mode of causation, as real or essential definitions (as opposed to nominal definitions) were considered to do according to authors contemporary to Clavius.[80] Hence, in the commentary on Euclid's *Elements*, the Jesuit professor not only suggested the complementarity of the two modes of definitions of the sphere, but also showed, as had done Peletier for the definition of the circle,[81] that the attributes of the sphere that are enunciated in the definition by property may be deduced from the knowledge of its mode of generation. The causal definition is, in other words, presented as the condition of the knowledge of the formal definition.

Therefore, in addition to setting forth the relation between the definitions and geometrical constructions (both abstract and concrete), genetic definitions would legitimate the definition by property in the eyes of the student. It would also offer the student an immediate access to the concept at hand, as well as to its mode of construction.[82] In a sense, Clavius considered that genetic definitions, through the experience they provide of the essential connection between geometrical objects and their properties, would allow for the type of induction that is necessary to gain knowledge of scientific principles according to Aristotle's *Posterior analytics*.[83] Yet, the abstract and universal character of geometrical objects would make this type of induction possible without requiring that the experience be repeated.[84]

8.7   Geometrical and Mechanical Processes in the Commentary on Sacrobosco's *Sphaera* and in the Commentary on Euclid

In the commentary on Sacrobosco's *Sphaera*, Clavius explained Euclid's genetic definition of the sphere through a concrete illustration. He invited the reader to imagine the production of a material sphere through the use of a lathe or a wooden or metallic device that would carve out a sphere in argile, or in any malleable and tractable matter, by rotating a semicircular arc around a pole embedded inside the material substrate.[85] As was shown

---

[80] De Angelis (1964, p. 82–98). See also V. De Risi's analysis of Leibniz's theory of definition (De Risi, 2016a, pp. 31–40).

[81] See *supra*, p. 80.

[82] This is comparable to the way Barrow would later describe the function of genetic definitions. See *supra*, n. 27, p. 5.

[83] Aristotle, *Posterior analytics* II.19, 100b2–4 (Tredennick 1989, pp. 260–261): "Clearly then it must be by induction that we acquire knowledge of the primary premises, because this is also the way in which general concepts are conveyed to us by sense-perception."

[84] This is quite similar to the role Barrow later attributed to motion in geometry, according to Dear (1995a, pp. 223–226).

[85] Clavius (1570, p. 21): "Est enim sensus, quod sphaera est tale solidum, quod ab arcu semicirculi, sua quidem diametro immobili, & fixa manente, una completa revolutione circumscribi intelligitur: Id autem Solidum circunscribi intelligitur, quod continue ab arcu circunducto tangitur. Ut si sumatur

above, Billingsley also included an illustration of a lathe in his commentary on Euclid's definition of the sphere (Fig. 5.7), where he explicitly referred to the tradition of Sacrobosco's *Sphaera*.[86]

This would again confirm that both Billingsley and Clavius related the generations of geometrical figures to physical or mechanical, rather than metaphysical, processes. This does not mean that Clavius actually used a mechanical device of the sort to teach the geometrical notion of sphere to his students (though he could have). It rather means that he established a relation between the abstract modes of generation of figures and mechanical processes, at least insofar as the latter are able to jog in the student's mind the idealised or imaginary representation of their geometrical counterpart. This shows therefore that the Jesuit professor regarded these concrete and instrumental images as a legitimate means for students to grasp the mode of causation of geometrical figures, as well as to perfectly and easily understand the relation between geometrical magnitudes and their essential properties. This interpretation would be corroborated by the representation of geometrical instruments which Clavius included in his commentary on the *Elements*.[87]

Generally speaking, Clavius' reference to instrumental procedures in his commentary on the *Elements* was part of an effort to teach, in addition to the proper intention and meaning of Euclid's text, the more practical aspects of geometry, for which he offered, as I already mentioned, an instrumental interpretation of the constructions presented in most of the problems of the first book in sections entitled *Praxis*.[88] Clavius also hinted at this approach in his commentary on Df. I.12, where he taught, in the text and through the appended illustration, how the set square could be used as a means of determining in

---

argilla, aut quaevis alia materia tractabilis, cui diameter aliqua pro materiae spissitudine inseratur, & ad huius diametri extremitates Semicirculi circumferentia utrinque applicata circunducatur, donec ad eum locum, ex quo dimoveri coepit, revertatur, tolletur omnis inaequalitas argillae, efficieturque figura sphaerica, sive rotunda. Tale igitur corpus rotundum a circunferentia semicirculi descriptum, Sphaera appellatur." Tartaglia presented a similar image in his commentary on Euclid's definition of the sphere in his Italian translation and commentary on the *Elements*, referring to the artisan's method for fabricating a sphere out of stone or of another matter. Tartaglia (1565, fol. 237r), Df. XI.10: "questa diffinitione ha insegnato alli artifici il modo di formare le palle di pietra, o d'altra materia, & che'l sia il vero el si fa che se uno artifice vol fare una palla di pietra che sia perfettamente al senso tonda lui forma prima un mezzo cerchio cioe giustando spesso quella forma secondo che va scarpellando & cosi pian piano la redusse a perfettione."

[86] See *supra*, pp. 157–159.

[87] See *supra*, pp. 200–201.

[88] See, for example, Clavius (1611–1612, I, p. 35), Prop. I.9: "*Praxis*. Dicto citius angulus quilibet rectilineus, ut *bac*, bifariam secabitur, hoc modo. Ex centro *a*, circino aliquo abscindantur rectae aequales *ad*, *ae*, cuiuscunque magnitudinis. Et circino non variato (posses tamen ipsum variare, si velles) ex centris *d*, & *e*, describantur duo arcus secantes sese in *f*. Recta igitur ducta *af*, secabit angulum *bac*, bifariam. Si enim ducerentur rectae *df*, *ef*, essent hae aequales, nempe semidiametri circulorum aequalium. Unde ut prius demonstrabitur, angulum *daf*, aequalem esse angulo *eaf*". See also Prop. I.10, p. 36; Prop. I.11, p. 36; Prop. I.12, p. 37; Prop. I. 22, p. 44; Prop. I. 23, p. 45; Prop. I. 31, p. 55.

practice whether an angle is right, obtuse or acute.[89] As Clavius himself expressed it in the commentary on the Prop. I.1, the practical and instrumental approach he applied in his commentary on the problems would aim to explain in a more straightforward manner the constructions which Euclid taught through a longer discourse.[90]

Therefore, in Clavius' mathematical teaching and interpretation of Euclid, the abstract and the concrete generations of figures are underlyingly conceived as connected, the latter corresponding to the material image of the former and the former corresponding to the idealised image of the latter. This would legitimate the material constructions carried out by students and their masters by the means of instruments in the course of studying and teaching Euclidean propositions, at least hypothetically. It would also allow us to directly interpret the operations authorised by Euclid's postulates as instrumental procedures, founding a geometry of the straightedge and compass, even if no reference to instruments was made in the *Elements*.

This would, furthermore, tend to identify the efficient cause of the geometrical figure (displayed in the *praedicatio causalis*) with the geometer himself (instead of the moving point or rotating semicircle), just as the artisan was presented as the efficient cause of the material sphere in the commentary on Sacrobosco's *Sphaera*. Indeed, in a pedagogical context, for instance, the disciple who is requested by his master to imagine a point producing a line through its flow or motion, as he is learning the first definitions of the *Elements*, corresponds to the primary agent of this production, the motion of the point being the tool through which he generates a line in his imagination.

Hence, in addition to the fact that Clavius did not refer to any metaphysical interpretation of the notion of flow of the point, he also did not take care to clearly distinguish the abstract generations of geometrical objects from their concrete counterparts, which are both presented as *descriptions*.[91] At most did he indicate, when dealing with definitions, that the description of geometrical figures takes place in the imagination of the mathematician.

---

[89] Clavius (1611–1612, I, p. 17), Df. I.12: "Facilius idem cognoscemus beneficio normae alicuius accurate fabricatae, qualem refert instrumentum *abc*, constans duabus regulis *ae*, *af*, ad angulum rectum in *a*, coniunctis. Nam si latus *ab*, huius normae, rectae, *ab*, applicetur, cadente puncto *a*, in punctum *a*; si quidem & normae latus *ac*, rectae *ac*, congruat, erit angulus *a*, rectus: si vero citra rectam *ac*, cadat normae latus *ac*, erit angulus *a*, obtusus: si denique latus normae *ac*, ultra rectam *ac*, cadat, acutus erit angulus, ut perspicuum est."

[90] Clavius (1611–1612, I, p. 29), Prop. I.1: "*Praxis*. Conabimur in singulis fere problematibus Euclidis tradere praxin quandam facilem, & brevem, qua effici possit id, quod Euclides pluribus verbis, atque lineis contendit construere; Idque in ijs praesertim observabimus, qua frequentiorem usum habent apud Mathematicos, & in quibus praxis compendium aliquod secum videtur affere."

[91] Compare Clavius (1611–1612, I, p. 35), Prop. I.9 (also quoted above, n. 88, p. 215): "Et circino non variato (posses tamen ipsum variare, si velles) ex centris *d*, & *e*, *describantur* duo arcus secantes sese in *f*." and Clavius (1611–1612, I, p. 18), Df. I.15: "Potest circulus etiam hac ratione *describi*. Circulus est figura plana, quae *describitur* à linea recta finita circa alterum punctum extremum quiescens circumducta, cum in eundem rursus locum restituta fuerit, unde moveri coeperat. Quae quidem *descriptio* persimilis est ei, qua ab Euclide sphaera *describitur* libri 11." (My emphasis.)

However, this insistence on the imaginary or abstract status of the considered motion would, it seems, not so much come from a will to distinguish the ontological status of geometrical and physical motions, as it was for Fine, than to set forth the mode through which mathematicians adequately apprehend geometrical concepts and properties.[92] This would be the case, in particular, for the counter-intuitive notions of breadthless length (Df. I.2) and depthless breadth (Df. I.5). These may only be apprehended as such by the means of the imagination, which would also allow us, in turn, to apprehend mechanical or concrete processes in an idealised manner.[93]

To this may also be added that, in the *Praxis* section of Prop. I.9, Clavius distinguished the description of the lines he presented in this context from the lines that are described in the context of "pure practice" (*nuda praxis*), through which only the lines necessary to the effective construction (as opposed to those required for the demonstration) are made apparent. As he wrote then, the multitude of lines (the straight lines and circles to be drawn in order to construct the sought object according to Euclid's text) may plunge the reader into darkness and bring about confusion ("linearum multitudo tenebras nobis offundat, pariatque confusionem").[94] Therefore, only the most necessary parts of the

---

[92]With regard to this, I would tend to disagree with the interpretation proposed by R. Feldhay (in Feldhay 1998) when she wrote that Clavius, although he promoted the link between mathematics and the physical world in his *Prolegomena* to his commentary on Euclid (Feldhay 1998, pp. 97–98), nevertheless attempted to preserve the separation between geometrical entities and motion, both as linked to physical matter and as considered in the context of geometry (*ibid.*, p. 125). One of the reasons that were then brought forth is Clavius' rebuttal of Peletier's rejection of superposition on account of its alleged mechanical status, claiming that Euclid had only intended for superposition to be done in the mind and not by actually moving triangles and angles. Yet, in Clavius' commentary, this argument aimed to address Peletier's confusion between problems and theorems, or between construction and rational assumption, rather than the admission of motion in geometry. On Clavius' response to Peletier in the debate on superposition, see Axworthy (2018).

[93]On Clavius' idealisation of practice, see Bos (2001, pp. 165–166). As noted also by H. Bos (Bos 2001, pp. 199–203 and 408–409) this attitude will be developed to a greater extent by Johannes Molther (1561–1618), who saw, in the constructions of pure geometry, imaginary and idealised versions of the instrumental processes of practical geometry, whereby he considered that Euclid's postulates necessarily implied motion and sense-perception. Molther (1619, p. 36): "Si quis autem existimet oportere puram Geometriam sola mentis actione, etiam secundum Postulata sua, exerceri; is quoque Regulae ac Circini materialis ideas (ἀφαιρέσει) Mathematica abstrahat & mente complectatur, ut in Phantasia per sensum interiorem Regulae ac Circini opera faciant. Sic enim proclive fuerit cogitando illud fingere, quod quomodo reipsa praestetur monstravimus." This passage is quoted, with an English translation, by Bos (2001, p. 200).

[94]Clavius (1611–1612, I, p. 35), Prop. I.9, *Praxis*: "Non descripsimus autem dictas lineas, ut nuda praxis haberetur: Id quod in aliis quoque praxibus, quoad eius fieri poterit, observabimus, ne linearum multitudo tenebras nobis offundat, pariatque confusionem."

[95]On the representation of compass arcs in geometrical diagrams in the early modern printed Euclidean tradition, see Lee (2018).

lines required by the construction are to be set forth as compass arcs or even as dotted lines.[95]

8.8 Geometrical and Mechanical Processes: The Quadratrix

If the imaginary motion of points and lines played, for Clavius, a foundational role for the constructive procedures required in Euclid's geometry and may sanction corresponding instrumental operations (at least as idealised procedures), not all generative motions were considered admissible depending on their use and function in geometry. Indeed, if the geometer could imagine any descriptions of lines through the motion of a point according to various paths, not all of the resulting curves may be mathematically determined, that is, measured precisely through rational means. Hence, some of the motions to which geometers may resort to solve certain mathematical problems would not be considered as legitimate in geometry, given that they result from a combination of motions whose ratio is not mathematically determinable. This was the case of the quadratrix, a curve used by Dinostratus (c. 390–c. 320 BC) to solve the quadrature of the circle and presented by Pappus, in Book IV of his *Mathematical collection*.[96] Clavius dedicated a small treatise to the quadratrix at the end of the sixth book of the second edition of his commentary on Euclid published in 1589,[97] that is, a year after Commandino edited and published Pappus' *Collection*.[98] The construction of the quadratrix involved, in its classical construction, the simultaneous motions of two adjacent sides of a square, one moving along a straight line,

---

[96] Pappus (Hultsch 1876–1878, pp. 250–259), IV.30–33, § 46–50.

[97] Clavius (1589, I, 894–918); Clavius (1611–1612, I, pp. 296–304). This treatise was later integrated in Clavius' *Geometria practica* published in 1604: Clavius (1604, pp. 359–370); Clavius (1611–1612, II, pp. 189–194). On Clavius' approach to the quadratrix, see Bos (2001, pp. 160–166).

[98] Bos (2001, 160) and Sasaki (2003, pp. 66–67).

[99] Pappus (1588b). Among the objections raised against the validity of the quadratrix as a geometrical means of squaring the circle, Sporus argued that, since the purpose of the quadratrix is to procure the knowledge of the ratio between the radius and the circumference of the circle, or between the straight and the circular, the fact that its construction precisely required the knowledge of the ratio between the two combined motions therefore caused it to be a *petitio principii*. Pappus (Hultsch 1876–1878, pp. 254–256), IV.31, § 46–47 and (Sefrin-Weis 2009, Prop. 31, pp. 132–133): "Sporus [...] is with good reason displeased with it, on account of the following observations. [...] For how is it possible when two points start from B, that they move, the one along the straight line to A, the other along the arc to D, and come to a halt at their respective end points at the same time, unless the ratio of the straight line AB to the arc BED is known beforehand? For the velocities of the motions must be in this ratio, also. Also, how do they think that they come to a halt simultaneously, when they use indeterminate velocities, except that it might happen sometime by chance; and how is that not absurd? [...] Without this ratio being given, however, one must not, trusting in the opinion of the men who invented the line, accept it, since it is rather mechanical". See also Bos (2001, pp. 40–43). *Cf.* Clavius (1611–1612, I, p. 296): "Sed quia duo isti motus uniformes, quorum unus per circumferentiam *db*, fit, & alter per lineas rectas *da*, *cb*, effici non possunt, nisi proportio habeatur

Fig. 8.2 Christoph Clavius,
Euclidis elementorum libri XV,
1607, p. 648. The quadratrix and
its mode of generation according
to Clavius. Courtesy Max Planck
Institute for the History of
Science, Berlin

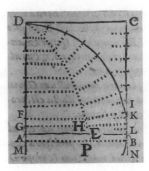

the other rotating around one of its fixed extremities and both meeting on the same side. The curve is drawn out by the points where the two moving lines intersect along their respective paths. Dinostratus' construction of the quadratrix was deemed non-geometrical by Pappus (and before him by Sporus[99]), because the ratio of the velocities of the combined straight and circular motions would not be accurately known, for which it was then presented as a mechanical rather than a geometrical mode of construction. Clavius, who acknowledged this difficulty, tried to offer a properly geometrical construction of this curve in his 1589 commentary on Euclid, using a pointwise construction (Fig. 8.2). Through this method, he evacuated the appeal to continuous motion, attempting to reconstitute the kinematic process in an indirect and static manner.[100]

As shown by H. Bos, Clavius did not however consider his construction as absolutely accurate, for which he later offered a more accurate alternative in his 1603 edition of

---

circularis lineae ad rectam, meritò à Pappo descriptio haec reprehenditur: quippe cum ignota adhuc sit ea proportio, & quae per hanc lineam investiganda proponatur."

[100] Clavius (1611–1612, I, p. 296): "Forte superiori anno incidi in librum 4. Pappi Alexandrini, ubi lineam quandam inflexam explicat, quam, ut ait, Dinostratus, & Nicomedes, & nonnulli iuniores excogitarunt ad circuli quadraturam, ideoque ab officio τετραγωνίζουσα ab eisdem appellata est, à nobis eadem de causa Quadratrix dicetur. Quanquàm autem praedicti auctores huiusmodi lineam conentur describere per duos motus imaginarios duarum rectarum, qua in re principium petunt, ut propterea à Pappo reiiciatur, tanquàm inutilis, & quae describi non possit; nos tamen eam sine illis motibus Geometricè describemus per inventionem quotvis punctorum, per quae duci debeat, quemadmodum in descriptionibus conicarum sectionum fieri solet." See also Clavius (1611–1612, I, p. 296), Clavius (1604, pp. 359–360) and Clavius (1611–1612, II, p. 189).

[101] Clavius (1611–1612, I, 297): "Verum puncta Quadraticis prope basem certius inveniemus, sine intersectionibus linearum hoc modo." See also Clavius (1604, p. 361) and Clavius (1611–1612, II, p. 190).

[102] H. Bos (Bos 2001, p. 164) noted that Clavius went from presenting his construction as geometrical to presenting it as geometrical "in a certain manner" (*geometrica quodammodo*): Clavius (1589, p. 897): "Haec igitur est descriptio lineae Quadratricis, quae Geometrica appellari potest." *Cf.* Clavius (1611–1612, I, p. 297): "Haec igitur est descriptio lineae Quadratricis, quae Geometrica *quodammodo* appellari potest." (My emphasis.) See also Clavius (1604, p. 362, 1611–1612, II, p. 191).

Euclid's *Elements* (and in his 1604 *Geometria practica*).[101] Although he then still presented his construction as more geometrical and as more accurate than the construction expounded by Pappus, he maintained certain reservations concerning the geometrical character of his construction of the quadratrix,[102] considering it as acceptable only in practice.[103] Because of the impossibility of rationally determining the combination of motions involved in the construction of the quadratrix, Descartes later placed this curve among mechanical (or non-geometrical) curves, along with other curves which result from a combination of motions that cannot be rationally determined.[104]

What this reveals, with respect to the issue at stake, is that, for Clavius, the mathematician is certainly free to imagine the generations of all kinds of lines and figures (as shown in his Df. I.4 through the example of mixed lines),[105] but that not all of the resulting magnitudes may be subject to a properly mathematical and rational determination. This shows also that a geometrical curve or figure, whether conceived as generated in the imagination or concretely, is not regarded as mathematically known until its mode of generation, that is, the motion that allows to generate it, is considered as mathematically known. This particular aspect, as we will see further, clearly resonates with the conception of geometry proposed by Descartes in the seventeenth century.[106]

8.9 Geometrical and Mechanical Processes: The Debate With Peletier on Superposition

In this regard, it is also important to mention Clavius' position on geometrical superposition, even if it does not directly concern the topic of generative motion. As was said earlier, Clavius was engaged in a debate on this issue with Peletier.[107] It may be recalled here that Peletier, in his 1557 commentary on Euclid's *Elements*, rejected superposition as a legitimate means of demonstrating the congruence of figures on account of its mechanical rather than geometrical character. In his response to Peletier, Clavius conceded that, if

---

[103] Clavius (1604, p. 359): "Haec [quadratrix] enim via licet ad Geometricè inveniendum punctum quoddam, nonnihil in ea desideretur, *accuratior* tamen est omnibus aliis, quas hactenus videre potui; ita ut *practicè* à scopo aberrare non possimus." (My emphasis.) See also Clavius (1611–1612, I, p. 189).

[104] On this issue, see Molland (1976), Mancosu (1996, pp. 71–79), Bos (2001, pp. 335–342), Mancosu (2008), Domski (2009) and Crippa (2014, pp. 68–72, 100–103, 221–228, 238–247 and 255–263). See also Panza (2011, p. 74–91) and *infra*, p. 254. Mancosu claimed (in Mancosu, 1996, pp. 74–76) that Clavius' treatment of the quadratrix was one of the sources of Descartes' reflection on pointwise constructions and on his own rejection of the quadratrix among geometrical curves in the *Géométrie*.

[105] See *supra*, p. 199.

[106] The relation between Clavius and Descartes is briefly considered in Chap. 10, pp. 252 sq.

[107] On the debate between Peletier and Clavius concerning superposition, see Axworthy (2018).

superposition were intended by Euclid as a constructive procedure and used within a problem as in Prop. I.2 or I.3 (mentioned by Peletier as illustrative examples in this contect), that is, as a procedure which is considered as actually effected on given geometrical figures, it would indeed be unsuitable to the geometrical evaluation of the congruence of figures. For it would not be based on any constructive postulate and would imply an empirical mode of determination of the equality of the compared magnitudes. However, as Clavius pointed out then, Euclid only intended superposition to be performed hypothetically ("*if* the triangle ABC be applied to the triangle DEF, and *if* the point A be placed on the point D and the straight line AB on DE. . ."[108]), all the more as superposition does not aim to construct or find any figures, but only to demonstrate the congruence of two already given figures. For this reason, as Clavius argued, superposition is only appealed to within theorems, whose purpose is to demonstrate properties or relations of geometrical figures (as classes of objects), and not to find or construct a particular object, which is the purpose of problems.

> [Mathematicians] do not want superposition to be done in fact (this would certainly be mechanical), but only to be done in thought and in the mind, which is the duty of the reason and of the intellect.[109]

Hence, for Clavius, the motion which may be imaginarily attributed to geometrical figures in the context of these theorems would only correspond to the intuition of the spatial relation between the two compared figures, but would not itself correspond to the source of the proof, that is, to the means by which the congruence of the two figures is established. This conclusion would rather be derived from the assumed equality of the parts of the triangles. This motion would therefore not need to be spatially determined so as to guarantee the equality and rigidity of the superposed figures and of their parts, as in Prop. I.2 and I.3, to which Peletier compared Prop. I.4 in order to justify his rejection of superposition. Indeed, even if the motion and the superposition of figures may be

---

[108] Euclid (Heath 1956, I, p. 247), Prop. I.4.

[109] Clavius (1611–1612, p. 121), Prop. III.16: "Neque enim volunt, re ipsa faciendam esse superpositionem, (hoc enim mechanicum quid esset) sed cogitatione tantum, ac mente, quod opus est rationis atque intellectus." This passage would indicate that, even if Clavius did not use a strict terminology in describing the mode according to which the generation of figures is apprehended by the mathematician, indifferently saying that it is imagined, conceived or intellectually embraced (see *supra*, pp. 203–204), he also identified then the mental faculties expressed here as *cogitatio, mens* and *intellectus* with a purely logical and non-intuitive cognitive activity, which would as such be regarded as separate from the imagination. The imagination would be here disregarded, since it corresponds specifically to the faculty which apprehends logical concepts in a spatial manner. Hence, it seems that, in the passages in which Clavius used *concipio, cogito* and *intelligo* as synonyms of *imaginor* (in the definitions of Book I), he was only making clear that the genetic processes of geometrical objects, which are appealed to as means of apprehending the essential properties of geometrical objects, should be understood as primarily and essentially occuring in the mind, as opposed to concretely.

performed in the imagination (or even concretely), these procedures would not play any part in the actual demonstration.

This particular example, as said, does not directly concern the type of geometrical motion which is investigated here. But it confirms that Clavius had a very clear conception of what kinematic processes were relevant to Euclid's problems and theorems, and of their ontological and epistemological status in this context. It furthermore confirms that the term "mechanical", when employed by Clavius as opposed to "geometrical", did not specifically refer to the use of instruments, as Clavius considered that instrumental procedures, when considered *in abstracto*, are able to express geometrically legitimate processes. This term then rather referred to the fact that the considered motion (whether understood as taking place imaginarily or concretely) is not mathematically known or rationally determined, but only depends on empirical judgment.

8.10   Geometrical Motion and the Affinity of Numbers and Magnitudes in the Commentary on Book II

As the other commentators considered here, Clavius appealed to a kinematic understanding of magnitudes in his commentary on Df. II.1. He did so in particular to legitimate the arithmetical interpretation of Book II,[110] which he instantiated through the numerical examples he appended to the propositions of Book II,[111] but also by inserting, in his commentary on Book IX, as had done Commandino, the arithmetical reinterpretation of the propositions of Book II.1–10 based on the work of Barlaam of Seminara.[112]

When dealing with Df. II.1, he explicitly compared the imaginary motion of the line-segment generating the surface of the parallelogram to the multiplication of a number by another.

> A parallelogram of this kind is produced by the imaginary motion of a line on another. For if it is conceived intellectually that the straight line AB is moved across downwards along the straight line AD, so that it always makes a right angle with AD, while the point A reaches the point D and the point B reaches the point C, the whole parallelogram ABCD will be described. The same thing will happen if AD is held to move across along the straight line AB, etc. For this reason, the parallelogram contained by two such straight lines is rightly said to be

---

[110] Malet (2006, pp. 69–70).

[111] Clavius (1611–1612, pp. 84–102).

[112] Clavius (1611–1612, I, p. 367): "Quoniam ad theorema sequens demonstrandum Theon quaedam assumit in numeris, quae demonstrata sunt de lineis libro secundo, perinde ac si eadem de numeris essent ostensa; non alienum instituto nostro duximus, nonnulla ex ijs, quae Geometrice ab Euclide libro 2. demonstrata sunt de lineis, hoc loco de numeris demonstrare, Quod idem & Barlaam monachum fecisse à nonnullis est traditum. Sequemur autem eundem ordinem, quem Euclidem in secundo libro tenuisse conspicimus." For these propositions, see Clavius (1611–1612, I, pp. 367–370).

rectangular. [. . .] But this understanding of the rectangular parallelogram as contained by two straight lines at right angles has a great affinity with the multiplication of one number by another. Just as, from the multiplication of 3 by 4, is produced the number 12, which is made in the form of a parallelogram, whereby it is said to be contained by 3 and 4, thus the parallelogram ABCD, which is encompassed by the two straight lines AB and BC, one of which is 3 palms long, and the other 4 palms long, similarly consists of 12 square-palms, which are certainly produced by the leading of the line AB, of 3 palms long, on the line BC, of 4 palms long, as indicated by the figure. And this is known to arithmeticians as well as to geometers and is demonstrated by Johannes Regiomontanus in the Prop. I.16 of his book *On triangles*. From this derives, according to what some say, that the rectangular parallelogram is generated by the leading of two lines, one on the other, about a right angle, just as the above-mentioned parallelogram is generated by leading the line AB on the line BC or (what is the same thing) by leading the line BC on the line AB. The same parallelogram is produced, whether the shorter line is led on the longer, or the longer on the shorter, just as the same number is also produced, whether the smaller number is led on the greater, or the greater on the smaller, as is shown by Euclid in Prop. VII.16. Indeed, this number 12 is produced by the multiplication of 3 by 4 as much as by the multiplication of 4 by 3.[113]

In this passage, Clavius clearly sets forth the commonness of the term *ductus* to geometers and to arithmeticians, being both used to designate the generation of magnitudes by motion and the generation of numbers by multiplication. He showed thereby the *affinity*, as he puts it, between these processes and therefore between numbers and magnitudes. But while Clavius proposed an explicit comparison between the modes of production of magnitudes and of numbers, which he sustained by a reference to Regiomontanus' *De triangulis*

---

[113]Clavius (1611–1612, I, pp. 82–83), Df. II.1: "quo ex motu imaginario unius lineae in alteram huiusmodi parallelogrammum conficitur. Si namque animo concipiatur recta *ab*, deorsum secundum rectam *ad*, moveri in transversum, ita ut semper angulum rectum cum *ad*, constituat, donec punctum *a*, ad punctum *d*, & punctum *b*, ad punctum *c*, perveniat, descriptum erit totum parallelogrammum *abcd*. Idem fiet, si *ad*, ponatur moveri in transversum secundum rectam, *ab*, etc. Quamobrem iure optimo sub talibus duabus lineis rectis contineri dicitur parallelogrammum rectangulum. [. . .] Habet autem comprehensio haec parallelogrammi rectanguli sub duabus rectis lineis angulum rectum continentibus, magnam affinitatem cum multiplicatione unius numeri in alterum. Sicut enim ex multiplicatione 3 in 4 producitur numerus 12 qui in formam parallelogrammi constituitur, unde & contineri dicitur sub 3 and 4. Ita quoque parallelogrammum *abcd*, comprehensum sub duabus rectis *ab*, *bc*, quarum illa sit 3 palmorum, haec autem 4 constat 12 palmis quadratis, qui quidem ex ductu lineae *ab*, 3 palmorum in lineam *bc*, 4 palmorum producuntur, ut figura indicat, notumque est Arithmeticis, atque Geometris, demonstraturque à Ioanne Regiomontano lib. 1. de triangulis, propos. 16. Hinc fit, ut nonnulli dicant, parallelogrammum rectangulum gigni *ex ductu* duarum linearum circa angulum rectum unius in alteram. Ut proxime antecedens parallelogrammum ex ductu lineae *ab*, in lineam *bc*, vel (quod idem est) ex ductu lineae *bc*, in lineam *ab*. Idem enim parallelogrammum procreatur, sive minor linea in maiorem, sive maior in minorem ducatur: quemadmodum etiam idem producitur numerus, sive minor numerus in maiorem, sive maior in minorem ducatur, ut ab Euclide demonstratur lib. 7 propos. 16. Tam enim ex multiplicatione 3 in 4 quam ex 4 in 3 producitur hic numerus 12."

[114]On the relation between numbers and magnitudes in Regiomontanus, see Bos (2001, pp. 136–138). On Regiomontanus and Clavius on this issue, see also Malet (2006).

omnimodis,[114] he distinguished, in this context, different levels of consideration of magnitudes and of their modes of generation. He thus placed, on one side, the motion through which the parallelogram is imagined to be generated according to its genetic definition, and which is then called *motus*, and, on the other side, the operative process through which the geometer is said to lead one side of the parallelogram along the other in order to reconstitute and measure the area of this figure. This latter process is then called *ductus* and directly compared to the operation of multiplication.

Indeed, when Clavius presented the rectangular parallelogram as produced by the transversal motion (*motus*) of one of its sides at the beginning of his commentary on Df. II.1 (in order to explain why Euclid stated that the parallelogram is contained by the two sides adjacent to the right angle), he remained within a strict geometrical framework, leaving aside any comparison with the multiplication of numbers. In this context, the illustration of the parallelogram follows the traditional Euclidean model of non-scaled lettered diagrams (Fig. 8.3). But when he later described the parallelogram as produced by the leading (*ductus*) of one side on the other, he directly compared this process to the multiplication of two specific numbers. For he established a correspondence between the multiplication of 3 by 4 (the two numbers taken here as an example), which are equal to 12, and the generation of the area of a parallelogram of sides equal to 3 by 4 units of length (here, palms, representing the width of a human hand), amounting to 12 square-units (Fig. 8.4).

The diagram on the right of the second illustration (Fig. 8.4), which depicts a rectangle subdivided into squares, is similar to the diagram presented by Commandino in his commentary on Df. II.1.[115] However, Clavius went further than Commandino in this respect by exhibiting also the relation between the generation of the parallelogram of 12 square-units (3 times 4 units of length) and the generation of a rectangular number through the multiplication of 3 times 4 point-units (consisting in 3 rows of 4 aligned points) in order to show the quantitative equivalence between the two. Moreover, while Commandino's diagram simply presented a lettered rectangle subdivided into 12 squares, Clavius added to his lettered rectangle the numerical values of the sides and of the total area. Hence, as other commentators considered here, Clavius explained the relation between the generation of the rectangular parallelogram and the multiplication of numbers through the notion of geometrical number. Yet, contrary to the commentators who appealed to a visual representation of a rectangular number (Fine and Billingsley),[116] the dots composing the geometrical number in Clavius' illustration are connected by line-segments in order to better show the correspondence between the geometrical number

---

[115] See *supra*, p. 192.

[116] See *supra*, p. 67 and 162.

[117] As seen above, Fine used this notion in this sense in the commentary on Df. I.30, but not in his commentary on Df. II.1. See *supra*, p. 69.

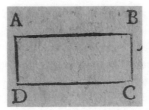

Fig. 8.3 Christoph Clavius, *Euclidis elementorum libri XV*, 1607, p. 164, Df. II.1. Diagram of the rectangular parallelogram to illustrate the generation of the figure by motion, which is not compared then to the generation of numbers by multiplication. Courtesy Max Planck Institute for the History of Science, Berlin

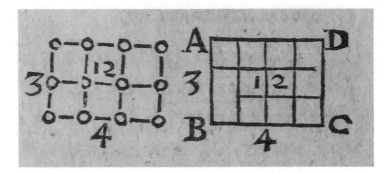

Fig. 8.4 Christoph Clavius, *Euclidis elementorum libri XV*, 1607, p. 165, Df. II.1. Diagrams illustrating the correspondence between the generation and the containing of rectangular numbers by the multiplication of two numbers and the generation and the containing of a rectangular plane surface by two line-segments touching at right-angles. As in Fine's commentary, Clavius compared the rectangular number to the geometrical rectangular parallelogram, although the latter is rather drawn here in the style of the parallelogram included in Commandino's commentary, with an indication of numerical values. Courtesy Max Planck Institute for the History of Science, Berlin

obtained by multiplication and the geometrical figure obtained by the transversal leading of the side of the parallelogram on its other side.

Furthermore, Clavius is the only commentator, out of those considered in this study, to have directly employed the term *ductus* (through the use of the verb *duco* or *ducatur*) in his commentary on the definitions of Book II[117] in order to designate the operation of multiplying numbers, according to the terminology used in medieval Latin arithmetical treatises.[118] He is thus the only one here to have explicitly connected the arithmetical and geometrical uses of the operative term *ductus* in this particular context and to have demonstrated to such a great extent the affinity or close connection between the generations

---

[118] See *supra*, p. 69.

of numbers and of magnitudes and thereby between arithmetical and geometrical operations. As in Peletier and Billingsley, the affinity between arithmetical and geometrical operations is also set forth by explicitly stating the commonness of the property of commutativity to arithmetical multiplication and to the generation of parallelogrammic areas by the *ductus* of one of their sides. The commonness of this property to geometry and to arithmetic, which allowed Clavius to fully justify the arithmetical and algebraic interpretation of Book II, is then founded on the correspondence between Df. II.1 and Prop. VII.16 of the *Elements*, which demonstrates the equality of the products of two numbers.[119] Therefore, as he expressed it a bit further in a manner quite similar to Billingsley, the propositions presented in Book II would be fundamental to teach the rules of algebra, as well as the operations of arithmetic as applied to surd numbers or irrational quantity, which one would never be able to attain through the sole study of integers or fractions.[120]

In the first definitions of Book I, Clavius did not mention the difficulties raised by the conception of a line produced by a point. Similarly, in the commentary on the definitions of Book II, no reservation was expressed concerning the comparison of the respective modes of generation of numbers and of magnitudes, as it was, on the contrary, in the other commentaries considered here. The production of the area of a parallelogram by the *ductus* of one of its sides along and according to the length of the other was then directly related to the arithmetical notion of *ductus*, or to the multiplication of a number by another, without raising doubts concerning the possibility for a line to produce a surface, or, by extension, for a point and a surface, to produce a line and a solid, respectively, through their motion. The affinity between discrete and continuous quantity, which is also marked by the rather traditional comparison between the unit, the point and the instant in relation to number, magnitude and time, respectively, in the commentary on Df. I.1,[121] would be founded on and represented by their common appeal to the notion of *ductus*. Now, as we have seen

---

[119] Euclid (Heiberg 1884, p. 222) and (Heath 1956, II, p. 316), Prop. VII.16: "If two numbers by multiplying one another make certain numbers, the numbers so produced will be equal to one another."

[120] Clavius (1611–1612, I, p. 83): "Sed iam ad propositiones secundi huius libri veniamus, in quibus sane operae pretium fuerit, multum laboris in eis exquisite intelligendis ponere, propter multiplicem earum usum cum in rebus Geometricis, tum in humanis commerciis. Nam ex nonnullis harum propositionum demonstrantur regulae illae admirabiles Algebrae, quibus vix credo in disciplinis humanis praestantius aliquid reperiri, quippe cum miracula quaedam numerorum (ut ita dicam) eruant tam abstrusa, ac recondita, ut facultas illa omnem captum humanum superare videatur, tanta nihilominus facilitate, atque voluptate, ut facilius videatur esse nihil. Ex aliis deinde propositionibus huius libri eliciuntur demonstrationes, quibus inter se adduntur, substrahuntur, multiplicantur, atque dividuntur numeri surdi, (quos dicunt) hoc est, qui nullo modo exprimi possunt; cuiusmodi sunt radices numerorum non quadratorum, aut non cubicorum, quae neque per Divinam potentiam in numeris possunt exhiberi, quod haec res contradictionem implicet, ut Philosophi, atque Theologi loquuntur."

[121] Clavius (1611–1612, I, p. 13), Df. I.1: "Denique in magnitudine id concipi debet esse punctum, quod in numero unitas, quodque in tempore instans."

earlier, this notion of *ductus* was also used by Peletier to specifically designate the generation of the surface (or solid) from the motion of a line (or surface) in the context of the definitions of Book I. It was furthermore also related to the notion of flow or motion of the point in the commentary on the postulates of Fine and Clavius. Its use in the context of Book II, and in arithmetic, would therefore seem to find its foundation, for the Jesuit professor, in the geometrical concept of generative motion or flow of the point, which was regarded in Book I as an idealised instrumental procedure.

* * *

Throughout his dense commentary on Euclid's *Elements*, Clavius did not hesitate to discuss the various modes of definition of geometrical objects and their epistemic and ontological implications. And, in his commentary on the definitions, he gave an important place to genetic definitions, not only as a didactic tool, but also as a means of connecting abstract and concrete generations of geometrical objects, as well as definitions and constructions.

Like Billingsley, Clavius made absolutely no reference to any metaphysical understanding of the notion of flow of the point (*fluxus*), which he used as a direct synonym of *motus* and which he straightforwardly conceived as a spatial process. In his commentary on the definition of the straight line, Clavius clearly showed that the spatial structure of the flow of the point, which indirectly points to the admission of an undetermined time-factor, governs the properties of the generated line and accounts accordingly for the attributes stated in the formal definition, exhibiting the essential relation of the *definiendum* to its properties. Also, as shown in his commentary on the definitions of the surface and of the solid, Clavius admitted no distinction between the mode of generation of the line (ῥύσις) and the modes of generation of other figures, whether obtained by transversal displacement or by rotation (φορά), with regard to their epistemological or ontological implications.

By appealing to the notion of flow of the point in his commentary on the first postulates (as had done Proclus, as well as Fine and Commandino), Clavius explicitly related this notion to the production of lines in the framework of geometrical constructions. Thus, for him, the imaginary generation of straight lines and circles played a role in the validation of the construction process, guaranteeing the connection between the lines produced in the framework of geometrical problems and the logical concepts of straight line and of circle. It is moreover, for the student, a means of obtaining knowledge of the defined magnitudes, since the deployment of the line through the motion of an indivisible point in the imagination would offer a first-hand experience of the essential relation between the line and its indivisibility according to breadth and depth, founding thereby the geometrical definition of the line itself. In this perspective, Clavius' pervasive use of genetic definitions displays his constructivist approach to the constitution of geometrical knowledge.

Moreover, in spite of the difference he established between abstract and concrete objects on the scale of exactness and separation from matter, Clavius had no difficulty in relating the modes of generation of geometrical figures with concrete and instrumental processes. While the latter could be used in a pedagogical context to teach the generations and properties of abstract geometrical objects, the former would, in turn, represent an idealised version of the latter.

On the issue of the logical and epistemological status of genetic definitions, Clavius did not, in his commentary on Euclid's *Elements*, establish an explicit distinction between definitions by property and genetic definitions, or definitions in the proper sense and descriptions, as was done by Foix-Candale and Billingsley. He did however display both definitions and referred to the notion of *descriptio* when dealing with genetic definitions and with the imaginary and concrete construction of geometrical objects in general. He also offered a clearer distinction between these two modes of definitions of geometrical objects in his 1570 commentary on Sacrobosco's *Sphaera*, as these were then presented as different types of definitions or *predicationes*. Hence, while Theodosius' definition of the sphere was designated as a *praedicatio formalis*, Euclid's definition was presented as a *praedicatio causalis*. In both texts, the two modes of definition are presented as complementary, though the genetic definition would have a stronger epistemic value, since it displays the process through which the defined magnitude is brought about, and is therefore necessary to legitimate the connection that is made in the definition by property between the *definiendum* and its attributes.

As was common to most of the commentaries considered here, Clavius appealed to a kinematic understanding of geometrical figures when commenting on the definitions of Book II and connected thereby the mode of generation of quadrilateral surfaces and the mode of generation of numbers by multiplication. In this context, he presented different interpretations of the generation of the rectangular parallelogram. Starting with the generation of a non-dimensioned rectangular parallelogram through the transversal motion of one of its sides, which is then designated as *motus* and treated as a geometrical notion only, he afterwards applied this process to the measurement of a parallelogram contained by sides of determined length, calling it *ductus*. He presented this process as the geometrical equivalent of the operation of multiplication in arithmetic, playing on the double meaning of the term *ductus* in Latin mathematical works. This process of *ductus*, which specifically came forth as an operative notion, contrary to the notion of *motus* presented in the same context, allowed Clavius to display the affinity between numbers and magnitudes, and thereby between arithmetic and geometry, which was then supported by the concept of geometrical number. Hence, no difficulty was explicitly raised in this context concerning the comparability of numbers and magnitudes, nor concerning the concept of point as the origin of the line. Generally speaking, Clavius did not take into consideration the difficulties raised by genetic definitions and motion within geometrical definitions, but actually made this a central element of his mathematical teaching, linking definitions and constructions, theory and practice, arithmetic and geometry.

Synthesis: Continuities and Transformations in the Status of Geometrical Motion and Genetic Definitions from Fine to Clavius

The general aim of this study was to investigate the place and status that was attributed to motion as a means of defining geometrical objects in the sixteenth-century Euclidean tradition, and to determine whether a significant change of attitude appeared throughout the considered period with regard to the admissibility of motion in geometry.

Considering only the sixteenth-century commentaries on Euclid's *Elements* that most significantly contributed to the question (along with other texts on the topic written by their authors), I therefore undertook to determine the extent to which genetic definitions were introduced within these commentaries, where they were introduced in these works and how they were formulated. This analysis revealed that not all of Euclid's commentators who made use of genetic definitions appealed to them to the same extent and formulated them in the same manner.

I also examined the ontological and epistemological status that was attributed to motion in geometry, in general and in the context of definitions more particularly. In this aim, I investigated the way in which the considered commentators characterised the difference and relation between genetic definitions and definitions by property. I sought furthermore to determine whether any reservations were then raised (explicitly or implicitly) concerning the legitimacy of genetic definitions, and, if so, whether any condition was stated for motion to be admitted in geometry. In relation to this, it was investigated whether, and how, the considered commentators received the conceptions of ancient mathematicians and philosophers concerning the status of motion and genetic definitions in geometry. This made it possible to discern different degrees of acceptance of motion as a genuine aspect of geometry, as well as different conceptions of the relation between geometrical and physical objects, as well as between the geometrical and the divine.

A. Axworthy, *Motion and Genetic Definitions in the Sixteenth-Century Euclidean Tradition*, Frontiers in the History of Science,
https://doi.org/10.1007/978-3-030-95817-6_9

I looked furthermore into how the use of genetic definitions contributed to the reassessment of the concept of continuous quantity and of its relation to discrete quantity in the Euclidean framework. Indeed, given the problems raised by the notion of a continuum as originating from an indivisible principle, the investigation of this issue offered complementary information on the way the considered authors positioned themselves with respect to ancient philosophical debates on the logical and mathematical validity of genetic definitions in geometry. It also provided us with an insight into the connections that existed between the wider admission of motion and genetic definitions in geometry and the changes that took place in the early modern era in the representation of geometrical science, its finality, objects and methods, as well as its relation to arithmetic.

The following pages offer a synthesis of the results of this study. This synthesis proposes a more general comparison of the various conceptions which were set forth on these issues by the investigated commentators of Euclid and aims to bring forward the most significant changes that took place over the sixteenth century with respect to the status of motion and genetic definitions in geometry. It will be followed by an epilogue which provides a brief overview of the later evolution of the question by looking at the case of Descartes and at his approach to geometrical motion in his own geometrical and philosophical work. I will attempt to show there how Descartes' attitude in this regard related to the tradition analysed here and went beyond it by developing a new geometrical science and philosophy of geometry in which the kinematic approach to geometrical objects held a central place.

9.1   Distribution and Terminology of Genetic Definitions in the Sixteenth-Century Euclidean Tradition

9.1.1   The Distribution of Genetic Definitions

Concerning the extent to which genetic definitions were added by the considered authors within their commentary on Euclid's *Elements*,[1] it has been established here that genetic definitions were most extensively introduced by Fine (with 16 noted occurrence), Billingsley (11 occurrences)—we may add to these 2 main occurrences in Dee's *Mathematicall praeface*—and Clavius (12 occurrences). Fewer occurrences were observed in the commentaries of Peletier (5 occurrences), Foix-Candale (5 occurrences) and Commandino (3 occurrences). It may however be noted that Fine and Peletier's commentaries covered only the first six books of the *Elements*. These commentaries could therefore have contained a greater number of genetic definitions, if they also had included Euclid's remaining books, especially Book XI.

---

[1] I have only counted here the genetic definitions presented in the commentaries on Euclid's *Elements*, excluding therefore those presented in the other sources considered throughout this study, such as Fine's *Geometria* or Peletier's *Louange de la Sciance*. A table in appendix (pp. 271–272) summarises the places where genetic definitions appeared in these commentaries on Euclid.

All commentators, except for Foix-Candale, included a genetic definition of the line in their commentaries on Euclid's Df. I.2. The genetic definition of the line also appeared when dealing with the point in Df. I.1 (in Fine and Peletier), with the extremities of lines in Df. I.3 (in Fine and Foix-Candale), with the straight line in Df. I.4 (in Fine, Peletier and Clavius), and with the extremities of surfaces in Df. I.6 (in Foix-Candale), but also when commenting on the first three Postulates (in Fine, Billingsley, Commandino and Clavius). The genetic definition of the line was also sometimes mentioned in the eleventh book, when commenting on the solid in Df. XI.1 (in Billingsley and Clavius) and on the sphere in Df. XI.14 (in Billingsley), where the relation and homology between the genetic definitions of the line, the surface and the solid were set forth.

However, the definition of the surface as generated by the motion of the line, as well as the definition of the solid as generated by the motion of the surface were encountered less often. When they were, it was rarely in a commentary on the definitions of the surface and on the solid (Df. I.5 and Df. XI.1, respectively); that was only the case in Fine for Df. I.5 and in Clavius for both Df. I.5 and Df. XI.1. The genetic definition of the surface was also mentioned when commenting on Df. I.4 (by Peletier) to introduce Df. I.5 and to show its homology with the genetic definition of the line; on Df. I.6 (by Fine and Foix-Candale); on the plane surface in Df. I.7 (by Clavius); on the boundary in Df. I.13 (by Fine and Foix-Candale, indirectly) and on Df. XI.14 (by Billingsley).

The genetic definition of the solid was otherwise presented when dealing with Df. I.7 and Df. I.13 (by Fine, indirectly); with the centre of the circle in Df. I.16 and with Df. XI.14 (by Billingsley).

Genetic definitions of more specific plane figures were also provided. A genetic definition of the circle was introduced when commenting on Df. I.15 (in Fine and Clavius) and Df. I.16 (in Peletier and Billingsley), where an indirect reference was sometimes made to Euclid's definition of the sphere (in Peletier and Clavius). A genetic definition of the square appeared in Fine's commentary, as he dealt with Df. I.30. A kinematic expression of the rectangular parallelogram was also provided in most of the analysed commentaries when dealing with Df. II.1.

Euclid's definition of the sphere (Df. XI.14) was already formulated as a genetic definition, as were the definitions of the cone (Df. XI.18) and of the cylinder (Df. XI.21). But what commentators did in this regard (i.e. Foix-Candale, Billingsley and Clavius) was to add a definition of the sphere by property, based on that found in Theodosius' *Spherics*. This genetic definition of the sphere was also referred to by Billingsley in his commentary on Df. XI.15 and Df. XI.16. No definitions by property of the cone and of the cylinder, independent from their genetic definitions, were however proposed in these commentaries.

Hence, genetic definitions were not used to the same extent by all the considered commentators. While some made an effort to systematically add a genetic definition to Euclid's definitions of the most fundamental objects of geometry (line, surface, solid) in order to show the synthesis of the different types of magnitude from the point (as in Fine and Clavius), others only added a genetic definition occasionally, as an alternative to Euclid's definition by property, most often to his definition of the line (as in Commandino).

This allowed the latter to implicitly point to the derivation of magnitudes from the point without giving too much importance to genetic definitions, which could have suggested that Euclid's list of definition was somehow incomplete or inadequate.

9.1.2   The Designation of Motion within Genetic Definitions

The terminology that was used to designate motion in these genetic definitions was also shown to vary from one author to the other, and sometimes from one passage to the next in the same commentary. The terms that were most commonly used to express the generation of the line, the surface and the solid are *fluxus* and *motus*. *Ductus* was also regularly encountered, though less often than *fluxus* and *motus* in the context of a genetic definition. In rare occasions, the term *progressus* was used.

These different terms were either distinguished in their meaning and function (as were *fluxus* and *ductus* in Peletier; or *fluxus* and *motus* in Commandino), or were used as direct synonyms (as in Fine and Clavius). *Progressus* was mainly used by Foix-Candale, but was also indirectly appealed to by Clavius, for instance in his commentary on Post. 1, through the use of the present participle *progrediens* as applied to the line. *Defluxus*, which has a similar meaning to *fluxus*, was also introduced by Fine. In Billingsley's commentary, which is the only commentary written in the vernacular out of those analysed here, we find the terms *moving*, *motion*, *draught*, *flowing*, *gliding* as applied to the generative motion of the point, line or surface. Dee also appealed to the terms *race* and *course*, in addition to *motion*, when applied to the generation of the line by a point.

The term *descriptio* and its related verbal forms were found in practically all of the considered commentaries, to designate either the type of definition that states the mode of generation of a geometrical object (as opposed to the definition that states its properties) or, more generally, the process by which the geometrical object is generated (concretely or imaginarily), as well as the result of this process. This term was also employed by Billingsley when speaking of the actual diagrams that were used in the book to illustrate Euclid's text.

Thus, setting aside the differences between Latin and vernacular languages (here, English), the way motion was referred to in the context of genetic definitions, within sixteenth-century commentaries on Euclid, was hardly uniform. This terminological diversity appeared not only among different authors, but also sometimes within the same text. Among the different terms that were used, *fluxus* (or *flow*, *flowing*) and *motus* (or *motion*, *moving*) were most often encountered, while *progressus*, which was only employed in the substantive by Foix-Candale, was the least common. The term *descriptio*, and its derivatives, was perhaps as common as *fluxus* and *motus*, but it had a larger scope than the latter inasmuch as it was not only used to designate the motion by which the geometrical object is produced (concretely or abstractly), but also the genetic definition itself.

9.2 The Ontological and Epistemological Status of Genetic Definitions

9.2.1 Ontology and Terminology

Considered in general, that is, independently from the specific discourses of the investigated authors, the different terms used to designate the motion of geometrical objects in sixteenth-century commentaries on the *Elements* carried different semantic, but also ontological implications. The term *fluxus*, *defluxus* or *flow* (i.e. ῥύσις in Greek) would initially suggest that the line flows, springs or emanates from the point as its source, without attributing any local form of motion to the point; the same could apply to the surface and the solid in relation to the line and the surface, respectively. On the other hand, the terms *motus*, *motion*, *movement*, or *moving* (i.e. φορά or κίνησις) would imply that the line, the surface or the solid is generated through the local displacement of a point, a line or a surface from one place to another, these objects producing magnitudes of a higher dimension by leaving behind traces of themselves in their motion. The terms *ductus* and *ductio* would be more straightforwardly associated with the operation of leading a geometrical object from one place to another by the action of an external agent in the context of a construction. This is also the case of the terms *draught* (used by Billingsley) and *descriptio*, which first suggest the procedure of drawing out a particular line or figure, in particular by instrumental means. As for the term *progressus*, it mainly suggests the generation of magnitude by a forward motion from a given place, as would *race* and *course*, though the latter would also evoke a local displacement in the stricter sense. The term *gliding* found in Billingsley would convey the notion of uniform and uninterrupted local motion, conforming to its empirical meaning as a smooth continuous displacement of a body through air or on water.

In the commentaries of Fine, Billingsley and Clavius, the terms *fluxus* and *motus* (or their English equivalent *flow/flowing* and *motion/moving*) were actually used as direct synonyms. Moreover, these authors indifferently used these two terms to speak of the generation of the line from a point or of other types of geometrical objects. This indicates that there was, for them, no strict distinction between these different generations, as for their status or function on a kinematic or on an ontological level, the rectilinear being compared to the circular and the flow of the point being treated as a local motion of the point.

Although in ancient Greek texts, different terms could be used to designate the motion of geometrical objects (φορά) and that of physical substances (κίνησις), the same term (*motus* or *motion* in English) was commonly used in Latin and in many vernacular languages to designate both. Hence, certain commentators of Euclid undertook to clearly mark the distinction, either by explaining the difference between the two, as did Fine, or by regularly characterising geometrical motion as abstract, imaginary or intelligible, as did Fine and Clavius.

In a different way, Peletier (who appealed to *fluxus*, *ductus* and *motus*) only used the term *fluxus* to designate the generation of the line from the point and used *ductus* to express the generation of the surface from the line, or *motus* (indirectly) to characterise the generation of the circle. These different uses of the terms *fluxus* and *ductus/motus* were shown to stem from Peletier's will to implicitly acknowledge their underlying metaphysical implications, according to his philosophical conception of the ontological status of mathematical objects. Yet, in a mathematical context, the flow of the point was interpreted by Peletier as a strictly local process, in the same way as the *ductus* or *motus* through which surfaces are conceived as generated.

Foix-Candale used both *progressus* (Df. I.3 and I.6) and *motus* (Df. I.13) to characterise the generations of geometrical objects. In this context, both terms aimed to convey the notion of expansion or growth of magnitude from itself, which indirectly referred to the essential geneses of intelligible geometrical figures, according to the Neoplatonic metaphysical interpretation of the term πρόοδος. Foix-Candale opposed this process to the motion introduced by Euclid in his definition of the sphere, which was then quite classically translated by the terms *ductio* and *circumductio*. On both an ontological and epistemological level, this *ductio* of the semicircle in the definition of the sphere was related to the motion (then also called *motus*) of the line-segment generating the rectangular parallelogram in his commentary on Df. II.1. Both processes corresponded indeed, for Foix-Candale, to *descriptiones*, that is, to accidental and extrinsic, rather than essential and intrinsic, accounts of the geneses of figures.

Commandino distinguished *fluxus* and *motus* to explain the meaning of the two first postulates, as did Proclus in the corresponding passage in his own commentary on Euclid, but he did not interpret this difference according to the ontological hierarchy that was intended by the Neoplatonic philosopher. While Proclus understood the flow of the point (ῥύσις) as an intelligible and non-spatial process, and the motion of the point (κίνησις) as the spatial and imaginary version of the latter, Commandino would have associated the first (*fluxus*) with the genetic definition of the line properly speaking, and the second (*motus*) with the actual process carried out by the geometer in the context of geometrical constructions.

As for the English terms that were also used by Billingsley and Dee, in addition to *flow*, *motion* or *draught*, namely, *gliding*, *race* or *course*, these were attributed, in the context of their commentary on Euclid, the same meaning as the former, even if the two latter may have been interpreted by Dee as derived from the ancient Greek notion of ῥύσις.

As said, *descriptio* was a wide-ranging term that was used to designate the genetic definition itself or the motion by which the geometrical object is held to be caused, both in the imagination or concretely. It also designated the particular object resulting from the imaginary or concrete drawing of the line or figure. As such, in comparison with the terms *fluxus* and *motus*, the motion that was designated by the term *descriptio* was more specifically related to the practice of the geometer.

Therefore, the various terms used in the sixteenth-century commentaries on Euclid to designate motion in genetic definitions were interpreted differently by their authors. While some authors attributed distinct meanings or at least distinct philosophical implications to these different terms, and to the different types of motion they expressed, they were generally taken as mutually synonymous by other commentators. Because *motus* could be used in Latin indifferently for a motion that is merely local, as in the case of geometrical objects, and for physical processes, such as the generation and corruption of material bodies, it was important for certain authors, and for Fine in particular, to distinguish the different levels of interpretation of this term.

9.2.2 Reservations and Justifications

In his commentary on Euclid, Fine explicitly acknowledged that a doubt could be raised concerning the admissibility of motion in geometry given that motion was defined by philosophers as an essential characteristic of physical substances and was traditionally used to distinguish mathematical objects from natural beings within the classification and hierarchisation of objects of knowledge. This distinction between physical and mathematical objects according to their degree of separation from matter and motion, which stemmed from Aristotle, is precisely what motivated some of the medieval commentators of Sacrobosco's *Sphaera* to question the suitability of Euclid's definition of the sphere in view of its appeal to motion. Fine admitted that the motion introduced by mathematicians merely takes place in the imagination and is only analogous to the *local* motion of physical substances, that is, not to motion as determined by final and material causes (i.e. the actualisation of the essence of a substance and the motion toward the natural place of a substance according to its material composition). Yet, he nevertheless showed that it is "in virtue of the variety of the lines and of the surfaces, and of their diverse motion or abstract flow" that "is drawn the various and practically infinite multitude of planes and solids".[2] In other words, without the imaginary motion through which geometrical objects are derived from each other according to the mathematician, these would be impossible to obtain and to know. This constructive approach to geometry thus fully legitimated the attribution of motion to geometrical objects. In this framework, the constant assertion that this motion is abstract or imaginary would suffice to avoid attacks from natural philosophers on account of the essential connection of motion with physical matter.

At the other end of the chronology, Clavius followed an approach similar to Fine's. Clavius indeed frequently appealed to genetic definitions in order to explain the properties and interrelations of geometrical objects, while explicitly and constantly stating the imaginary or abstract status of the motion attributed thereby to geometrical objects. But although he asserted the abstract nature of geometrical objects, he did not, contrary to Fine, formulate any discourse to restrict the ontological status of geometrical motion. Rather, he openly displayed the connection between physical and mathematical motion by pointing to the homology between the generation of magnitudes in the imagination and the production of their concrete counterpart by instrumental means. Hence, the generative

---

[2]Fine (1536, p. 2), Df. I.7. See *supra*, n. 26, p. 45.

motions of geometrical objects came forth, in Clavius' commentary, as idealised instrumental procedures. These, in turn, are presented as objects susceptible to be studied geometrically. But beside idealised instrumental processes, the imaginary motion of geometrical objects would allow us to conceive, for Clavius, an infinite variety of curves, straight, circular, mixed and parallel, just as it was, for Fine, a means of producing an immense number of different figures, plane and solid. The capacity of the imagination to derive any type of geometrical objects by different types of motion revealed therefore, for the Jesuit professor, the unlimited potential of the mathematician's mind rather than its wandering into falsehood, even if its conceptions cannot all be rationally and mathematically determined, as shown by the case of the quadratrix.[3]

Between Fine and Clavius, Billingsley clearly asserted the ontological separation between geometrical objects and material realities, notably as geometrical objects such as the point, the line or the surface cannot be found in nature and may only be displayed, as defined by the geometer, in the imagination. Yet, he also admitted their close connection, due to their common reliance on spatiality and motion. In this regard, he explicitly compared the generative motion of geometrical objects to instrumental processes, as would Clavius a few years later. In Billingsley's case, this clearly resonates with the important place held, in late sixteenth-century England, by practical geometry, which focused on hands-on knowledge and instrumental practice and which would have contributed to assert the legitimacy of genetic definitions. In Billingsley's commentary, this practical approach to geometry was embodied by his appeal to paper figures which the reader was asked to cut out (sometimes after being redrawn), fold and paste within and beside the book in order to acquire a better understanding of Euclid's abstract stereometrical concepts. Thus, according to Billingsley's commentary on the *Elements*, the introduction of motion and of genetic definitions in geometry would not only be fully legitimate, but would be crucial both to the justification of the scientific value of geometry (through the notion of causal definition) and to the unification of theoretical and practical geometry.

In a very different manner, the implicit restrictions Peletier established regarding the admissibility of motion and of genetic definitions in geometry were determined both by his conception of the ontological status of mathematical objects, which he regarded as divine as for their essence and origin, and by the epistemic limitations he imposed on mathematical knowledge as a result of the irreducible gap between the human cognitive faculties and the transcendent nature of the divine. Peletier considered indeed the human mind as unable to grasp the true geneses of geometrical objects, just as their true properties and relations as transcendent entities. The mode of generation which the mathematician attributes to the line, the surface and the solid would then have a limited value as a means of conveying the essential mode of being, of causation and the mutual relations of these objects, as well as their relationship with numbers and physical substances. Hence, for Peletier, geometrical

---

[3] See *supra*, pp. 217–219.

objects, such as we conceive and know them, would only correspond to conjectural representations. It nevertheless remains that these conjectural conceptions of the generations and properties of geometrical objects would, for him, be perfectly admissible as a means of founding a coherent geometrical science as defined by human criteria of intelligibility.

Notwithstanding, when defining geometrical objects and their generations, the mathematician, for Peletier, would still be required to attempt, as much as it is possible, to describe them in the way that most suits their true origin and nature. This is very likely why he maintained in his commentary a terminological distinction between the generation of the line from the point and the generation of other geometrical objects from the line and the surface, the first being properly designated as a flow (*fluxus*) and the latter as a leading or a motion (*ductus* and *motus*). This distinction, even if irrelevant on a mathematical level, would have enabled him to implicitly hint at the ancient conception that posited the primordial and simultaneous emanation of all quantities and substances from a unique indivisible source, according to the metaphysical conception he presented in his philosophical and literary writings. The geometrical concept of point, and the role it is attributed by mathematicians in the generation the line, would allow the geometer to express this metaphysical notion in a form most understandable to the human mind, that is, as a local process.

Foix-Candale's position, in this respect, was quite similar to that of Peletier. He asserted indeed the transcendent origin and essence of mathematical objects, which was corroborated by his assertion of the divine nature of the geometrical sphere and of the mathematical unit in his commentary on the *Poimandres*. He considered also that one kind of derivation of geometrical objects (i.e. through the expansion of magnitude) is more suited than another (i.e. through the translation of an object of lower dimension) to express this intelligible origin and to reveal the essential properties of geometrical figures. The definition of the sphere provided in Book XI of the *Elements*, which defines this figure as generated through the local rotation of a semicircle—the same would apply to the cone and of the cylinder, which were defined as generated by the rotation of a triangle and a rectangle, respectively—would therefore only possess an operative value, referring to the practice of the geometer, and not to the proper mode of being of this figure. Foix-Candale thus regarded the mode of generation of lines, surfaces and solids through the translation of a point, a line or a surface, respectively, as incompatible with a truly rational and mathematical exposition of geometry, as he believed the *Elements* to be. On the other hand, the definition of the sphere provided by Theodosius, which stated the spatial properties of the sphere and suggested the outward motion of the sphere's magnitude from its centre, would be more suited to express the true essence of the sphere. The same would go for his own definitions of magnitudes in terms of *progressus*.

In Dee's *Mathematicall praeface*, no doubt nor restriction was brought forth regarding the legitimacy of genetic definitions, or of motion in general, in geometry. On the contrary, the motion of geometrical objects, and the motion of the point in particular, was presented as essential to geometry, and to mathematics more generally. As shown by the *Monas hieroglyphica*, the generative motion of geometrical objects would moreover reveal the interrelations of all the causes (mathematical, physical, divine and magical) that are at play in the constitution of the universe. Dee nevertheless implictly admitted a difference between its interpretation in the context of natural, theological or esoteric philosophy, on the one hand, and in mathematics, on the other hand, the latter relating strictly to the properties of the divisible and quasi-material magnitudes formed in the imagination. It remains that these were then considered as images of higher and indivisible substances, which themselves are deprived of spatiality and motion, following Proclus' doctrine of geometrical objects. In this regard, just as Proclus, Dee regarded motion as a perfectly legitimate attribute of geometrical objects, when they are considered as objects of the imagination. The imagination would indeed endow geometrical points, lines and figures with the intelligible matter necessary to investigate their mathematical properties and mutual relations.

Commandino did not either express any reservation concerning the use of motion and of genetic definitions in geometry, which is partly due to his following Proclus for many parts of his commentary on Euclid's first book. It was notably shown here that he textually took up the distinction between flow and motion present in Proclus' commentary on the third postulate. Yet, this did not mean that he followed Proclus in attributing to the flow of the point a metaphysical and non-spatial meaning, which the motion of the point in the imagination would aim to reproduce in a spatial manner. Rather, Commandino considered the imagination as a "drawing board" through which the geometer may study the properties of geometrical objects after having abstracted them from the consideration of physical magnitudes. As said, he used the terminological disjunction proposed by Proclus between ῥύσις and κίνησις (or between *fluxus* and *motus*) as a distinction between an intellectual or logical apprehension of the object of the definition and its actual deployment by the geometer in the context of constructions.

* * *

Thus, within these commentaries, two main positions were held concerning the ontological status of geometrical motion and the admissibility of genetic definitions in geometry. While Fine, Billingsley and Clavius focused on the difference (and connection) between geometrical and physical motion, Peletier and Foix-Candale focused rather on the difference (and connection) between the generations of geometrical objects as apprehended by the mathematician and the true geneses of geometrical objects that transcends human apprehension.

The positions held by Dee and Commandino may be placed somewhere between these two main standpoints. Dee asserted indeed the divine origin of geometrical objects, but related their generations to the physical world as well as to the metaphysical realm. Commandino, as for him, did not assert the divine nature of mathematical objects and pointed to their separation from physical matter. But, at the same time, he took up Proclus' distinction between the non-spatial procession of the suprasensible figures and the spatial generation of geometrical objects and reinterpreted it into a distinction between the purely logical apprehension of geometrical concepts and their unravelling in the geometer's imagination.

Now, concerning the admissibility of motion as means of definition, Fine, Billingsley and Clavius all considered genetic definitions as fully legitimate in geometry. This is not only determined by the extent to which they used genetic definitions in their commentaries on the *Elements*, but also by the fact that they explicitly placed motion among the objects of the geometer, at least in the cases of Fine and Billingsley. In this respect, Dee and Commandino presented a rather similar attitude, even if they did not make an extensive use of genetic definitions, which is understandable in Dee's case, as he did not himself publish a commentary on the *Elements*. In fact, Dee presented motion as fundamental to the constitution of quantity and related it to all the types of cause at play in the universe. In Commandino's case, motion was not only presented as relevant to define geometrical objects, but as necessary for the geometer to deploy geometrical notions spatially, playing therefore a foundational role with respect to the geometer's constructions.

By contrast, Peletier and Foix-Candale emitted restrictions concerning the admissibility of genetic definitions in geometry, notably because these would not be able to truly express the proper geneses of geometrical objects, given the ontologically transcending nature of the latter. Genetic definitions would, in other words, only offer a conjectural representation of the true mode of causation of geometrical objects. But while, for Peletier, these would be admissible to constitute a humanly accessible science of geometry, for Foix-Candale, the type of genetic definitions commonly used by mathematicians, which present geometrical objects as derived from the translation or rotation of a point or an object of lower dimension, would be unsuited to a truly scientific approach to geometry. These would, for him, be too closely related to the mechanical means through which mathematicians concretely produce geometrical objects and which convey an empirical rather than a properly rational and mathematical mode of definition of these objects.

9.2.3 Genetic Definitions and Definitions by Property

The difference and relation between genetic definitions and definitions by property in geometry was most explicitly set forth by commentators when considering Euclid's definition of the sphere (Df. XI.14), which was often compared to the non-genetic definition of the sphere proposed in Theodosius' *Spherics*. This was the case in the commentaries of Foix-Candale and Billingsley, as well as in that of Clavius (though less directly), and allowed these commentators to shed light on the status and function they attributed to genetic definitions in general.

For Foix-Candale, the genetic definition of the sphere proposed by Euclid would be epistemically inferior to its definition by property, since it only presents its *descriptio*. To him, the *descriptio* would express an accidental or extrinsic mode of generation of the defined object, teaching the practical precepts according to which geometers construct the objects which are enunciated in the definitions by property. This would also apply to the definitions of the line, the surface and the solid as resulting from the translation of a point, a line and a surface, respectively. The type of genetic definition that would, on the other hand, express the essential mode of generation of a figure is one that would present the figure as caused by the expansion of its magnitude from its centre or from itself until it is stopped by its boundary or extremity, exhibiting thereby its spatial properties. This is what would express Theodosius' definition of the sphere, but also Foix-Candale's own genetic definitions of the line, the surface and the figure. Concerning the relation between these different types of definition, Foix-Candale thought that the kind of genetic definitions provided by Euclid in Book XI of the *Elements* express the practical precepts according to which geometers construct the objects which are enunciated by the definitions by property.

In Billingsley's commentary, the genetic definition or description of the sphere would state its proper cause and mode of production. The same would apply to the genetic definitions of the line, the surface and the solid in general, to which he explicitly related Euclid's definition of the sphere. Although a description would not, to Billingsley, state the essential properties of the figure and would not correspond to a definition in the strict sense, it would be truly relevant to the mathematician's knowledge of geometrical objects. It would indeed correspond to a "causal definition", in the sense that it would display the efficient cause through which geometrical objects are brought about. This notion of causal definition points to the epistemic importance that would be granted to genetic definitions in geometry in the following century. Pointing to the existence of a causal form of knowledge within mathematics, this notion of "causal definition" also resonated with the debate on the scientific status of mathematical demonstrations that was taking place at the time in Italy.

In Clavius' commentary on Euclid, the genetic definition of the sphere was not distinguished from a proper definition, but it was distinguished from a definition by property in its epistemic and pedagogical function. This difference of function was set forth most explicitly in his commentary on Sacrobosco's *Sphaera*, where Clavius designated the definition by property as a *praedicatio formalis* and the genetic definition, as a *praedicatio causalis* (evoking Billingsley's notion of causal definition). Clavius' commentary on the definitions of the line, the surface and the solid confirmed that "causal" and "formal" definitions have an equal epistemic relevance within geometry and complement each other. But while the latter would simply state the essential properties that define geometrical objects, the former would display the reason for the essential relation between geometrical objects and their properties. When represented by and within the imagination, the concept defined within the *praedicatio causalis* would indeed offer a first-hand experience of the essential connection that exists between the *definiendum* and its attributes.

Similar information was set forth by Fine and Peletier concerning the two definitions of the circle, which were analogous in this regard to the two definitions of the sphere by Euclid and Theodosius. While Fine tended to reserve the status of *definitio* to the definition by property, he did however (like Peletier and Clavius) consider the genetic definition as crucial to offer a proper understanding of the essential properties of the circle. More generally, Fine considered that genetic definitions had the function of unravelling the synthetic and constructive movement of geometrical knowledge, which starts from the indivisible point and ends with the most complex solids, enabling us to conceive a quasi-infinite number of different geometrical figures. This idea was also clearly set forth through Clavius' approach to genetic definitions in Book I and XI, and mostly in Df. I.4, where he showed that the fact of attributing motion to geometrical objects allows us to imagine the construction of any sort of lines, even those whose quantitative properties cannot be mathematically determined. Peletier also presented genetic definitions as displaying the synthesis of the three dimensions from the point, though this representation, given its conjectural status, only consisted, to him, in a representation that is conform to a humanly apprehensible constitution of geometry.

In Commandino's commentary, the distinction between definition by property and definition by genesis was only indirectly set forth. But the fact that he placed the genetic definition of the line among other possible definitions of the line, following Proclus' commentary on Euclid's *Elements*, tends to indicate that he did consider genetic definitions as proper definitions.

Thus, although the difference between definitions by property and genetic definitions was acknowledged by nearly all of the considered commentators and was interpreted along the same lines, they did not all attribute the same status and function to genetic definitions within the discourse of the geometer and in the constitution of geometry. This difference led certain authors (such as Fine and Foix-Candale) to dismiss genetic definitions as definitions in the proper sense. In one case, it was because motion was not part of the essential attributes of geometrical objects according to their definitions by property. In the other, it was on account of the type of motion commonly admitted in classical genetic definitions (such as Euclid's definition of the sphere). In the remaining cases (e.g. Billingsley or Clavius), genetic definitions were merely considered as a special type of definitions, and in particular, as causal definitions.

9.3   Geometrical Motion and the Mode of Composition of Magnitudes

9.3.1   The Composition of the Continuum

One of the issues that was raised in Antiquity concerning the validity of genetic definitions was the manner in which they expressed the mode of composition of magnitudes. Indeed, the notion of line as derived from the flow of a point raised the long-debated question of the possibility for a line to be caused by a point and of the manner in which this should be considered to take place, given that no aggregation of contiguous indivisible elements could be held to constitute a continuous quantity, according to Aristotle. In itself, this issue does not directly concern the legitimacy of the attribution of motion to geometrical objects, but touches on the logical possibility of defining the line as caused by the motion of point, and, by extension, to define the surface as caused by the motion of a line, and the solid as caused by the motion of a surface. It relates also, more generally, to the very nature of continuous quantity, which was held to encompass magnitude, but also time and motion itself. Hence, one of the functions that was attributed to genetic definitions in the commentaries analysed here was the elucidation of the mode of composition of continuous quantity or, in other words, the mereological relation between the point and the line or between magnitudes of immediately higher and lower dimensions.

With regard to this issue, Peletier and Billingsley followed ancient authors such as Eratosthenes and Theon of Smyrna in asserting that a line can both originate from a point and be properly continuous insofar as it is constituted by the *flow* or *motion* of the point and not by an addition or aggregation of indivisible points or units, as in the case of number. In other words, the fact that motion itself consists in a continuous quantity would make it possible for the moving point to cause a properly continuous line and, from there, any continuous magnitude.

By choosing to define the line, the surface and geometrical figures in general, as generated by the stretching or expansion of its own magnitude, Foix-Candale adopted an approach that is not dissimilar to that of Peletier and Billingsley. For it would solve the problems raised by the notion of magnitude as derived from an indivisible principle. Yet, contrary to them, he did so by avoiding, rather than by using, the notion of flow of the point, given that it is the magnitude of the line or of the surface that would then be spatially extending. Therefore, even if Foix-Candale did refer to the point as the origin of continuous quantity (in what may be considered as a philosophical rather than a mathematical statement), the way he described the mode of generation of geometrical objects allowed him to entirely leave aside the question of the mereological relation between geometrical objects of different dimensions.

John Dee, like Peletier and Billingsley, took into account the problems raised by the notion of line as derived from the point and also distinguished the relation of the line to the point and the relation of number to the unit by appealing to a kinematic undertanding of magnitude. Yet, his attitude toward this issue revealed to be quite different from that of his French predecessor and of his collaborator. In his *Mathematicall praeface*, Dee asserted that while number, or discrete quantity, is divisible into indivisible elements such as points, it is not constituted nor generated by points, since the arithmetical unit is not endowed with position and cannot be attributed any motion. Conversely, while magnitude does not consist of points, it is generated by the motion of a point. As a result, only magnitude may be held as derived from a process of generation or synthesis from an indivisible principle. Number, which would then correspond to the "Number Numbred" present in the human soul, would not have any generative power but only be subjectable to an analysis into its constitutive elements.

Fine's approach to this issue consisted in defining the generative flow of the point as a process of infinite multiplication of the point by itself, which evoked certain fourteenth-century indivisibilist interpretations of the continuum. By assuming thereby that the line may be conceived as composed of an infinite number of partless elements, Fine could account for the infinite divisibility of the line while hinting at the connection between the respective constitutions of continuous and discrete quantities. Such as presented in his *Arithmetica practica*, this could be justified at a higher level by the numerical order of the divine Creation. It could thus offer a legitimation, on a philosophical level, for the arithmetical interpretation of certain definitions and propositions found in Euclid's geo-metrical books, in particular within Book II.

Hence, the specific issue of the role of genetic definitions in the elucidation of the composition of magnitude may offer us a certain representation of the transition from the premodern to the early modern era in the history of mathematics. On one hand, Peletier, Dee and Billingsley clearly insisted on the distinction between the compositions of discrete and continuous quantity, following Aristotle's doctrine of quantity. The same attitude was set forth in a mathematical context by Foix-Candale. And Fine's representation of the flow of the point in terms of multiplication, even if more modern than the latter position by certain aspects, pointed to the middle ages by evoking the fourteenth-century debates between divisibilists and indivisibilists. On the other hand, the fact that Commandino and Clavius did not address this issue in their commentary on Euclid's first definitions would hint at the progressive erosion of the distinction between discrete and continuous quantity that culminated in the early modern era with the seventeenth-century developments of analytic geometry and infinitesimal calculus.

9.3.2   Genetic Definitions and the Relation Between Arithmetic and Geometry

Practically all the commentators considered here added a genetic definition of the rectan-gular parallelogram, or at least offered a kinematic expression of the mode of generation of this figure, in their commentary on Df. II.1. This aimed to explain what Euclid meant when he stated that rectangular parallelograms are contained by the two straight lines surrounding their right angle and how this determined the quantitative relation between the sides and the area of this figure. By assuming that the rectangular parallelogram is generated through the transversal motion of one of its sides along its adjacent side while remaining at right angles, Euclid's commentators were able to account for the quantitative relation between the two sides surrounding any right angle of the parallelogram and the area they enclose. In most cases (i.e. in Fine, Peletier, Billingsley and Clavius), the explanation of this definition was the occasion to show the correspondence between the generation of the rectangular parallelogram through motion and the multiplication of two numbers in arithmetic, both implying a property of commutativity. In Foix-Candale and Commandino, this correspondence was indirectly set forth through the notion of measure-ment. The rectangular area was then understood as measured by the motion of one of its sides along the other. In Commandino, the notion of measurement was also suggested by a diagram of a rectangle subdivided into square-units and by a reference to Regiomontanus' *De triangulis omnimodis*.

In Peletier and Billingsley's cases, the comparison between the generation of the rectangular parallelogram through the transversal motion of a line and the multiplication of numbers was mainly presented as a didactic tool to help the reader understand the meaning of Euclid's Df. II.1. In Billingsley, this comparison also aimed to justify the arithmetical interpretation of Euclid's second book, since, as he expressed it, the properties and generations of geometrical figures may be useful to understand certain properties of numbers, and of arithmetical and algebraic operations.

Fine and Clavius both showed the connection between the arithmetical process of multiplication and the generation of magnitudes through motion by their common designation as *ductus*. Indeed, the term *ductus* was regularly used in premodern arithmetic to designate the operation of multiplying a number by another (or by itself). Moreover, the relation between the geometrical and arithmetical generations of quantities was visually expressed in the commentaries of Fine and Clavius through the illustration of a rectangular figurate number, as it was also the case in Billingsley's commentary. In Clavius' case, the relation between rectangular numbers and rectangular surfaces was explicitly said to reveal the affinity between numbers and magnitudes.

In Dee's *Mathematicall praeface*, the connection between numbers and magnitudes relied both on the arithmetical foundation of the divine Creation, as in Fine's *Arithmetica practica*, and on the treatment of numbers within practical and applied mathematics, where the unit is taken as divisible and where numbers take on the properties of magnitudes, as spatial and dimensional objects.

Therefore, although certain sixteenth-century commentators of Euclid such as Peletier and Billingsley used motion to distinguish the mode of generation of magnitudes from that of numbers when considering the relation of the indivisible point to the line in their commentary on Book I, they most often did not hesitate to use motion to compare and connect magnitudes (mainly represented by quadrilaterals) and numbers when dealing with Book II. This would be because, even without assuming the actual homogeneity between discrete and continuous quantities, these authors admitted that the moving line has, to the surface it generates, a quantitative relation which the flowing point may never have to the line, since the point is not itself a quantity.

Looking at Peletier's case in particular, it appears that although he clearly distinguished, in his commentary on the *Elements*, the modes of composition of numbers and of magnitudes on a mathematical level, he nevertheless asserted, in his scientific poetry, the coincidence of numbers and magnitudes and of their respective modes of generation on a metaphysical level. As he explained in the commentary on Euclid's Df. I.16, our comprehension of the relationship between different types of geometrical objects is restricted by our inadequate comprehension of the properties and geneses of mathematical objects given their divine origin and condition. The same would go for our comprehension of the true relation between numbers and magnitudes, which is revealed by our ignorance of the exact value of certain ratios and by the need for the mathematician to reduce the parts of lines to lines (and so on for the surface and the solid) in order to avoid confusion.[4]

---

[4]Peletier (1557, p. 2), Df. I.1: "Sed quia Magnitudinum partes, naturam totius denominationemque retinent, partes enim Linearum, lineae sunt: Superficierum, superficies: & Corporum, corpora: *alioqui vaga & confusa esset rerum substantia.*" (My emphasis.) See *supra* n. 81, p. 96.

Hence, while nearly all the considered authors compared the mode of generation of magnitudes to that of numbers at the beginning of their commentary on Euclid's Book II, they did not attribute the same degree of legitimacy to this comparison. Peletier and Billingsley, who clearly distinguished the constitution of the line by the flow of a point from the constitution of number by addition or multiplication of units, and would as such have rejected the notion of surface as constituted by the aggregation of lines, both presented this comparison as a useful tool to understand the definitions and propositions of Book II. This allowed them to legitimately appeal to numerical examples to explain certain properties of magnitudes without asserting the homogeneity of discrete and continuous quantities in a mathematical context (though it existed for Peletier on a metaphysical level). A similar attitude is to be found in Dee's *Mathematicall praeface*, insofar as he distinguished the modes of composition of numbers and magnitude in theory, but allowed their comparability in a practical context.

Foix-Candale and Commandino, as for them, only suggested the link between the generation of the parallelogram and the multiplication of two numbers in an indirect manner. These authors therefore seem to have found useful to show the connection between the compositions of numbers and magnitudes as a practical or pedagogical tool, without having to openly challenge the separation Euclid established between arithmetic and geometry in the *Elements*. Fine was also part of those authors who asserted the separation between numbers and magnitudes in principle, but his position on this issue is less clear insofar as he also compared the flow of the point to the multiplication of the point and connected arithmetic and geometry through their common use of the term *ductus*, as did Clavius much later. Clavius clearly differed from his predecessors in the way he openly asserted the connection or *affinity* between numbers and magnitudes. In this framework, he not only left aside the traditional distinction between the compositions of discrete and continuous quantity, but also clearly displayed the correspondence between abstract geometrical figures, geometrical numbers and dimensioned geometrical figures.

9.4 A Changing Approach to Motion and to Genetic Definitions from Fine to Clavius

The commentaries on the *Elements* which I have investigated here, and which were selected because they contain a significant number of additional genetic definitions, do not allow us to discern a linear evolution or progression in the approach to motion as a means of defining geometrical objects within the sixteenth-century Euclidean tradition. Yet, their analysis, beyond the great diversity of positions they presented concerning the status and uses of geometrical motion, did reveal certain general tendencies.

9.4.1 Changes in the Epistemological Status of Genetic Definitions

Among the features that were the most stable throughout the considered period, we may retain that the attribution of motion to geometrical objects in the context of genetic definitions was practically always considered as mathematically relevant, even if issues were raised concerning the ontological status of this motion and the function of genetic definitions within geometry. As it happens, the categorisation of genetic definitions as *descriptiones*, by opposition to definitions in the proper sense, rarely contributed to reduce their relevance for the study of geometrical objects. When it did, as in Foix-Candale's commentary, these were replaced by other types of genetic definitions, which would be more proper to convey the true essences of the defined objects.

Where a distinct change may be discerned within this tradition is in the role given to genetic definitions in the constitution of geometrical knowledge and of its place in the teaching of geometry. Certainly, Fine, in the 1530s, already attributed a strong place to genetic definitions in the comprehension of the properties and relations of geometrical objects. Yet, this role was emphasised and extended by Billingsley and more so by Clavius, in comparison with their immediate predecessors. They presented indeed genetic definitions as exhibiting the causes of geometrical objects at a time when the ability of mathematicians to propose a properly causal knowledge of their objects was being challenged.

The importance Fine, Clavius, but also Billingsley, attributed to genetic definitions in geometry, in addition to their constructive conception of geometry, was also related to the primarily pedagogical aim of their commentaries, which would have incited them to give a key role to notions that facilitate the intuition of geometrical notions for students or readers unaccustomed to the abstractness of ancient Greek geometrical concepts. In this regard, what was made clear by Clavius, more conspicuously than in Fine and Billingsley, is that genetic definitions, and their deployment in the imagination, would allow students to obtain a first-hand experience of the causal relation between geometrical objects and their essential properties.

9.4.2 Changes in the Ontological Status of Motion within Genetic Definitions

From Fine to Clavius, a significant change is to be found in the attitude of commentators concerning the ontological status of geometrical motion. In particular, the need for a clear distinction between physical and mathematical motion, which was mainly due to the dominance of Aristotelian ontology in late medieval and early Renaissance scientific discourse, was disappearing toward the end of the sixteenth century. Indeed, Billingsley and Clavius, contrary to Fine, did not question the legitimacy for geometrical motion to be properly designated as "motion", even if they recognised the ontological difference between physical and mathematical objects. At most did they feel the need to state the imaginary character of geometrical motion. But, for both authors, this would be as much a means of clarifying the ontological status of geometrical motion as it would be of emphasising the fact that this motion is not subjected to the imperfection of physical matter, for example to distinguish abstract lines or figures from those that are concretely produced by instrumental means.

Thus, the notion that the generative motion attributed to geometrical objects, especially the flow of the point, expressed an ancient metaphysical conception, which presented the world as flowing or emanating from a single indivisible divine principle, also tended to disappear. If Fine only referred to the Pythagorean understanding of the generation of magnitudes in indirect manner, it was clearly admitted by Peletier, Foix-Candale and Dee. However, it was totally absent from the commentaries of Billingsley and Clavius, as well as from the commentary of Commandino, whose quotation of Proclus did not imply the admission of Neoplatonic conceptions concerning the ontological status of geometrical objects. What we saw appearing instead toward the end of the sixteenth century, especially in the commentaries of Billingsley and Clavius, is a stronger emphasis on the relation and homology between the generative motions of geometrical objects and instrumental processes.

Indeed, if Fine, just as Billingsley and Clavius, related genetic definitions to the geometer's constructions, notably through the connections he made between the genetic definitions of the straight line and of the circle and the first three postulates, he did not present these as idealised instrumental procedures. Through Billingsley's use of the notion of description, genetic definitions also came to be implicitly connected with the engraving of the diagrams that accompanied Euclid's text within its various editions. This approach was furthermore supported by Dee in his annotations to Billingsley's commentary on Euclid, as he explained how to interpret the geometer's diagrams and asserted at this occasion the importance of practical and applied mathematics for theoretical mathematics.[5]

To a certain extent, the connection that was made in this context between genetic definitions and the concrete production of geometrical objects hints at the shift that was progressively taking place from the sixteenth to the seventeenth century in the respective functions and status of the point and of the line in the pictorial arts. As was shown by C. Fowler,[6] the line replaced the point as the beginning and foundation of artistic practice, which followed the stronger emphasis placed in this context on bodies in motion, or bodies as viewed from a variety of perspectives, rather than on bodies considered as a set of spatially fixed points. In the treatises on pictorial representation of Leon-Battista Alberti (1404–1472), Albrecht Dürer (1471–1528), as well as Piero della Francesca (1415–1492), the point corresponded to the principle and limit by which the space and proportions of viewed bodies were both demarcated and measured. But it later became valued as the

---

[5]Dee (Billingsley 1570, p. 386v), Prop. XII.18. See also *supra*, p. 174–175.
[6]Fowler (2017).

generative principle of the line, the line being then conceived as the proper beginning and limit of bodies in motion, and as a representation of the mobile point of view of the artist.

In a geometrical context, this position also anticipated the seventeenth-century approach to geometry and to geometrical objects as expressing and as represented by instrumental processes, an approach which was most canonically developed by Descartes, as we will see further.[7]

At any rate, the approach adopted by Foix-Candale (that is, his avoidance of the association between genetic definitions and constructive procedures and, as a general rule, his dismissal of the comparison between instrumental and geometrical procedures) was progressively abandoned.

9.4.3 Changes in the Notion of Magnitude

When considering the way genetic definitions were used to express the mode of composition of magnitudes, it appeared that the ancient philosophical issues raised by the generation of the line from the point tended to be given less importance toward the end of the sixteenth century. This is shown by the cases of Commandino and Clavius, who did not tackle the question in their commentaries on Euclid's definitions. This attitude actually contrasted with the positions of their predecessors, from Fine to Dee, whether these presented a position closer to indivisibilism (as Fine) or to continuism (as Peletier, Foix-Candale and Billingsley, as well as Dee, in a slightly different manner).

The fact that the issue of the composition of the continuum was becoming less relevant toward the end of the considered period paralleled the progressive tendency to assert the connection between numbers and magnitudes, as shown most clearly in the commentary of Clavius. As such, if most of the considered authors, when commenting on Euclid's Df. II.1, directly compared the generation of rectangular parallelograms by the transversal motion of a line-segment with the generation of numbers by multiplication (with the exception of Foix-Candale and Commandino), with Clavius, the *affinity* between the arithmetical and geometrical operations is asserted without restrictions. His attitude in this regard demonstrated a greater freedom in the treatment of the concept of quantity, eroding the separation Euclid had established between numbers and magnitudes in the *Elements*. By contrast, his predecessors generally appealed to this comparison in a more cautious manner, by stating its pedagogical function or by assuming this connection as effective and conform to truth only on a metaphysical level. Among these, however, Billingsley joined Clavius in openly asserting that geometry is crucial to the comprehension of certain properties of arithmetical and algebraic operations and in demonstrating this through an arithmetical treatment of Euclid's book.

* * *

---

[7] See *infra*, p. 251 sq.

Hence, throughout the sixteenth-century Euclidean tradition, the motion that was referred to in the context of geometrical definitions progressively freed itself from its metaphysical understanding and was more directly connected to mechanical and physical processes. In this context, it is likely that the reading of Proclus' *Commentary on the first book of the Elements*, which was widely circulated in the later part of the sixteenth century and which legitimated the use of genetic definitions in geometry, enabled interpretors of Euclid to feel more entitled to introduce genetic definitions in their commentary on the *Elements*. The development and extensive circulation of practical geometry in print throughout Europe over the century, to which commentators of Euclid such as Fine, Peletier and Clavius contributed,[8] may also have played a part in this process. Indeed, early modern practical geometry treatises generally defended a hands-on approach to geometry, in which instruments held a central place, and sometimes reinterpreted some of Euclid's geometrical principles and propositions according to this perspective.[9] This practical approach to geometry, and to Euclidean geometry in particular, would certainly have helped justify the direct comparison between the abstract derivation of geometrical objects through the motion of points, lines and surfaces, on the one hand, and instrumental processes, on the other, within sixteenth-century commentaries on the *Elements*. Furthermore, the measuring procedures taught in medieval and Renaissance practical geometry treatises would also have contributed to legitimate the arithmetical treatment of magnitudes. In effect, lines, plane and solid figures were then directly interpreted as lengths, areas and volumes and attributed thereby a numerical expression.

This evolution points to the developments that would take place in seventeenth-century geometry and epistemology of mathematics, within which motion was to be regarded as an essential instrument for the definition and study of geometrical objects and helped legitimate the application of geometry to the investigation of mechanical and physical phenomena. In particular, the fact that the interpretation of geometrical motion was to be disengaged from ancient philosophical conceptions—both those that ontologically distinguished geometrical motion from the motion of physical substances and those that related it to a non-spatial and transcendent process—would have given mathematicians more freedom when appealing to motion in their scientific and pedagogical practices, as well as in their reinterpretation of ancient mathematics. This left more place for progress in the investigation of the properties of figures and of their relationship with numbers, on the one hand, and with physical quantities and processes, on the other. Through this transformation, motion would not only become an inherent part of the definition and treatment of geometrical objects, playing a foundational role in the investigation of their properties and relations, but would also help bind together theory and practice, as well as rational and empirical geometrical knowledge. It would also contribute to connecting geometry with arithmetic and with the physico-mechanical sciences, as well as to adapting the textual canon of Euclid's *Elements* to new pedagogical and scientific developments.

---

[8] Fine (1532, fol. 50r sq.); Peletier (1572), Peletier (1573) and Clavius (1604).

[9] These aspects are notably visible in the practical geometry of Fine, Peletier and Clavius.

Later Developments in the Seventeenth Century: A Cartesian Epilogue

10

I have, at several occasions in the course of this study, pointed out the importance that was attributed to motion and to the kinematic treatment of geometrical objects in seventeenth-century geometry, in comparison with the place it was given in premodern geometry.[1] Although an investigation of these seventeenth-century developments exceeds the scope of this work, notably as such an investigation has already been carried out in a number of studies,[2] it will be useful to outline here certain aspects of this later treatment of the question and the way it relates (directly or indirectly) to the discussions which have been analysed in the last eight chapters. Rather than offering a summary of the various ways in which the question was dealt with over this later period, which would be redundant with the brief survey of notable treatments that was proposed in the introduction, I have chosen to focus on one particular author, René Descartes, by offering a brief consideration of his treatment of geometrical motion in his mathematical and philosophical thought.

Descartes represents indeed a figure of crucial importance when considering the aftermath of the sixteenth-century conceptions and uses of motion and genetic definitions in Euclidean geometry. He provided a redefinition of classical geometry that was centered on the generation of lines through motion, especially by instrumental means, and that connected furthermore a kinematic and an arithmetical treatment of geometrical objects. Moreover, Descartes, who was born in 1596 and who died in 1650, is situated at the junction between two periods in the Western histories of mathematics and of philosophy, canonically instantiating the transition between the premodern and the early modern era in both domains. Moreover, although it is difficult to assess with certainty the influences of

---

[1] See *supra*, pp. 5–7, 219 and 248–249.

[2] A list of references to relevant studies is provided *supra*, n. 24–35, pp. 5–7.

Descartes' earlier mathematical thought, his studies at the Jesuit Collège de la Flèche from about 1607 to 1615[3] would have provided him access to the sixteenth-century Euclidean tradition through his reading of Clavius' commentary on the *Elements*,[4] in which the Jesuit professor compiled the most notable contributions of his predecessors. Descartes also knew Clavius' *Geometria practica*[5] and his *Algebra*,[6] respectively published in 1604 and 1608. C. Sasaki has notably shown that Clavius' *Algebra* exerted a certain influence, albeit indirectly, on the formation of Descartes' mathematical thought.[7]

Without making here any claims, or at least any definite claims, regarding the influence of Clavius or of any of the investigated authors on the young Descartes,[8] it is relevant to look at the way Descartes dealt with certain issues considered within the tradition examined here and which were central to his own contributions to mathematics, to philosophy and to the science of nature. Such issues concern the relation between genetic definitions and instrumental processes, the part played by the kinematic understanding of magnitudes in connecting discrete and continuous quantity, and the role of genetic definitions in the reconstitution of the divine creation of the universe. The analysis of these questions, which only brushes over certain conceptual convergences and divergences between Descartes and the authors considered here as an invitation to study the matter in greater depth, aims to show how the French mathematician and philosopher contributed to expand the role and importance of the kinematic treatment of magnitudes in geometry and was able to offer thereby an ontological, epistemological as well as metaphysical foundation to the mathematisation of nature.

10.1   Genetic Definitions and Instrumental Processes

We have seen here that for certain sixteenth-century commentators of Euclid, notably for Billingsley and most of all Clavius, the motion entailed by genetic definitions was connected to instrumental processes and came to be interpreted as idealised instrumental

---

[3] Sasaki (2003, chap. 1). On mathematical education in Jesuit colleges during the early modern era, see Romano (1999, Part 1).

[4] Costabel (1983) and Sasaki (2003, pp. 2–3 and 45–48).

[5] Clavius (1604). On Descartes' knowledge of this work, see Gilson (1976, pp. 181–183). H. Bos (in Bos 1990) has presented some of the connections between Descartes' geometry and Clavius' *Geometria practica*.

[6] Clavius (1608). On Descartes' knowledge of this work, see Sasaki (2003, pp. 47–48 and 72–93). See also Gilson (1976, pp. 193–194) and *infra*, p. 261.

[7] Sasaki (2003, pp. 47–48 and 72–93).

[8] To make such claims would require a separate analysis, which exceeds the scope of the research I have undertaken for this study. In any case, the fact that Descartes read some of Clavius' mathematical works, notably his commentary on Euclid and his *Algebra* (Sasaki 2003, pp. 45–48) is not sufficient *per se* to demonstrate that Descartes drew on Clavius' interpretation of Euclidean geometry.

procedures. However, these authors, along with Oronce Fine, clearly asserted the distinction between geometrical and material objects, and between their respective modes of generation on an ontological level.

In his earlier programme for the transformation of geometry,[9] as well as in his 1637 *Géométrie*,[10] Descartes not only admitted a kinematic approach to geometrical objects,[11] and to curves in particular, but he made no clear distinction between instrumental and abstract generations of curves on an ontological and epistemological level in the context of his geometry, regarding the latter as idealisations of the former.[12] For Descartes, a curve was regarded as properly geometrical when it was precise, exact and could therefore be rendered intelligible through rational means. And this would not depend, for him, on the intrinsically abstract nature of the motion through which it is generated, but rather on this generative motion's ability to be quantitatively determined or measured with accuracy, in other words, mathematically known, as are the circular and rectilinear motions generable by the means of the compass and straightedge, respectively.[13] Such criteria would also be valid for curves that resulted from a plurality of motions as long as the ratio between the different motions could be accurately measured or mathematically determined, as are the curves produced by linkage instruments,[14] which are subordinated to one simple motion.[15]

---

[9]That is, the programme he outlined for geometry in 1619 within his letter to Isaac Beeckman (1588–1637) from 26 March 1619 (Descartes, AT 10, 1986a, pp. 154–160), in which he sets forth the main precepts of an art through which all problems dealing with quantity (discrete and continuous) may be solved. On this programme and the role Beeckman played in its earlier development, see Shea (1991, pp. 9–13 and 35–67). On Descartes' earlier mathematical thought, and on his mathematical knowledge and research before 1637, see also Costabel (1983) and Rabouin (2010) and Rabouin (2018).

[10]Descartes (1637) and (transl. Smith 1954). On the structure, aim and specificities of Descartes' *Géométrie*, see Bos (1990). D. Rabouin (Rabouin 2018) has shown that it is far from obvious that a direct continuity may be established between the programme outlined in the 1619 letter to Isaac Beeckman and the program developed in the *Géométrie* with respect to the mathematical knowledge and practice on which they rely.

[11]On the place of motion in Descartes' geometrical and philosophical thought in general, see, for instance, Nikulin (2002, in part. pp. 210–234) and Domski (2009).

[12]Bos (1981, p. 320) and Bos (2001, pp. 338–339).

[13]Descartes (1637, p. 316) and (Smith 1954, p. 43): "it seems very clear to me that if we make the usual assumption that geometry is precise and exact, while mechanics is not; and if we think of geometry as the science which furnishes a general knowledge of the measurement of all bodies, then we have no more right to exclude the more complex curves than the simpler ones, provided they can be conceived of as described by a continuous motion or by several successive motions, each motion being completely determined by those which precede; for in this way an exact knowledge of the magnitude of each is always obtainable."

[14]Descartes himself designed such instruments. On these instruments, see Serfati (1993), Bos (2001, pp. 237–245) and Mancosu (1996, pp. 72–73).

[15]Descartes (AT 10, 1986a, p. 157): "For continuous quantity I hope to prove that, similarly, certain problems can be solved by using only straight or circular lines, that some problems require other

Through Descartes' algebraic treatment of magnitudes in his 1637 *Géométrie* and the resulting classification of curves, geometrical curves also came to be characterised by the fact that they may be expressed by algebraic equations.[16]

By opposition, the curves that would result from two distinct motions that do not have a determined and measurable ratio to each other and which may not be expressed by algebraic equations, such as the quadratrix or the spiral,[17] were considered irrational and unintelligible, lacking the precision and exactness required in mathematics.[18] Such curves were therefore regarded as pertaining *only* to the domain of mechanics, wherefore they were designated as mechanical curves.[19] The designation of such curves as "mechanical" was grounded in the previously mentioned discourse held by Pappus, in the *Mathematical collection*, concerning the mode of generation of the quadratrix.[20]

The distinction between geometrical and mechanical curves was central to Descartes' geometry.[21] It established the boundaries of geometry and the criteria for the characterisation, and thereby for the definition, of curves, which then conveyed the properties of continuous magnitudes in general.[22] In this context, it is the motion by which a curve was conceived to be generated, and so its "specification by genesis" rather

---

curves for their solution, but still curves which arise from one single motion and which therefore can be traced by the new compasses, which I consider to be no less certain and geometrical than the usual compasses by which circles are traced". This translation is drawn from Bos (2001, p. 232).

[16] Descartes (1637, p. 319) and (Smith 1954, p. 48): "I think the best way to group together all such curves and then classify them in order, is by recognizing the fact that all points of those curves which we may call 'geometric', that is, those which admit of precise and exact measurement, must bear a definite relation to all points of a straight line, and that this relation must be expressed by means of a single equation."

[17] On the generation of these curves, see pp. 24 and 217–219.

[18] Descartes (1637, p. 317) and (Smith 1954, p. 44): "the spiral, the quadratrix, and similar curves [...] really do belong only to mechanics, and are not among those curves that I think should be included here, since they must be conceived of as described by two separate movements whose relation does not admit of exact determination."

[19] In Descartes' above-mentioned letter to Beeckman, such curves were also called "imaginary" (*imaginariae*). Descartes (AT 10, 1986a, p. 157), transl. Bos (2001, p. 232): "and, finally, that other problems can only be solved by curved lines generated by separate motions not subordinate to one another; *certainly such curves are imaginary only*; the well known quadratrix line is of that kind." (My emphasis.)

[20] As was shown earlier, Clavius dealt with the quadratrix in the second edition of his commentary on Euclid and in his *Geometria practica*. See *supra*, pp. 211 and 218.

[21] On Descartes' classification of curves, see Molland (1976), Bos (1981), Bos (1988, pp. 15–19), Bos (1990), Bos (1996), Bos (2001, pp. 335–342 and 355–359), Mancosu (1996, pp. 71–82) and Mancosu (2008). As shown by R. Rashed (in Rashed 2005), earlier classification of curves according to the motions used to generate them are to be found in the geometrical work of medieval Arabic mathematicians, such as al-Sijzi.

[22] Jullien (1996, pp. 26–32 and 36–37) and Nikulin (2002, pp. 228–229).

than its "specification by property"[23] (in which consists the algebraic expression of curves), that was held as most fundamental and most relevant to define the curve.[24] Hence, as was asserted by H. Bos,[25] while Descartes made the kinematic understanding of geometrical objects a foundational element of his geometry, corresponding to a proper means of definition or representation, the algebraic expression of curves only represented a tool to classify them,[26] even when he gave a larger place to algebra in his 1637 *Géométrie* in comparison with his earlier geometrical work.[27]

Furthermore, as curves came to be classified, and therefore defined, according to the motion by which they are generated,[28] the curve and its generative motion tended to be presented as conceptually identical, at least as inextricably connected. For the properties of the motion by which a curve is described would apply to the curve itself, and vice versa. Moreover, as Bos pointed out, knowing the means by which a curve is traced would be

---

[23] I take up here the terminology used in Molland (1976).

[24] Molland (1976, p. 23).

[25] Bos (1981, p. 331).

[26] This is the interpretation defended in Molland (1976, p. 42), Bos (1981, pp. 322–324 and 331), Bos (1988, pp. 16–17), Bos (1990), Serfati (1993), Domski (2009) and Crippa (2017). This position was challenged by E. Giusti (in Giusti 1987, p. 429), who considered, on the contrary, that the identification of the curve with its algebraic expression was central to Descartes' geometry and that the kinematic processes through which they are constructed only played a secondary role. Against this conception, M. Domski (Domski 2009) showed that the kinematic construction and characterisation of curves was central not only to Descartes' geometrical work, but also to his philosophical doctrine, eventually replacing the concept of method which previously connected the different aspects of his earlier scientific and philosophical thought. Moreover, as noted by Bos (Bos 1981, pp. 323–324) and Mancosu (Mancosu, 1996, p. 73), the fact for all algebraic equations to define a geometrical curve was only implicitly assumed by Descartes. V. Jullien (Jullien 1996, pp. 51–67), who discussed these different interpretations in detail, showed that, while the kinematic (and properly geometrical) mode of specification of curves allows us to know these for what they are, that is, geometrical objects, the algebraic mode of specification enables us to set these in order (see *infra*, n. 28, p. 255), thereby asserting the interdependence of these two modes of apprehension of curves in Descartes' *Géométrie*.

[27] On the development of Descartes' geometrical programme and on his approach to algebra at the various stages of its development, see Bos (2001, pp. 285–287 and 352–354), Serfati (1993), Guicciardini (2009, pp. 38–47) and Rabouin (2010).

[28] Within the genre of geometrical curves however, the various classes of curves were distinguished according to the degree of simplicity of their equations: Descartes (1637, p. 319) and (Smith 1954, p. 48): "If this equation contains no term of higher degree than the rectangle of two unknown quantities, or the square of one, the curve belongs to the first and simplest class, which contains only the circle, the parabola, the hyperbola, and the ellipse: but when the equation contains one or more terms of the third or fourth degree in one or both of the two unknown quantities (for it requires two unknown quantities to express the relation between two points) the curve belongs to the second class; and if the equation contains a term of the fifth or sixth degree in either or both of the unknown quantities the curve belongs to the third class, and so on indefinitely."

equivalent, for Descartes, to knowing the curve itself.[29] Since Descartes placed instrumental processes and abstract generations of geometrical objects on a comparable level (epistemologically and ontologically), the latter corresponding to idealisations of the former, the fact of knowing the instrumental process by which a curve is traced would then allow us to know the properties of the curve itself. Thus, in addition to bringing together abstract and instrumental generations of geometrical objects, Descartes promoted a stricter identification between geometrical objects themselves and the processes through which they are generated, an identification already underlyingly present in Clavius' commentary on the *Elements*, as shown in particular by his commentary on Euclid's Df. I.4.[30]

But while sixteenth-century commentators of Euclid mostly presented genetic definitions (whether or not they were related to instrumental processes) as complementary to, yet as distinct from, definitions by property (which represented the primary and most valid mode of definition of geometrical objects), Descartes saw the kinematic characterisation of curves as the most relevant means of defining a curve and of determining its essential properties. It is indeed the quality of the motion that would generate a curve that allowed it to be designated as "geometrical" or "mechanical".

It remains that, as we have seen, some of these sixteenth-century commentators of Euclid placed great emphasis on genetic definitions in geometry and asserted moreover their properly mathematical character. Some of these, and Clavius in particular, also explicitly pointed to the relation between genetic definitions of geometrical objects and constructive instrumental procedures. This would have certainly contributed to provide an epistemic context that was favourable to the development of Descartes' approach to geometry.

10.2   Geometrical and Mechanical Processes

It may be recalled here that the distinction between "geometrical" and "mechanical" in the context of geometry had a long history from Pappus to Descartes and, within the tradition investigated here, was brought forth by Peletier,[31] Foix-Candale and Clavius when discussing the status of geometrical superposition in Euclid's *Elements*.[32] I have already

---

[29] Descartes (1637, p. 307) and Descartes (Smith 1954, p. 22): "it is required to *discover and trace* the curve" ("il est aussy requis de *connoistre, &* de *tracer* la ligne") and Descartes (1637, p. 319) and (Smith 1954, p. 48): "I could give here several other ways of *tracing and conceiving* a series of curved lines" ("Je pourrois mettre icy plusieurs autres moyens pour *tracer & conçevoir* des lignes courbes"). (My emphasis.) Bos (1981, pp. 307–308); Serfati (1993) and Jullien (1996, p. 49).

[30] See *supra*, pp. 198–199.

[31] Although this is not properly relevant here, it may be noted that Descartes knew and took up certain notational elements from Peletier's algebra (Peletier 1554). On this issue, see Costabel (1983).

[32] See *supra*, n. 10, p. 76.

mentioned that the motion supposedly involved in geometrical superposition differs in its function and status from that implied by the generation of lines and figures, since superposition, such as used in Euclid's *Elements*, aims to demonstrate the congruence of two geometrical objects and not to generate any magnitude. Yet, it may be interesting here to briefly compare the way "geometrical" was opposed to "mechanical" by these sixteenth-century authors in this context.

As I have shown elsewhere,[33] Peletier considered superposition as mechanical and rejected it as non-geometrical in view of the spatially and quantitatively indeterminate character of this motion. Such a motion, which is not legitimated by any constructive postulate, would not guarantee the equality of the moved lines and angles and would therefore not enable the mathematician to rationally demonstrate the congruence of the superposed figures. Foix-Candale, as for him, directly rejected this procedure as mechanical because of its allegedly instrumental character, an interpretation that was not made explicit in Peletier's argumentation against superposition.[34] Nevertheless, both Peletier and Foix-Candale considered that the superposition of figures (whether or not it was interpreted as instrumentally performed) allowed for an empirical rather than a rational, and thus properly mathematical, demonstration of the congruence of figures. Hence, "mechanical" ultimately meant for both authors "inexact" and "imprecise", which relates to the meaning given to this term by Descartes in the framework of his classification of curves. However, the fact that Foix-Candale considered superposition as mechanical because of its alleged instrumental character clearly distinguishes his interpretation of the term "mechanical" from that of Descartes, since the latter considered that all curves, including those that are geometrical, are traceable by instrumental means.

For that matter, another crucial difference between Descartes and Foix-Candale is the fact that the latter, contrary to the former, distinguished extrinsic and intrinsic generations of geometrical objects, that is, the motion by which the geometer is held to construct a given magnitude and the generative motion that can be considered as proper to it by essence. Peletier held in that regard a position similar to that of Foix-Candale, since he made a difference between the metaphysical and the mathematical contexts when considering the mode of generation and the ontological status of geometrical objects. Yet, he did not consider the extrinsic generation of geometrical objects as improper to define it in a geometrical context.

As for Clavius, who did not distinguish extrinsic and intrinsic generations of geometrical objects in his commentary on Euclid,[35] his understanding of the term "mechanical" was related to that of Descartes insofar as he partly interpreted it according to Pappus' discussion of the quadratrix.[36] Clavius agreed indeed with Pappus (and, through him,

---

[33] Axworthy (2018). See also *supra*, n. 24, p. 101 and n. 109, p. 220 for Foix-Candale and Clavius.

[34] See *supra* n. 24, p. 101.

[35] What Clavius did distinguish was the concrete and abstract production of lines and figures.

[36] On Clavius' treatment of the quadratrix, see Bos (2001, pp. 159–166). See also *supra*, pp. 217–219. On Descartes and the quadratrix, see Molland (1976); Mancosu (1996, pp. 74–77); Bos (1981) and

with Sporus) that the motion involved in the generation of the quadratrix cannot be measured accurately or be mathematically determined, wherefore he adopted a pointwise construction of this curve. He regarded this mode of construction as more accurate and more geometrical than its construction through continuous motion (at least in the earlier versions of his discourse on this issue).[37]

However, as was stated by Bos, Clavius' treatment of the issue implicitly revealed that he applied in theoretical geometry a notion of precision (and of imprecision) drawn from geometrical practice (i.e. the practical execution of geometrical constructions).[38] Descartes, on the other hand, considered neither modes of construction (pointwise or by motion) as properly geometrical in the case of the quadratrix. To him, pointwise constructions would in fact only be admissible in geometry insofar as they allow us to construct arbitrary points of the curve and to reconstitute thereby the entire curve and the continuous motion through which it is generated, which was not the case of the pointwise construction of the quadratrix.[39] Hence, as was shown by Molland and Bos, only a continuous motion was fundamentally regarded as a genuinely geometrical mode of construction of curves for Descartes.[40]

As for Clavius' discussion of geometrical superposition (which first appeared in the same edition of the *Elements* as his treatise on the quadratrix) and his interpretation of the term "mechanical" in this context, it suffices to note that he simply rejected Peletier's assertion of the mechanical character of this demonstrative procedure. Indeed, as he pointed out then, since the propositions in which superposition is used (Prop. I.4, I.8 and III.24) correspond to theorems and not to problems, this procedure would merely be used to rationally demonstrate the congruence of the two figures by hypothetically assuming their

---

Serfati (1993). H. Bos (in Bos 2001, p. 245, n. 29) wrote that Descartes likely learned about the quadratrix from Clavius. Descartes could have read about it either in Clavius' commentary on the *Elements* (from the second edition) or in his *Geometria practica*. The former would be confirmed by a letter to Marin Mersenne (1588–1648) from November 13, 1629 (Descartes, AT 1, 1987, pp. 70–71), transl. Mancosu (1996, 78): "[...] the helix [...] is a line that is not accepted in geometry any more than that which is called quadratrix, since the former can be used to square the circle and to divide the angle in all sorts of equal parts as precisely as the latter can, and has many other uses as you will be able to see in Clavius' commentary to Euclid's *Elements*." On this issue, see Mancosu (1996, p. 78) and Sasaki (2003, pp. 4 and 70–71).

[37] Bos (2001, pp. 163–165).

[38] Bos (2001, p. 166). However, it must be noted that Bos (in Bos 1990, p. 356) did not consider that Clavius, in the *Geometria practica*, expected the reader to actually perform the construction with the straightedge and compass, though he would expect it to be executed mentally.

[39] Vuillemin (1960, pp. 83–87), Molland (1976, pp. 41–42), Bos (1981, pp. 303 and 315–319) and Domski (2009). In this case, geometrical would not be understood by opposition to mechanical, where it would mean precise and exact, but rather by opposition to algebraic, where it is essentially related to the proper mode of generation of curves, and of geometrical objects in general, whether abstract or concrete. On the other modes of construction considered by Descartes, notably through strings, see Bos (1981) and Bos (2001, pp. 346–349).

[40] Molland (1976, pp. 41–42) and Bos (1981, pp. 303 and 331).

superposition, side to side and angle to angle. This assumption only consists in a statement of the equality and therefore of the congruence of their parts. No superposition, and so no motion of the compared figures, is required to be done *in fact* ("cum *re ipsa* translatio nulla facta sit"), as would the procedures involved in a geometrical construction (even if performed in the imagination).[41] However, Clavius did admit that, if superposition were used as a constructive procedure in the context of a problem, it would be improper to demonstrate the congruence of geometrical objects, as it would not be founded on any constructive postulate and would not be able to rationally demonstrate the congruence of figures, actually depending on empirical judgment. It would, in this case, not be admissible in geometry.[42] Thus, in both discussions, Clavius connected the term "mechanical" with inexactness, imprecision and lack of mathematical intelligibility in the context of constructions, which tends to convey the meaning Descartes later attributed to this term, though it will then be applied to a different set of objects, methods and conception of geometrical exactness.[43]

10.3 Geometrical Motion and the Connection Between Arithmetic and Geometry

In most of the sixteenth-century commentaries on the *Elements* which were analysed here, motion and genetic definitions played a role in connecting arithmetic and geometry when dealing with Df. II.1. In this framework, the transversal motion of the line generating a quadrilateral was nearly systematically compared to the arithmetical operation of multiplication. In doing so, some of these commentators asserted the existence of a more general connection between the generation of geometrical objects and arithmetical and algebraic operations. These hinted thereby at the possibility of interpreting magnitudes arithmetically and algebraically in the context of Euclid's geometry, challenging the separation between arithmetic and geometry that was inherent to the structure of the *Elements*. Clavius stood out again within this tradition by the extent to which he affirmed the connection, or "affinity" as he put it, between arithmetic and geometry.

---

[41] Clavius (1586, p. 343) and Clavius (1589, pp. 368–369), Prop. III.16. See also Axworthy (2018, 28–29).

[42] Clavius (1586, pp. 342–343) and Clavius (1589, p. 368), Prop. III.16: "Hic certe Peletarium iure carpere potuissem, si id mihi fuisset propositum, ut falso criminatur; maxime in eo, quòd eadem ratione usui fore existimavit superpositionem in demonstrandis problematibus, ac theorematibus. Nam non satis intellexisse videtur, quo pacto Geometrae superpositionem illam usurpent. Neque enim volunt, *re ipsa* faciendam esse figurarum superpositionem, (*hoc enim mechanicum quid esset*) sed cogitatione tantum, ac mente, quod opus est rationis atque intellectus." (My emphasis.)

[43] The history of this notion was retraced by H. Bos (in Bos 2001). A further assessment of this notion, in relation to the development of Cartesian geometry specifically, is provided in Panza (2011).

Yet, none of the authors considered here took the groundbreaking step taken by Descartes, when he devised his rules for applying arithmetical operations[44] to magnitudes in his 1637 *Géométrie* by looking at the multiplication of lines as resulting in a new line-segment and not in a quadrangular surface.[45] This approach, which was based on the admission of a unit segment[46] and on the theory of proportions,[47] made it possible for Descartes to overcome the limitations brought by dimensions and by the requirement of homogeneity in the representation and treatment of continuous quantity. It therefore allowed him to establish a more direct and more general connection between the constructions of magnitudes and the arithmetical operations employed in algebra, through which he was able to solve a much greater range of geometrical problems compared to his predecessors.[48] This connection between geometrical constructions and algebraic operations, which was conveyed by the correspondence defined in the *Géométrie* between geometrical curves and algebraic equations, did not however aim to translate an ontological

---

[44] Addition, substraction, multiplication, division and root extraction.

[45] Descartes (1637, pp. 297–299) and (Smith 1954, pp. 2–7). Vuillemin (1960, pp. 92–93), Bos (1996), Bos (2001, pp. 293–298), Mancosu (2008, pp. 111–113), Guicciardini (2009, pp. 38–40) and Crippa (2017). On the progressive development of this technique in Descartes' geometrical thought, see Bos (2001, pp. 264–298). Yet, it must be noted that the geometrical representation of the operation of multiplication as resulting in a quadrilateral was part of Viète's *logistica speciosa*, which has been described as the birth place of geometrical analysis. Viète was able to go further than the third dimension by admitting higher levels or degrees of interpretation of the three dimensions or types of magnitudes. On Viète's *logistica speciosa* and its contribution to the unification of arithmetic and geometry, see Bos (2001, pp. 125 and 147–151). Descartes however hardly applied Viète's method in the *Géométrie* (Bos 2001, pp. 300–301).

[46] As expounded by D. Crippa (in Crippa 2017, pp. 1250–1254), this unit segment is to be conceived according to the mode of the arithmetical unit and not according to the mode of a determinate segment used to measure lengths (as it was in the context of practical geometry). This conception therefore allowed for the direct application of arithmetical operations to line-segments. However, such as used by Descartes, the actual segment that is taken as a unit may change according to the geometrical problem that is to be solved, that is, according to the specific relation that exists between the lines considered in the problem at hand.

[47] Crippa (2017).

[48] As shown by R. Rashed (in Rashed 2005), the application of algebra to the resolution of geometrical problems is not an innovation brought by Descartes, as it represents a practice already in use among Arabic mathematicians from the tenth century (with al-Khayyām in particular) to solve plane and solid geometrical problems, i.e. problems solvable by the means of straight lines and circles and by the means of conic sections, respectively (as per Pappus' classification of problems in the *Collection* IV.30, § 57–59, Hultsch 1876–1878, p. 270). Yet, Descartes' modernity in this regard consisted in going further than his predecessors in the resolution of geometrical problems and in using algebra also to solve linear problems, i.e. problems solvable by the means of more complex lines, such as the spiral, the quadratrix or the conchoid. On this aspect, see also Bos (1996) and Rabouin (2018).

identity between numbers and magnitudes, but rather the possibility to apply to both domains a common set of operational rules.[49]

If Clavius may have indirectly contributed to the formation of Descartes' new approach to geometry, it would altogether be through his extensive use of genetic definitions in his commentary on Euclid's definitions, through the way he used them to connect instrumental and abstract generations of lines, and through the link he established between arithmetic and geometry. His assertion of the connection between arithmetic and geometry was not only instantiated by his commentary on Euclid's Df. II.1, as well as by his own arithmetical treatment of the propositions of Book II (also in Book IX through the work of Barlaam),[50] but also by his *Algebra*. C. Sasaki has shown indeed that Clavius' *Algebra*, which was read by Descartes, must have played a role in the earlier formation of the French philosopher's approach to mathematics.[51] This work was, admittedly, a product of the medieval and Renaissance tradition of cossist algebra.[52] However, in this treatise, which featured several problems drawn from the *Arithmetic* of Diophantus of Alexandria (200/214–284/298),[53] the Jesuit professor aimed to elevate algebra to the status of a theoretical science, as are arithmetic and geometry, leaving aside its representation as a tool of commercial arithmetic.[54] In this context, he furthermore defined algebra as independent from any specific type of quantity (discrete or continuous, abstract or concrete), presenting it as a general art of solving problems that stretches across all mathematical sciences.[55] This approach is illustrated in this context by his use of geometrical diagrams to solve certain algebraic problems,[56] as well as by the crucial role he attributed to Euclid's Book II of the *Elements*

---

[49] Crippa (2017, pp. 1253–1254).

[50] See *supra*, n. 112, p. 221.

[51] Sasaki (2003, pp. 72 and 92–93).

[52] Sasaki (2003, pp. 74–77). It has been shown that, in his earlier work, Descartes used cossic notations. On this topic, see Costabel (1983, pp. 644–654). C. Sasaki (in Sasaki 2003, p. 101) and D. Rabouin (in Rabouin 2010, pp. 436–437) noted certain similarities between the notations used by Descartes and some of those used by Clavius in his *Algebra*.

[53] Heath (1910) and Sasaki (2003, pp. 74–79).

[54] Clavius (1608, p. 1): "Et dignitate summa, & amplissimarum laudum praeconijs nulli postrema ea ars est, quam recepto vocabulo Algebram nostrates appellarunt." Sasaki (2003, pp. 75–76).

[55] Clavius (1608, p. 2): "Quàm igitur scopus iste late vagatur, qui nec genus ullum numerorum, nec ullius magnitudinis diversitatem, à se alienam putat? ut non numerorum modo latebras omnes detegat, sed uniuscuiusque etiam molis finitam magnitudinem, sonorum metrum, ponderum momenta, mensurarumque certos terminos assequatur; neque ulla Arithmeticae quaestio subijciatur, quam non veluti suam agnoscat Algebra, atque expediat. Tam multas, tam varias, tam obscuras, tam difficiles Mathematicae partes una Algebra pertractat universas." Sasaki (2003, pp. 72–73). See also Crippa (2017, p. 1247).

[56] Clavius (1608, p. 344): "Ex hoc aenigmate facile intelligitur, eum, qui quaestiones per Algebram solvere vult, debere optime esse exercitatum in Geometriae scientia, ut cap. l. diximus. Hoc enim aenigma ab eo, qui Geometriam ignorat, vix, aut nullo modo solvetur, ut patet." Sasaki (2003, pp. 79–81).

in demonstrating basic operations of algebra and arithmetic as applied to square and cubic roots.[57] This geometrical approach to algebra clearly embodied what he designated, in his commentary on Euclid, as the *affinity* between numbers and magnitudes, and which was supported by his kinematic understanding of geometrical objects.

10.4   The Constructive Approach to Geometry of Clavius and Descartes

Clavius' constructive approach to geometry may (as said) also have exerted a certain influence on Descartes, as for the epistemic function he attributed to the tracing out of the curve.[58] As W. Shea expressed it, the fact that Descartes, in the *Géométrie*, continued to give precedence to the tracing out of the curve over its algebraic determination as a mode of representation and definition would be founded on "his traditional conception of geometry as the *construction* of a problem, and his belief that only when the intersection of curves was traced out by one continuous motion, and thereby rendered visible to the physical eye or to the imagination, could we have a *clear* and *distinct* conception of the geometrical solution".[59] Now, the fact for a continuous motion to render a curve visible and thereby knowable clearly resonates with Clavius' words as he wrote that the mathematician, in order to set the line *before our eyes*, imagines that the point moves according to a given trajectory and that the resulting line corresponds to the trace left by the point in its motion. The infinite variety of traces that may be produced and apprehended in this manner is made clear through Clavius' commentary on Df. I.4, where the point is attributed a diversity of possible trajectories (straight, curved, mixed and even parallel) and where the motion is presented as the cause of both the line itself and of the knowledge of the line's nature and properties.[60] If Clavius did not present, in this context, some of these curves as more or less intelligible than others,[61] he did make a distinction in this direction when considering

---

[57] See *supra*, n. 112, p. 221.

[58] This is also suggested in Dear (1995a, pp. 220–221). See also Dear (1995b, pp. 55–56).

[59] Shea (1991, p. 67). The emphasis is proper to the original text.

[60] See *supra*, p. 199. This association between tracing and knowing is made clear in Clavius' commentary on Df. XI.1 (see *supra*, p. 201): "In order to allow us *to correctly understand the line*, mathematicians enjoin us to imagine a given point as moved from one place to another [...] But to allow us *to perceive the surface*, they advise us to imagine a line as transversally moved [...] In this way also, in order *to place before our eyes the body or solid*, that is, the quantity endowed with three dimensions, they recommend us to conceive a surface uniformly transversally raised or moved." (My emphasis.)

[61] Descartes rejected, for instance, the intelligibility and therefore the geometrical nature of mixed lines, part curved and part straight, given that he refuted the possibility of knowing the relation between the straight and the circular: Descartes (1637, p. 340) and (Smith 1954, p. 91): "geometry should not include lines that are like strings, in that they are sometimes straight and sometimes curved, since the ratios between straight and curved lines are not known, and I believe cannot be

curves as solutions of problems. Indeed, when dealing with the quadratrix, Clavius conceded that the motion through which a curve is traced out renders it intelligible only insofar as it is quantitatively determinable or measurable.

Clavius however clearly differed from Descartes inasmuch as he still defended a conception of geometry as a science of figures and as an axiomatic science.[62] In fact, for the Jesuit professor, the generative process would serve the construction of a line or figure, which, although abstract and representing a class of objects, is mainly considered in itself and in relation to its parts. Descartes, on the other hand, conceived geometry as dealing with relations between continuous quantities.[63] In this framework, the motion tracing the curve, which is fundamental to Descartes' geometry, visually expresses a set of properties pertaining to continuous quantity in general, which can therefore be expressed algebraically.[64] These properties, fundamentally, correspond to the objects of the *mathesis universalis*, which are none other than ratios, proportions and relations.[65] As expressed by V. Jullien, the fact that Descartes focused on curves rather than on figures would be due to the fact that plane and solid figures are too complex to reveal the more general order of continuous quantity, an order that is to be ultimately investigated and determined by the intellect in a purely rational manner.[66] Moreover, the construction of curves is, in Descartes' geometrical thought, inextricably linked to the image of geometry that was provided in the *Géométrie* as an art or science of solving geometrical problems.[67] In this context, the construction of curves would provide us with a knowledge of the properties and relations of continuous quantities that goes far beyond that which is delivered by the construction of a square or a rectangle.

---

discovered by human minds, and therefore no conclusion based upon such ratios can be accepted as rigorous and exact." On this issue, see Bos (1981, pp. 324–325) and Mancosu (1996, pp. 77–79).

[62] The difference between the structure of Descartes' geometry and the Euclidean model of geometry as a deductive science, that is, which would derive propositions about geometrical objects from axioms, is presented in Bos (1990, pp. 352–353).

[63] Nikulin (2002, pp. 209–210 and 126).

[64] Jullien (1996, pp. 36 and 40) and Bos (1996).

[65] Jullien (1996, pp. 28–32 and 41–42), Nikulin (2002, pp. 124–126) and Rabouin (2018). More generally, on Descartes' concept of *mathesis universalis* and its relation to algebra, see Klein (1968, pp. 197–211). As D. Crippa has convincingly shown (in Crippa 2017, pp. 1255–1256), this science of relations nevertheless remained ontologically determined, since Descartes, in the *Géométrie*, did not go as far as to consider the relations expressed by algebraic equations in general (as would later do Frans van Schooten (1615–1660)), but only as referring to specific kinds of mathematical objects, and most of all to geometrical curves.

[66] Jullien (1996, pp. 31 and 34–38). See also Klein (1968, pp. 197–211) and Nikulin (2002, p. 197).

[67] Bos (1990, pp. 351–353 and 356) and Bos (1996).

10.5   Genetic Definitions and the Reconstruction of the World in Descartes' *Le Monde*

Concerning the metaphysical interpretation of genetic definitions, I have shown that commentators such as Peletier and Dee, as well as Foix-Candale less directly, appealed to a kinematic understanding of geometrical objects in order to explain the mode according to which the universe was created. Indeed, these authors compared the generation of the line from the point to the primordial procession of the universe from a unitarian divine principle and attributed to the geometrical point itself (then assimilated to the arithmetical unit and to the physical atom) an active role in the constitution of the world.[68]

Now, the way Descartes used and characterised geometrical motion in a philosophical context may also bear some connection to this metaphysical interpretation of genetic definitions. For Descartes, in his unpublished treatise *On the World* (*Le monde*) written between 1629 and 1633,[69] established a correspondence between the motion geometers attribute to points, lines and surfaces in order to generate geometrical magnitudes and the motion that God would have placed within the universe at the moment of its creation. Through this conception, he was notably able to philosophically justify the application of geometry, and of mathematics in general, to the investigation of the physical world.

Descartes' aim, in this treatise, was to present a universe that would be intellectually conceivable and therefore physically possible, since he believed that God can create anything that is conceivable.[70] Now, to him, a universe in which change and motion are defined only in terms of spatial transport or translation, as is the motion attributed to geometrical objects by mathematicians, would be more intellectually conceivable and physically admissible than the universe devised by the Scholastics, in which change was rather defined in terms of act and potentiality.[71]

According to this mechanistic conception of the universe, God would have placed in the cosmos, and in each of its components, a determined quantity of motion, whose perpetual presence, cohesion and operation in the world would be guaranteed by its continuous

---

[68] See *supra*, pp. 236–238.

[69] Descartes (AT 11, 1986b) and (transl. Gaukroger 1998).

[70] Descartes (AT 11, 1986b, p. 36); (Gaukroger 1998, p. 24): "And my purpose [...] is not to explain the things that are in fact in the actual world, but only to make up as I please a world in which there is nothing that the dullest minds cannot conceive, and which nevertheless could not be created exactly the way I have imagined it. [...] Indeed, since everything I propose here can be imagined distinctly, it is certain that even if there were nothing of this sort in the old world, God can nevertheless create it in a new one; for it is certain that He can create everything we imagine."

[71] Descartes (AT 11, 1986b, p. 39) and (Gaukroger 1998, p. 26): "They themselves admit that the nature of their motion is very little understood. And trying to make it more intelligible, they have still not been able to explain it more clearly than in these terms: *Motus est actus entis in potentia, prout in potentia est*. These terms are so obscure to me that I am compelled to leave them in Latin because I cannot interpret them."

conservation by the divine Creator.[72] In this framework, if the motion which geometers[73] attribute to geometrical objects best represents the motion God installed in the universe, it is because it is, for Descartes, the only type of motion that may be clearly and distinctly understood by the human intellect, founding thereby the higher intelligibility (and therefore possibility) of this cosmogonic narrative.[74]

For M. Domski, the function Descartes attributed to the generative motion of geometrical objects in his hypothetical recreation of the cosmos—that is, to make the order and mode of causation of the physical world intelligible—would be conceptually related to the function he assigned to the continuous motion by which curves are generated in the *Géométrie*. For the motion or the tracing out of the line is then not only what enables us to represent and define a curve, but also what allows us to know it.[75] This metaphysical conception would ultimately justify the fundamental role Descartes attributed to the kinematic treatment of curves in his geometry. As was mentioned above, this representation also underlies Descartes' geometrical approach to the physical world, which was founded on his assimilation of geometrical and physical objects insofar as they share the properties of being extended and locally movable.[76]

Thus, the kinematic treatment of geometrical objects played a central part in Descartes' scientific and philosophical thought, since it enabled him to unify not only the science of continuous quantity, as well as mathematics as a whole, but also mathematics, physics and

---

[72] On Descartes' representation of the universe, and the role played by spatially-defined motion within it, see Shea (1991, pp. 251–277) and Gaukroger (1995, pp. 237–256). See also Dear (2001, Chap. 5).

[73] Descartes may perhaps have been thinking of the genetic definitions provided by Clavius in his commentary on Euclid.

[74] Descartes (AT 11, 1986b, p. 39) and (Gaukroger 1998, p. 26): "By contrast [with the nature of the motion defined by the scholastics], the nature of the motion that I mean to speak of here is so easily known that even geometers, who among all men are the most concerned to conceive the things they study very distinctly, have judged it simpler and more intelligible than the nature of surfaces and lines, as is shown by the fact that they explain 'line' as the motion of a point and 'surface' as the motion of a line." It is interesting to note that the distinction between motion as defined in terms of power and act and motion as merely defined in terms of spatial transport parallels the distinction that was presented by Fine between the motion as understood by natural philosophers and motion as understood by mathematicians (see *supra*, pp. 54–55).

[75] Domski (2009, pp. 128–129). See also Vuillemin (1960, pp. 93–94).

[76] On this assimilation and its role in the foundation of the Cartesian geometrical approach to physical science, see, for instance, Nikulin (2002, pp. 115–121). The assimilation of the geometrical and the physical through their common spatial properties does not mean that Descartes did not admit any difference between geometrical and physical things, since the material substrate of the latter makes it impossible for them to perfectly instantiate geometrical lines, surfaces and solids (Nikulin 2002, pp. 115–120). However, the possibility of treating physical things geometrically would be founded, for Descartes, on the fact that these share with geometrical objects their most essential properties, that is, extension and movability. This would enable the imagination and the mind to dismiss the discrepancies between physical and geometrical objects and to investigate the former in a geometrical manner.

theology. Through his geometrical treatment of extended things (physical and geometrical) and his algebraic treatment of curves, he pointed furthermore to the possibility of expressing physical phenomena in algebraic terms. In this regard, Descartes' identification between the motion of abstract geometrical objects and idealised instrumental processes, which left no room for a distinction between intrinsic and extrinsic generations as applied to geometrical objects, found its full justification in his mechanistic representation of the universe, in which all motions are conceived as properly spatial and externally caused, being conserved by God at each instant.[77]

It should be added that Descartes' own imaginary reconstruction of the world on the basis of a merely spatial conception of motion is connected with his constructivist epistemology, according to which the world can only be properly known insofar as it is mentally recreated, as were curves in his geometrical work.[78] In this context, the physical reality of the reconstructed world is not relevant here, since its truth is founded on the fact that it is conceivable, and therefore possible by essence. Its logical necessity is indeed considered as properly grounded in principles (i.e. the genetic definitions of geometrical objects) that are clear and distinct to the mind and on its compatibility with the phenomenal existence of the universe.[79]

Descartes' representation of God, the universe and the science of nature, is, admittedly, quite different from that which was proposed by the above-mentioned sixteenth-century authors, who followed the ancient representations of the cosmos and of nature found in the Platonic, Aristotelian, as well as Stoic and Hermetic doctrines. These, for instance, all maintained to some extent the Aristotelian notion of natural substances as composed of matter and form, even if they also adhered to the Platonic theory of transcendent forms, as well as to elements of ancient atomism.[80] Their representation of the universe, and the role played by motion within it, therefore clearly differed from Descartes' mechanical universe, in which only motion defined in terms of transport was considered relevant and which openly challenged the scholastic physics of forms and qualities.[81]

---

[77] Vuillemin (1960, pp. 89–90).

[78] Nikulin (2002, pp. 211–223).

[79] As God is, for Descartes, the foundation of all things, real and possible, contingent and necessary, the determination of the essence prevails over that of the existence within the scientific investigation according to the principles of Cartesian epistemology. On this distinction and relation between essence and existence as applied to geometrical concepts in Descartes, see Nikulin (2002, pp. 118 and 160–161).

[80] Descartes' philosophical doctrine did however relate by certain aspects to ancient Platonism and atomism.

[81] See *supra*, n. 71, p. 264. To this may be added the opposition between the geocentric cosmological model retained by most sixteenth-century mathematicians, including those investigated in this study, and the heliocentric model defended by Descartes. His support of the heliocentric model in *Le monde* led him to renounce publishing his work after hearing of the condemnation of Galileo.

Also, Peletier, as he described the flow of the point as a principle of the Creation in the *Louange de la Sciance*, attributed to it a non-spatial character, to which may be related Foix-Candale's non-spatial procession of the multiple from God in the *Pimandre*. Contrary to these, Descartes did not depict the generative flow of the point itself as a transcendent and divine unitarian principle of the Creation, nor did he conceive it ever as a non-spatial process. In fact, it is precisely in virtue of its spatial character that the geometrical notion of flow of the point would be, for Descartes, suitable to represent the motion God placed in the physical universe and to ultimately legitimate the application of geometry to the study of physical phenomena.[82]

10.6   Genetic Definitions and the Intelligibility of Geometry

Nevertheless, the higher intelligibility Descartes attributed to mathematics, and to geometry in particular, which is founded on the clarity and evidence of geometrical notions and on their ability to directly express the universal notions of order and measure,[83] indirectly evokes parts of Peletier's conception of geometry. In the preface to his commentary on Euclid, Peletier wrote indeed that order and proportion, which are considered in geometry, are the things that are the most useful to human life on account of the fact that the power of geometry lies in everything.[84] Thus, the study of geometry, by giving us the occasion to mentally recreate the world, would progressively enable us to unravel the order installed by

---

[82] The difference between the philosophy of geometry of Neoplatonists (in Plotinus and Proclus) and that of Descartes was investigated by D. Nikulin (Nikulin 2002).

[83] This is the case in his reconstruction of the world's creation in *Le monde*, but also in his works on the elaboration of the method, in the *Regulae and directionem ingenii* and the *Discours de la méthode*. On these aspects of Descartes' philosophy of geometry, see Jullien (1996, pp. 19–24 and 42–47) and Nikulin (2002, pp. 106–107).

[84] Peletier (1557, sig. 4r–v): "Nihil enim in rebus humanis ferè aliud est quod expediat aut iuvet, praeter ordinem & proportionem: id est, in omnibus moderationem. Ubique igitur latet vis quaedam Geometriae". The notions of "order" and "proportion", as they are then said to correspond to "moderation" or "measure" in everything (*in omnibus moderatio*), appear to have a mostly moral and practical connotation. Yet, Peletier also wrote, in his commentary on Prop. V.1, that geometry entirely consists in proportions (Peletier 1557, p. 120: "Geometria enim quantacunque est, tota in Proportionibus est: neque aliud quicquam spectat, quàm ut Lineas Lineis, Superficies Superficiebus, & Corpora Corporibus componat & comparet."), which then shows the connection, in his thought, of the moral and mathematical understanding of the term *proportio*.

God in the universe.[85] This stance, which echoes the philosophical thought of Cusanus, was also later presented by Kepler.[86]

The assertion of the higher intelligibility of mathematics was also present in another way in the attitude of Clavius, who defended, in the *Prolegomena* to his commentary on Euclid, the mathematicians' ability to settle any argument (by opposition to the endless debates of philosophers and logicians).[87] E. Kessler has in fact explicitly presented Clavius as an intermediary between Proclus and Descartes in the development of the concept of method.[88] As he compared the nature of problems in geometry and in dialectic,[89] Clavius asserted indeed the higher certainty of the knowledge procured by a geometrical problem over that provided by a problem in dialectic. The reason for this would be that, in a geometrical problem, we are required to construct our objects and are therefore able to

---

[85] Peletier (1557, sig. 4r–v): "In qua meditatione quanto maiores progressus fecerimus, tanto propiùs ad Deum accedere videmur. Ac quemadmodum Mens illa aeterna, praeteritorum meminit, praesentia cernit, futura perspicit, simul verò omnia amplectitur & moderatur: ita praeclarus Geometriae artifex suas cogitationes in unum collatas, ad rem suam convertit, et suum quendam Mundum universa speculatione intuetur". See also *ibid.* (sig. 4v): "Geometricae positiones, quae operas auxiliarias inter se praestant, omnia in rerum natura mutuis alternisque subsidijs niti & consistere declarant. Quinetiam amicitiae ipsius iura, in Figurarum similitudine, quarum colligationem Diameter efficit, conspicua sunt. Ad summam, haec imago & facies Geometrica eiusmodi est, ut in ea. Mundi quandam *theorian* possis agnoscere". See *supra*, p. 94 and Axworthy (2013).

[86] On this position in Kepler and its link to Descartes' constructivism, see Nikulin (2002, pp. 212–213).

[87] Clavius (1611–1612, I, p. 5): "Demonstrant enim omnia, de quibus suscipiunt disputationem, firmissimis rationibus, confirmantque, ita ut vere scientiam in auditoris animo gignant, omnemque prorsus dubitationem tollant: Id quod alijs scientijs vix tribuere possumus, cum in eis saepenumero intellectus multitudine opinionum, ac sententiarum varietate in veritate conclusionum iudicanda suspensus haereat, atque incertus. Huius rei fidem aperte faciunt tot Peripateticorum sectae, (ut alios interim Philosophos silentio involvam) quae ab Aristotele, veluti rami è trunco aliquo, exortae, adeo & inter se, & nonnumquam à fonte ipso Aristotele dissident, ut prorsus ignores, quidnam sibi velit Aristoteles, num de nominibus, an de rebus potius disputationem instituat. Hinc fit, ut pars interpretes Graecos, pars Latinos, alij Arabes, alij Nominales, alij denique Reales, quos vocant, (qui omnes tamen Peripateticos se esse gloriantus) tanquam ductores sequantur." On the epistemological status of mathematics according to Clavius, see, for instance, Kessler (1995) and Higashi (2018, pp. 356–360).

[88] Kessler (1995).

[89] Clavius (1611–1612, I, p. 8): "Dictum est autem hoc genus demonstrationum Problema, ad similitudinem problematis Dialectici. [...] Est tamen discrimen non parvum inter Dialecticorum & Mathematicorum problema. Nam in problemate Dialectico utravis pars contradictionis suscepta confirmatur tantum probabiliter, ita ut intellectus cuiusque ambigat, utranam illius pars vera sit. In Mathematico vero, quamcunque quis partem elegerit, eam firma demonstratione, ita ut nihil omnino dubij sit reliquum, comprobabit. Si enim Geometra statuat ex puncto quolibet lineae rectæ propositæ lineam perpendicularem educere, efficiet utique hoc ipsum ratione constanti, & evidenti".

obtain a complete knowledge of their properties and relations with the construction.[90] By contrast, in dialectic, the demonstration of a problem is only based on probable arguments.

Thus, although geometry, and the place held by motion in the geometer's approach to his objects, was defined by Descartes within a system of representations that is quite different from those of his predecessors, we find in their comparison certain continuities, many of which are significantly stronger with Clavius than with any other author considered in this study. These continuities are mainly based on the way motion, and the kinematic treatment of geometrical objects, allowed Descartes and some of his sixteenth-century predecessors to connect intellection, imagination and the senses, theory and practice, arithmetic and geometry, but also mathematics and nature. Yet, these connections, as we have seen, are neither direct nor absolute, as Descartes went by many aspects beyond the conceptions and treatment of geometrical motion proposed by the sixteenth-century commentators of Euclid.

* * *

It is not the place here to explore in greater depth the relations and differences between Descartes and the investigated authors with respect to the interpretation and uses of motion and genetic definitions in geometry, nor to go further into the analysis of later developments on these questions. But the provided elements will hopefully suffice to illustrate the way Descartes and other seventeenth-century thinkers continued and went beyond the discussions presented in the sixteenth-century Euclidean tradition on the topic. Admittedly, Descartes' approach to geometrical motion, though crucial to the treatment of the question in the early modern era, was itself surpassed by other conceptions and applications of geometrical motion in mathematics, in the physico-mathematical sciences and in the philosophy of mathematics, as those brought forth by Leibniz and Newton, with the admission of transcendental curves within geometry[91] and the consideration of all types of mechanical processes as mathematically determinable.[92] But the way Descartes dealt with geometrical motion in his mathematical and philosophical work, and the way his conceptions on these matters related to and differed from those of sixteenth-century commentators of Euclid, point to the changes that would later take place concerning the role and status of motion in geometry. This manifests, in other words, the growing importance of motion for the practice and representation of geometry, as for the mathematisation of the physical sciences, in the early modern era.

---

[90] Kessler (1995, pp. 300–302).

[91] On this topic, see Bos (1988).

[92] On Newton and the way he surpassed Descartes' views on the matters, see Guicciardini (2009, pp. 299–305 and 313–315).

Appendix

This table summarises the references made to genetic definitions in the commentaries on Euclid's *Elements* considered in this study.[1] The list of occurrences counts both genetic definitions that were added by the investigated commentators to Euclid's definitions in their exposition of the *Elements*, and the genetic definitions that were already present in Euclid's text and to which the commentators added a non-genetic definition. This is here only the case of Df. I.14, that is, the definition of the sphere. This table indicates, for each definition, the geometrical objects for which a genetic definition was given or implicitly referred to. The latter are given in parentheses.

Euclid	Fine	Peletier	Foix-Candale	Dee–Billingsley	Commandino	Clavius
Preface	line		(point as principle of magnitude)	line		
Df. I.1	line	line	(point as principle of magnitude)			
Df. I.2	line	line		line	line	line
Df. I.3	line	(line)	magnitude; line			
Df. I.4	line; surface	line; surface				line
Df. I.5	surface					surface
Df. I.6	surface		line; surface			
Df. I.7	solid; (plane figures; solid figures)					surface
Df. I.13	(line; surface; solid)		magnitude			

(continued)

---

[1] The references made to genetic definitions outside the framework of the commentaries on Euclid (in Fine's *Geometria*, Peletier's *Louange de la Sciance* or Dee's *Monas hieroglyphica*) were not taken into account here.

© The Author(s), under exclusive license to Springer Nature Switzerland AG 2021
A. Axworthy, *Motion and Genetic Definitions in the Sixteenth-Century Euclidean Tradition*, Frontiers in the History of Science,
https://doi.org/10.1007/978-3-030-95817-6

Euclid	Fine	Peletier	Foix-Candale	Dee–Billingsley	Commandino	Clavius
Df. I.15	circle					circle; (sphere)
Df. I.16	circle	circle; (sphere)		line; surface; solid; circle		
Df. I.30	square					
Post. 1	line					line
Post. 2	line			line		line
Post. 3	circle			circle	line; circle	circle
Df. II.1	rectangular parallelogram	rectangular parallelogram	rectangular parallelogram	rectangular parallelogram	rectangular parallelogram	rectangular parallelogram
Df. XI.1				*Preface of book XI*: line; (surface; solid)		line; surface; solid
Df. XI.2						solid
Df. XI.14			[Df. XI.12]: sphere	[Df. XI.12]: line; surface; solid; sphere		sphere
Df. XI.15				[Df. XI.13]: sphere		
Df. XI.16				[Df. XI. 14]: sphere		
Df. XI.17						sphere

References

Al-Khayyām, ʿUmar. 1851. *L'algèbre d'Omar Al-Khayyāmī, publiée, traduite et accompagnée d'extraits de manuscrits inédits*, transl. and comm. F. Woepcke. Paris: B. Duprat.

Al-Sijzī, Abū. 1851. Treatise on the trisection of the rectilinear angle. In *L'Algèbre d'Omar al-Khayyāmī*, transl. F. Woepcke. Paris: B. Duprat.

Andreae, Johannes Valentinus. 1614. *Collectaneorum mathematicorum*. Tübingen: J. A. Cellius.

Apollonius of Perga. 1891. *Apollonii Pergaei quae Graece exstant cum commentariis antiquis*, ed. J. L. Heiberg, vol. 1. Leipzig: B.G. Teubner.

———. 1990. *Conics. Books V to VII: The Arabic translation of the lost Greek original in the version of the Banū Mūsa*, ed., transl. and comm. G.J. Toomer. New York/Berlin: Springer.

———. 1998. *Conics. Books I-III*, transl. D. Densmore. Sante Fe: Green Lion Press.

———. 2009. *Apollonius de Pergé. Coniques. Tome 4: Livres VI et VII. Commentaire historique et mathématique*, ed. and transl. R. Rashed. Berlin/New York: De Gruyter.

Aquinas, Thomas. 1989. *Expositio libri Posteriorum. Editio altera retractata*, ed. R.-A. Gauthier. Rome/Paris. Commissio Leonina, Vrin.

Archibald, Raymond Clare. 1950. The first translation of Euclid's *Elements* into English and its source. *The American Mathematical Monthly* 57 (7): 443–452.

Archimedes of Syracuse. 1881. Archimedis opera omnia cum commentariis Eutocii, vol. II. ed. and transl. J.L. Heiberg. Leipzig: B.G. Teubner.

———. 1897. *The works of Archimedes*, transl. T.L. Heath. Cambridge: Cambridge University Press.

Aristotle. 1989. *Posterior analytics*, transl. H. Tredennick. Cambridge, MA: Harvard University Press.

———. 1995. *The complete works of Aristotle*, transl. J. Barnes. Princeton: Princeton University Press.

Armstrong, Arthur H. 1995. Plotinus. In *The Cambridge history of later Greek and early medieval philosophy, part III*, ed. A.H. Armstrong, 236–268. Cambridge: Cambridge University Press.

Arnaud, Sophie. 2005. *Ratio et Oratio: la voix de la Nature dans l'oeuvre de Jacques Peletier du Mans (1517–1582)*. Paris: Honoré Champion.

Arthur, Richard T.W. 2021. *Leibniz on time, space, and relativity*. Oxford: Oxford University Press.

Axworthy, Angela. 2004. *Le Statut des disciplines mathématiques au XVIᵉ siècle au regard des préfaces aux Éléments d'Euclide de Niccolò Tartaglia et de Christoph Clavius*. MA Dissertation, Université François-Rabelais – Centre d'Études Supérieures de la Renaissance, Tours.

———. 2013. *The ontological status of geometrical objects in the commentary on the Elements of Euclid of Jacques Peletier du Mans (1517–1582)*. Fondation Maison des Sciences de l'Homme, Working Paper 41.

© The Author(s), under exclusive license to Springer Nature Switzerland AG 2021
A. Axworthy, *Motion and Genetic Definitions in the Sixteenth-Century Euclidean Tradition*, Frontiers in the History of Science,
https://doi.org/10.1007/978-3-030-95817-6

————. 2016. *Le Mathématicien renaissant et son savoir. Le statut des mathématiques selon Oronce Fine*. Paris: Classiques Garnier.

————. 2017. La notion géométrique de flux du point à la Renaissance et dans le commentaire des *Éléments* de Jacques Peletier du Mans. In *Mélanges offerts à Joël Biard*, ed. C. Grellard, 453–464. Paris: Vrin.

————. 2018. The debate between Peletier and Clavius on superposition. *Historia mathematica* 45 (1): 1–38.

————. 2020. Oronce Fine and Sacrobosco: From the edition of the *Tractatus de sphaera* (1516) to the *Cosmographia* (1532). In *De sphaera of Johannes de Sacrobosco in the early modern period*, ed. M. Valleriani, 185–264. Springer.

Bacon, Francis. 2000. *The New Organon*, introd. L. Jardine and transl. M. Silverthorne. Cambridge: Cambridge University Press.

Baldi, B. 1998. *Le Vite de' matematici*, ed. and comm. E. Nenci. Milan: Francoangeli.

Baldwin, Robert. 2006. John Dee's interest in the application of nautical science, mathematics and law to English naval affairs. In *John Dee: Interdisciplinary studies in English Renaissance thought*, ed. S. Clucas, 97–130. Springer.

Banks, Kathryn. 2007. Space and light: ficinian neoplatonism and Jacques Peletier Du Mans's 'Amour des Amours'. *Bibliothèque d'Humanisme et Renaissance* 69 (1): 83–101.

Barany, Michael J. 2010. Translating Euclid's diagrams into English, 1551–1571. In *Philosophical aspects of symbolic reasoning in early modern mathematics*, ed. A. Heeffer and M. Van Dyck, 125–163. London: College Publications.

Barker, Peter, and Kathleen Crowther. 2013. Training the intelligent eye: Understanding illustrations in early modern astronomy texts. *Isis* 104: 429–470.

Barozzi, Francesco. 1560. *Procli Diadochi Lycii philosophi Platonici ac mathematici probatissimi in primum Euclidis Elementorum librum commentariorum ad universam mathematicam disciplinam principium eruditionis tradentium libri IIII. A Francisco Barocio patritio Veneto summa opera, cura, ac diligentia cunctis mendis expurgati: scholiis, et figuris, quae in graeco codice omnes desiderabantur aucti: primum jam Romanae linguae venustate donati, & nunc recèns editi. Cum catalogo deorum, & virorum illustrium, atque autorum: elencho librorum, qui vel ab autore, vel ab interprete citati sunt: & indice locupleti notabilium omnium in opere contentorum*. Padua: G. Percacino.

Barrow, Isaac. 1734. *The usefulness of mathematical learning explained and demonstrated, being mathematical lectures read in the publick schools at the University of Cambridge*. London: S. Austen. [repr. London: Cass Publishing Company, 1970].

————. 1751. *Euclide's Elements. The whole fifteen books, compendiously demonstrated: With Archimedes's theorems of the sphere and cylinder, investigated by the method of indivisibles. Also, Euclide's data, and a brief treatise of regular solids*. London: W. & J. Mount.

Berger de Xivrey, Jules. 1843. *Recueil des lettres missives de Henri IV. Tome 1: 1562–1584*. Paris: Imprimerie Royale.

Bernhardt, Jean. 1978. Infini, substance et attribut: Sur le Spinozisme. *Cahiers Spinoza* 2: 53–92.

Bérulle, Pierre (de). 1644. *Les Oeuvres de l'éminentissime et reverendissime Pierre Cardinal de Berulle,… augmentées de divers opuscules de controverse et de piété, avec plusieurs lettres: et enrichies de sommaires et de tables par les soins du R. P. François Bourgoing*. Paris: A. Estienne and S. Huré.

Billingsley, Henry. 1570. *The Elements of Geometrie of the most auncient Philosopher Euclide of Megara Faithfully (now first) translated into the Englishe toung, by H. Billingsley, Citizen of London. Whereunto are annexed certaine scholies, Annotations, and Inventions, of the best Mathematiciens, both of time past, and in this our age. With a very fruitfull Praeface made by M. I. Dee, specifying the chiefe Mathematicall Sciences, what they are, and wherunto commodious: Where, also, are disclosed certaine new Secrets Mathematicall and Mechanicall, until these our daies, greatly missed*. London: J. Daye.

Billingsley, Dale. 1993. Authority in early editions of Euclid's *Elements*. *Fifteenth Century Studies* 20: 1–14.

Bodin, Jean. 1597. *Le Theatre de la nature universelle*. Paris: Jean Pillehotte.

Boethius, Anicius Manilius Severinus. 1867. *Anicii Manilii Torquati Severini Boetii De institutione arithmetica libri duo, De institutione musica libri quinque. Accedit geometria quae fertur Boetii*, ed. G. Friedlein. Leipzig: B.G. Teubner.

———. 1983. *Boethian number theory: A translation of the De institutione arithmetica*, transl. M. Masi. Amsterdam: Rodopi.

———. 2002. *Boèce. Institution arithmétique*, ed. and transl. J.-Y. Guillaumin. Paris: Les Belles Lettres.

Bos, Henk J.M. 1981. On the representation of curves in Descartes' *Géométrie*. *Archive for History of Exact Sciences* 24: 295–338.

———. 1988. Tractional motion and the legitimation of transcendental curves. *Centaurus* 31: 9–62.

———. 1990. The Structure of Descartes' *Géométrie*. In *Descartes: il metodo e I saggi*, ed. G. Belgioso et al., 349–369. Florence: A. Paoletti.

———. 1996. Tradition and modernity in early modern mathematics: Viète, Descartes and Fermat. In *L'Europe mathematique, histoires, mythes, identités*, ed. C. Goldstein, K. Chemla, J. Gray, and J. Ritter, 183–204. Paris: Éditions de la Maison des Sciences de l'Homme.

———. 2001. *Redefining geometrical exactness: Descartes' transformation of the early modern concept of construction*. New York: Springer.

Boudet, Jean-Patrice. 2007. Charles V, Gervais Chrétien et les manuscrits scientifiques du collège de Maître Gervais. *Médiévales* 52: 15–38.

Breger, Herbert. 1991. Der mechanistische Denkstil in der Mathematik des 17. Jahrhunderts. In *Gottfried Wilhelm Leibniz im philosophischen Diskurs über Geometrie und Erfahrung*, ed. H. Hecht, 15–46. Berlin: Akademie Verlag.

Brigaglia, Aldo. 2012. Apollonius de Pergé. Coniques. Tome 4: Livres VI et VII. Commentaire historique et mathématique. Édition et traduction du texte arabe by Roshdi Rashed. Book review. *Aestimatio* 9: 241–260.

Brioist, Jean-Jacques. 2009. Oronce Fine et cartographical methods. In *The worlds of Oronce Fine. Mathematics, instruments and print in Renaissance France*, ed. A. Marr, 137–155. Donington: S.Tyas.

Brun, Robert. 1934. Un illustrateur méconnu. Oronce Finé. *Arts et Métiers graphiques* 41: 51–57.

———. 1966. Maquettes d'éditions d'Oronce Finé. In *Studia bibliographica in honorem Herman de La Fontaine Verwey*, ed. S. van der Woude, 36–42. Amsterdam: M. Hertzberger.

Busard, H.L.L. 2005. *Campanus of Novara and Euclid's elements*. Stuttgart: Steiner.

Calhoun, Alison. 2017. Montaigne's Swerve: The geometry of parallels in the essays and other writings. *Neophilologus* 101: 351–365.

Campanus de Novara. 1482. *Praeclarissimus liber elementorum Euclidis perspicassimi: in arte geometriae incipit foelicissime*. Venice: Erhardt Ratdolt.

Campbell, Andrew. 2012. The reception of John Dee's *Monas hieroglyphica* in early modern Italy: The case of Paolo Antonio Foscarini (c. 1562–1616). *Studies in History and Philosophy of Science* 43: 519–529.

Casalini, Cristiano, and Claude Pavur. 2016. *Jesuit Pedagogy 1540–1616. A Reader*. Chestnut Hill: Institute of Jesuit Sources – Boston College.

Champier, Symphorien. 1507. *Symphoriani Champerii Liber de quadruplici vita. Theologia Asclepii Hermetis Trismegisti discipuli cum commentariis eiusdem Symphoriani*. Lyon: E. Gueynard and J. Huguetan.

Chase, Michael. 2008. Albert le Grand sur la dérivation des formes géométriques: Un témoignage de l'influence de Simplicius par le biais des Arabes? In *Proceedings of the Conference "Damascius et le parcours syrien du néoplatonisme"* (Damas, 27–28 October 2008). Damas: Presses de l'IFPO.

Cifoletti, Giovanna. 1992. *Mathematics and rhetoric: Peletier and Gosselin and the making of the French algebraic tradition*. PhD Dissertation, Princeton University, Princeton, NJ.

———. 1995. La question de l'algèbre. Mathématiques et rhétorique des hommes de droit dans la France du XVIe siècle. *Annales. Histoire, Sciences Sociales* 50 (6): 1385–1416.

———. 2009. Oronce Fine's legacy in the French algebraic tradition: Peletier, Ramus and Gosselin. In *The worlds of Oronce Fine. Mathematics, instruments and print in Renaissance France*, ed. A. Marr, 114–136. Donington: S. Tyas.

Claessens, Guy. 2009. Clavius, Proclus, and the limits of interpretation: Snapshot-idealization versus projectionism. *History of Science* 47: 317–336.

———. 2015. The drawing board of imagination: Federico commandino and John Philoponus. *Journal of the History of Ideas* 76 (4): 499–515.

Clavius, Christoph. 1570. *Sphaeram Ioannis de Sacrobosco commentarius*. Rome: Helianus.

———. 1574. *Euclidis elementorum libri XV: Accessit XVI. De solidorum regularium comparatione: Omnes perspicuis demonstrationibus accuratisque scholijs illustrati*. Rome: V. Accolti. [repr. B. Grassi: 1589; Cologne, B. Ciotti: 1591; Rome, L. Zanetti: 1603; Frankfurt, N. Hoffmann: 1607; Cologne, M. Cholinus: 1607; Mainz, Antonius Hierat: 1611–1612; Frankfurt, Jonas Rosa: 1644; 1654].

———. 1581. *Gnomonices libri octo*. Rome: F. Zanetto.

———. 1583. *Epitome Arithmeticae Practicae*. Rome: D. Basa.

———. 1586. *Theodosii Tripolitæ Sphæricorum libri III a Christophoro Clavio Bambergensis Societatis Iesu perspicuis demonstrationibus ac scholijs illustrati. Item eiusdem Christophori Clavii Sinus, lineæ tangentes et secantes, triangula rectilinea atque sphærica*. Rome: D. Basa.

———. 1588. *Novi calendarii romani apologia*. Rome: Santi & C.

———. 1589. *Euclidis elementorum libri XV: Accessit XVI. De solidorum regularium comparatione: omnes perspicuis demonstrationibus accuratisque scholijs illustrati*. Rome: B. Grassi.

———. 1593. *Astrolabium*. Rome: Bartolomeo Grassi.

———. 1603. *Romani calendarii a Gregorio XIII P.M. restituti explicatio*. Rome: L. Zannetti.

———. 1604. *Geometria practica*. Rome: L. Zannetti.

———. 1607. *Tabulae sinuum, tangentium et secantium, ad partes radij 10,000,000. & ad scrupula prima Quadrantis, et ad earum praxin brevis introductio*. Mainz: J. Albinus.

———. 1608. *Algebra*. Rome: B. Zannetti.

———. 1611–1612. *Opera omnia*. Mainz: A. Hierat.

Cléro, Jean-Pierre, and Évelyne LeRest. 1981. *La Naissance du calcul infinitésimal au XVIIe siècle*. Paris: CNRS.

Clulee, Nicholas H. 1977. Astrology, magic, and optics: Facets of John Dee's early natural philosophy. *Renaissance Quarterly* 30 (4): 632–680.

———. 1988. *John Dee's natural philosophy: Between science and religion*. London, New York: Routledge.

———. 2005. The *Monas hieroglyphica* and the alchemical thread of John Dee's career. *Ambix* 52: 197–215.

Commandino, Federico. 1558. *Federici Commandini Urbinatis in Planisphaerium Ptolemaei commentarius*. Venice: A. Manuzio.

———. 1562. *Claudii Ptolemaei liber de analemmate, A Federico Commandino Urbinate instauratus, & commentariis illustratus, Qui nunc primum eius opera e tenebris in lucem prodit. Eiusdem Federici Commandini liber de Horologiorum descriptione*. Rome: P. Manuzio.

———. 1565a. *Archimedis de iis quae vehuntur in aqua libri duo*. Bologna: A. Benacio.

———. 1565b. *Federici Commandini Urbinatis. Liber de centro gravitatis solidorum.* Bologna: A. Benacci.

———. 1566. *Apollonii Pergaei conicorum libri quattuor. Unà cum Pappi Alexandrini Lemmatibus, et commentariis Eutocii Ascalonitae. Sereni Antinsensis philosophi Libri duo nunc primim in lucem editi. Quae omnia nuper Federicus Commandinus Urbinas mendis quamplurimis expurgata e Graeco convertit, & commentariis illustravit.* Bologna: A. Benacio.

———. 1572a. *Euclidis elementorum: libri XV. Unà cum scholiis antiquis.* Pesaro: C. Francischino.

———. 1572b. *Aristarchi de magnitudinibus et distantiis solis et lunae liber cum Pappi Alex. explicationibus. A Federico Commandino Urbinate in latinum conversus, ac commentariis illustratus.* Pesaro: C. Francischino.

———. 1575a. *De gli elementi d'Euclide libri quindici. Con gli scholii antichi. Tradotti prima in lingua latina de M. Federico Commandino da Urbino, & con commentarii illustrati, et hora d'ordine dell'istesso trasportati nella nostra vulgare, & da lui riveduti.* Urbino: D. Frisolino.

———. 1575b. *Heronis Alexandrini Spiritalium Liber. A Federico Commandino Urbinate, ex Graeco, nuper in Latinum conversus.* Urbino: D. Frisolino.

———. 1588a. *Archimedis opera nonnulla. Quae hoc libro continentur: Circuli dimensio. De lineis spiralibus. Quadratura paraboles. De conoidibus, & sphaeroidibus. De arenae numero.* Venice: P. Manuzio.

———. 1588b. *Pappi Alexandrini mathematicae collectiones a Federico Commandino urbinate in Latinum conversæ, et commentariis illustratae.* Pesaro: G. Concordia.

Commandino, Federico, and John Dee. 1570. *De superficierum divisionibus liber Machometo Bagdedino ascriptus. Nunc primum Ioannis Dee Londinensis, & Federici Commandini Urbinatis opera in lucem editus. Federici Commandini de eadem re libellus.* Pesaro: G. Concordia.

Corry, Leo. 2013. Geometry and arithmetic in the medieval traditions of Euclid's Elements: A view from Book II. *Archive for History of Exact Sciences* 67 (6): 637–705.

Cosentino, Giuseppe. 1970. Le matematiche nella "Ratio studiorum" della Compagnia di Gesù. *Miscellanea Storica Ligure, nuova serie* II (2): 169–213.

———. 1971. L'insegnamento delle matematiche nei collegi gesuitici dell'Italia settentrionale. Nota introduttiva. *Physis* 13: 205–217.

———. 1973. Le matematiche nei collegi gesuitici dell'Italia settentrionale. *Physis* 12: 212–231.

Costabel, Pierre. 1983. L'initiation mathématique de Descartes. *Archives de Philosophie* 46 (4): 637–646.

Counet, Jean-Michel. 2005. Mathematics and the divine in Nicholas of Cusa. In *Mathematics and the divine: A historical study*, ed. T. Koetsier and L. Bergmans, 273–290. Amsterdam: Elsevier.

Crippa, Davide. 2014. *Impossibility results: From geometry to analysis. A study in early modern conceptions of impossibility.* PhD dissertation, Université Paris–Diderot – Sorbonne Paris Cité, Paris.

———. 2017. Descartes on the unification of arithmetic, algebra and geometry via the theory of proportions. *Revista Portuguesa de Filosofia* 73 (Fasc. 3/4): 1239–1258.

Cusa, Nicholas (de). 1972. *Nicolai de Cusa Opera omnia*, ed. J. Koch and K. Bormann, vol. III. Hamburg: F. Meiner.

D'Ailly, Pierre, and Pedro Sanchez Ciruelo. 1498. *Uberrimum sphere mundi commentum intersertis etiam questionibus domini Petri de Aliaco.* Paris: G. Marchant and J. Petit.

Dagens, Jean. 1951. Le commentaire du *Pimandre* de François de Candale. In *Mélanges d'histoire littéraire offerts à Daniel Mornet*, 21–26. Paris: Nizet.

———. 1961. Hermétisme et cabale en France de Lefèvre d'Étaples à Bossuet. *Revue de littérature comparée* 35 (1): 5–16.

Dainville, François (de). 1970. How did Oronce Fine draw his large map of France? *Imago Mundi* 24: 49–55.

De Angelis, Enrico. 1964. *Il Metodo geometrico nella filosofia del seicento*. Pisa: Università degli studi – Istituto di filosofia.

De Gandt, François. 1995. *Force and geometry in Newton's principia*, transl. Curtis Wilson. Princeton: Princeton University Press.

De Léon-Jones, Karen. 2006. John Dee and the Kabbalah. In *John Dee: Interdisciplinary studies in English Renaissance thought*, ed. S. Clucas, 143–158. Dordrecht: Springer.

De Morgan, Augustus. 1837. *Notices of English mathematical and astronomical writers between the Norman conquest and the year 1600*. Companion to the Almanac for 1837.

De Pace, Anna. 1993. *Les Matematiche e il mondo: Ricerche su un dibattito in Italia nella seconda metà del Cinquecento*. Milano: Francoangeli.

De Risi, Vincenzo. 2007. *Geometry and monadology: Leibniz's analysis situs and philosophy of space*. Basel, Boston, Berlin: Birkhäuser.

———. 2016a. *Leibniz on the parallel postulate and the foundations of geometry. The unpublished manuscripts*. Cham: Springer.

———. 2016b. The development of Euclidean axiomatics. The systems of principles and the foundations of mathematics in editions of the *elements* in the early modern age. *Archive for History of Exact Sciences* 70: 591–676.

Dear, Peter. 1995a. *Discipline & experience: The mathematical way in the scientific revolution*. Chicago: University of Chicago Press.

———. 1995b. Mersenne's suggestion: Cartesian meditation and the mathematical model of knowledge in the seventeenth century. In *Descartes and his contemporaries: Objections and replies*, ed. R. Ariew and M. Grene, 44–62. Chicago: University of Chicago Press.

———. 2001. *Revolutionizing the sciences: European knowledge and its ambition, 1500–1700*. Princeton: Princeton University Press.

Dee, John. 1558. *Propaedeumata Aphoristica Ioannis Dee, Londinensis, de Praestantioribus quibusdam naturae virtutibus*. London: H. Sutton. [repr. R. Wolfe, 1568].

———. 1564. *Monas hieroglyphica*. Antwerp: G. Sylvius.

———. 1577. *General and rare memorials pertayning to the perfect arte of navigation*. London: J. Daye.

———. 1842. *The private diary of Dr. John Dee: And the catalogue of his library of manuscripts, from the original manuscripts in the Ashmolean Museum at Oxford, and Trinity College Library*, ed. J. O. Halliwell-Phillipps. London: J. B. Nichols and son.

———. 1851. The compendious rehearsall of John Dee. In *Autobiographical tracts of Dr. John Dee*, ed. J. Crossley. Manchester: Chetham Society.

Debus, Allen, and John Dee. 1975. *The Mathematicall praeface to the Elements of geometrie of Euclid of Megara (1570)*. New York: Science History Publications.

Del Monte, Guidobaldo. 1577. *Mechanicorum liber*. Pesaro: G. Concordia.

Demerson, Guy. 1975. Dialectique de l'amour et Amour des amours chez Peletier du Mans. In *Actes du colloque Renaissance-classicisme du Maine*, 263–282. Paris: Nizet.

Demoss, David and Devereux Daniel. 1988. Essence existence and nominal definition in Aristotle's posterior analytics II 8-10. Phronesis 33(1–3): 133–154.

Descartes, René. 1637. *Discours de la méthode pour bien conduire sa raison et chercher la vérité dans les sciences. Plus la Dioptrique. Les Meteores. Et la Geometrie*. Leiden: J. Maire.

———. 1954. *The Geometry of René Descartes with a facsimile of the first edition, 1637*, transl. D.E. Smith and M.L. Latham. New York: Dover Publications.

———. 1986a. Physico-mathematica, Compendium musicae, Regulae ad directionem ingenii, Recherche de la vérité, Supplément à la correspondance. In *Oeuvres de Descartes X*, ed. C. Adam and P. Tannery. Paris: Vrin.

―――. 1986b. Le monde. Description du corps humain. Passions de l'âme. Anatomica. Varia. In *Oeuvres de Descartes XI*, ed. C. Adam and P. Tannery. Paris: Vrin.

―――. 1987. Correspondance: avril 1622 – février 1638. In *Oeuvres de Descartes I*, ed. C. Adam and P. Tannery. Paris: Vrin.

―――. 1998. The World and other writings, ed. and transl. S. Gaukroger. Cambridge: Cambridge University Press.

Destombes, Marcel. 1971. Oronce Fine et son globe céleste de 1553. In *Actes du XIIe Congrès international d'histoire des sciences (Paris, 1968)*, 41–50. Paris: A. Blanchard.

Dhombres, Jean. 2006. La mise à jour des mathématiques par les professeurs royaux. In *Histoire du Collège de France*, ed. A. Tuilier, vol. I, 377–420. Paris: Fayard.

Dillon, John M. 2014. Dionysius the areopagite. In *Interpreting Proclus: From Antiquity to the Renaissance*, ed. S. Gersh, 111–124. Cambridge: Cambridge University Press.

Diogenes Laërtius. 1925. *Lives of the eminent philosophers*, vol. II, transl. R. Drew Hicks. London/ New York: W. Heinemann, G.P. Putnam's Sons.

Djebbar, Ahmed. 2001. Omar Khayyām, Epître sur l'explication des prémisses problématiques du livre d'Euclide. *Farhang* 14 (39–40): 79–136.

Domski, Mary. 2009. The intelligibility of motion and construction: Descartes' early mathematics and metaphysics, 1619–1637. *Studies in History and Philosophy of Science. Part A* 40: 119–130.

Du Préau, Gabriel. 1549. *Deux livres de Mercure Trismegiste Hermes tres ancien Theologien, & excellant Philosophe, l'un de la puissance et sapience de Dieu, l'autre de la volonté de Dieu, avecq'un Dialogue de Loys Lazarel, poëte Chrestien intitulé le Bassin d'Hermes*. Paris: E. Groulleau.

Dunn, Richard. 2006. John Dee and astrology in Elizabethan England. In *John Dee: Interdisciplinary studies in English Renaissance thought*, ed. S. Clucas, 85–94. Springer.

Dupèbe, Jean. 1999. *Astrologie, religion et médecine à Paris. Antoine Mizault (ca. 1512–1578)*. PhD Dissertation, Université Paris-Nanterre, Paris.

Duplessis-Mornay, Philippe (de). 1585. *De la vérité de la religion chrestienne, contre les athées, épicuriens, payens, juifs, mahumédistes et autres infidèles*. Paris: C. Micard.

Dye, Guillaume, and Bernard Vitrac. 2009. Against the *Geometers* by sextus empiricus: Sources, targets, structure. *Phronesis* 54: 155–203.

Eagleton, Catherine. 2009. Oronce Fine's sundials: The sources and influences of 'De Solaribus Horologijs'. In *The worlds of Oronce Fine. Mathematics, instruments and print in Renaissance France*, ed. A. Marr, 83–99. Donington: S. Tyas.

Easton, Joy B. 2008. Dee, John. In *Complete dictionary of scientific biography*, ed. Charles C. Gillispie, vol. 4, 5–6. New York: Charles Scribner's Sons.

Errard, Jean. 1598. *Les Six premiers Livres des Elémens d'Euclide: traduicts et commentez par J. Errard de Bar-le-Duc, ingénieur du treschrestien Roy de France et de Navarre*. Paris: G. Auvray.

Faivre, Antoine. 2008. Foix-Candale, François. In *Dictionary of Gnosis & Western Esotericism*. Leiden: Brill.

Fauvel, John. 1987. *The Greek study of curves. Topics in the history of mathematics, Unit 4*. Milton Keynes: The Open University Press.

Feldhay, Rivka. 1998. The use and abuse of mathematical entities: Galileo and the Jesuits revisited. In *The Cambridge Companion to Galileo*, ed. P.K. Machamer, 80–145. Cambridge: Cambridge University Press.

Ficino, Marsilio. 1471. *Mercurii Trismegisti liber De potestate et sapientia dei, e Graeco in Latinum traductus a Marsilio Ficino Florentino ad Cosmum Medicem patriae patrem, Pimander incipit*. Treviso: G. de Lisa.

Fine, Oronce. 1516. *Mundialis sphere opusculum Joannis de Sacrobusto*. Paris: R. Chaudière. [Repr. 1519, 1524, 1527 and 1538.].

———. 1532. *Protomathesis: Opus varium, ac scitu non minus utile quàm iucundum, nunc primum in lucem foeliciter emissum*. Paris: G. Morrhe.

———. 1536. *In sex priores libros Geometricorum elementorum Euclidis Megarensis demonstrationes*. Paris: S. de Colines.

———. 1542. *Orontij Finei Delphinatis, Regii mathematicarum professoris, De mundi sphaera, sive cosmographia, primáve astronomiae parte, Libri V: Inaudita methodo ab authore renovati, proprijsque tum commentarijs et figuris, tum demonstrationibus et tabulis recens illustrati*. Paris: S. de Colines.

———. 1544. *In sex priores libros Geometricorum elementorum Euclidis Megarensis demonstrationes, recèns auctae, & emendatae*. Paris: S. de Colines.

———. 1551a. *L'Esphere du monde, proprement ditte cosmographie, composee nouvellement en françois, et divisee en cinq livres, comprenans la premiere partie de l'astronomie, et les principes universels de la geographie et hydrographie. Avec une epistre, touchant la dignité, perfection et utilité des sciences mathematiques. Par Oronce Fine, natif du Daulphiné, lecteur Mathematicien du tres-chrestien Roy de France en L'université de Paris*. Paris: M. de Vascosan.

———. 1551b. *De speculo ustorio, ignem ad propositam distantiam generante, Liber unicus, ex quo duarum linearum semper appropinquantium, & nunquam concurrentium colligitur demonstratio*. Paris: M. de Vascosan.

———. 1551c. *In sex priores libros Geometricorum elementorum Euclidis Megarensis demonstrationes, recèns auctae, & emendatae*. Paris: Regnault Chaudière.

———. 1587. *Opere di Orontio Fineo del Delfinato divise in cinque Parti, Aritmetica, Geometria, Cosmografia, e Orivoli. Tradotte da Cosimo Bartoli, Gentilhuomo, e Academico Fiorentino. Et gli Specchi, Tradotti dal Cavalier Ercole Bottrigaro, Gentilhuomo Bolognese, Nuovamente poste in luce*, transl. C. Bartoli. Venice: F. Franceschi.

Foix-Candale, François (de). 1566. *Elementa Geometrica, Libris XV. Ad Germanam Geometriae Intelligentiam è diversis lapsibus temporis iniuria contractis restituta, ad impletis praeter maiorum spem, quae hactenus deerant, solidorum regularium conferentiis ac inscriptionibus. His acceßit decimus sextus liber, de solidorum regularium sibi invicem inscriptorum collationibus*. Paris: J. Royer.

———. 1574a. *Mercurii Trismegisti Pimandras utraque lingua restitutus, D. Francisci Flussatis Candallae industria*. Bordeaux: S. Millanges.

———. 1574b. *Le Pimandre de Mercure Trismegiste nouvellement traduict de l'exemplaire grec restitué en la langue françoyse par François Monsieur de Foys dela famille de Candalle*. Bordeaux: S. Millanges.

———. 1578. *Elementa Geometrica, Libris XV. Ad Germanam Geometriae Intelligentiam è diversis lapsibus temporis iniuria contractis restituta, ad impletis praeter maiorum spem, quae hactenus deerant, solidorum regularium conferentiis ac inscriptionibus. Accessit decimussextus liber, de solidorum regularium sibi invicem inscriptorum collationibus. Novissimè collati sunt decimusseptimus & decimusoctavus, priori editione quodammodo polliciti, de componendorum, inscribendorum, & conferendorum compositorum solidorum inventis, ordine & numero absoluti*. Paris: J. du Puys.

———. 1579. *Le Pimandre de Mercure Trismegiste de la philosophie Chretienne, Cognoissance du verbe divin, et de l'excellence des œuvres de Dieu, traduit de l'exemplaire grec, avec la collation de tresamples commentaires, par François Monsieur de Foix, de la famille de Candalle*. Bordeaux: S. Millanges.

Forcadel, Pierre. 1564. *Les Six premiers livres des Elements d'Euclide, traduicts et commentez par Pierre Forcadel de Bezies, Lecteur ordinaire du Roy es Mathematiques en l'université de Paris*. Paris: J. de Marnef and G. Cavellat.

Forshaw, Peter J. 2005. The early alchemical reception of John Dee's 'Monas Hieroglyphica'. *Ambix* 52 (3): 247–269.

Fowler, Caroline O. 2017. The point and its line: An early modern history of movement. *Visual culture and mathematics in the early modern period*, ed. I. Alexander-Skipnes. New York: Routledge.

French, Peter J. 1987. *John Dee: The World of the Elizabethan Magus*. London: Routledge.

Fried, Michael N., and Sabetai Unguru. 2001. *Apollonius of Perga's Conica. Text, context, subtext*. Leiden: Brill.

Friedman, Michael. 2018. *A history of folding in mathematics mathematizing the margins*. Basel: Birkhäuser.

Gagné, Jean. 1969. Du quadrivium aux scientiae mediae. In *Arts libéraux et philosophie au Moyen âge. Actes du IVe Congrès international de philosophie médiévale*, 975–986. Paris/Montréal: Institut d'études médiévales, Vrin.

Gallois, Lucien. 1918. Un géographe dauphinois: Oronce Fine et le Dauphiné sur sa carte de France de 1525. *Recueil des travaux de l'institut de géographie alpine* 6: 1–25.

Gatto, Romano. 1994. *Tra Scienza e immaginazione. Le matematiche presso il collegio gesuitico napoletano (1552–1670 ca.)*. Florence: Olschki.

———. 2006. Christoph Clavius' *Ordo Servandus in addiscendis disciplinis Mathematicis* and the teaching of mathematics in Jesuit colleges at the beginning of the modern era. *Science & Education* 15: 235–258.

Gaukroger, Stephen. 1995. *Descartes: An intellectual biography*. Oxford: Clarendon Press.

Giacomotto-Charra, Violaine. 2012. Le commentaire au Pimandre de François de Foix-Candale: l'image d'une reine-philosophe en question. *Albineana, Cahiers d'Aubigné* 24: 207–224.

Gilbert, Neal W. 1960. *Renaissance concepts of method*. New York: Columbia University Press.

Gilly, Carlos, and Cis van Heertum, eds. 2002. *Magia, Alchimia, Scienza dal '400 al '700: L'influsso di Ermete Trismegisto/Magic, Alchemy and Science 15th–18th Centuries: The Influence of Hermes Trismegistus*. Florence: Centro Di.

Gilson, Etienne, ed. 1976. *René Descartes: Discours de la méthode*. Paris: Vrin.

Giorgio, Francesco. 1525. *De Harmonia Mundi totius Cantica tria*. Venice: B. Vitali.

Giusti, Enrico. 1987. La Géométrie di Descartes tra numeri i grandezze. *Giornale critico della filosofia italiana* VII: 409–432.

Goldstein, Marie. 1972. The historical development of group theoretical ideas in connection with Euclid's axiom of congruence. *Notre Dame Journal of Formal Logic* 13: 331–349.

Goulding, Robert. 2010. *Defending Hypatia. Ramus, Savile, and the Renaissance rediscovery of mathematical history*. Dordrecht: Springer.

Gracilis, Stephanus (Étienne). 1557. *Euclidis Elementorum libri XV quibus, cum ad omnem mathematicae scientiae partem, tum ad quamlibet geometriæ tractationem facilis comparatur aditus*. Paris: G. Cavellat.

Grosseteste, Robert. 2011. *De luce*, ed. C. Panti. Pisa: Edizioni Plus, Pisa University Press.

Gueroult, Martial. 1974. *Spinoza II: L'âme*. Hildesheim: Georg Holms Verlag.

Guicciardini, Niccolò. 2009. *Isaac Newton on mathematical certainty and method*. Cambridge: MIT Press.

Halliwell, James O., ed. 1839. *Rara mathematica*. London: J. W. Parker.

Halsted, George Bruce. 1879. Note on the first English Euclid. *American Journal of Mathematics* 2: 46–48.

Harari, Orna. 2003. The concept of existence and the role of constructions in Euclid's *Elements*. *Archive for the History of Exact Science* 57: 1–23.

Harkness, Deborah. 2007. *The Jewel house: Elizabethan London and the scientific revolution*. New Haven: Yale University Press.

Harrie, Jeanne E. 1975. *François Foix de Candale and the hermetic tradition in sixteenth-century France*. PhD dissertation, University of California, Riverside.

———. 1978. Duplessis-Mornay, Foix-Candale and the hermetic religion of the world. *Renaissance Quarterly* 31 (4): 499–514.

Heath, Thomas L. 1910. *Diophantus of Alexandria. A study in the history of Greek algebra*. Cambridge: Cambridge University Press.

———. 1956. *The thirteen books of Euclid's Elements*. New York: Dover Publications.

———. 1981a. *A history of Greek mathematics. From Thales to Euclid*. Vol. I. New York: Dover.

———. 1981b. *A history of Greek mathematics. From Aristarchus to Diophantus*. Vol. II. New York: Dover.

Heiberg, Johan Ludvig. 1883. *Euclidis opera omnia*. Vol. I. Leipzig: B.G. Teubner.

———. 1884. *Euclidis opera omnia*. Vol. II. Leipzig: B.G. Teubner.

———. 1885. *Euclidis opera omnia*. Vol. IV. Leipzig: B.G. Teubner.

Hero of Alexandria. 1903. *Heronis Alexandrini opera quae supersunt omnia, III*, ed. H. Schöne. Stuttgart: B.G. Teubner.

———. 1974. *Heronis Alexandrini opera quae supersunt omnia IV*, ed. J.L. Heiberg. Stuttgart: B.G. Teubner.

Higashi, Shin. 2007. Giorgio Valla et Alessandro Piccolomini. Quelques aspects de la réception de Proclus à la Renaissance. *Bulletin of Liberal Arts Education Center* 27: 31–52.

———. 2018. *Penser les mathématiques au XVIe siècle*. Paris: Classiques Garnier.

Hillard, Denise, and Emmanuel Poulle. 1971. Oronce Finé et l'horloge planétaire de la Bibliothèque Sainte-Geneviève. *Bibliothèque d'Humanisme et de Renaissance* 33: 311–349.

Hobbes, Thomas. 1839. *The English works of Thomas Hobbes of Malmesbury*, ed. W. Molesworth. London: J. Bohn.

Holtzmann, Wilhelm (Xylander). 1562. *Die Sechs Erste Bücher Euclidis vom anfang oder grund der Geometri*. Basel: J. Oporinus.

Homann, Frederick A. 1983. Christopher Clavius and the Renaissance of Euclidean Geometry. *Archivum Historicum Societatis Iesu* 52: 233–246.

Huffman, Carl A. 2005. *Archytas of Tarentum: Pythagorean, philosopher and mathematician king*. Cambridge: Cambridge University Press.

Iamblichus. 1894. *Iamblichi in Nicomachi arithmeticam introductionem liber*, ed. H. Pistelli. Leipzig: B.G. Teubner.

———. 2014. *Nicomachi arithmeticam*, ed., transl. and comm. N. Vinel. Pisa/Rome: Fabrizio Serra editore.

Ighbariah, Ahmad, and Roy Wagner. 2018. Ibn al-Haytham's revision of the Euclidean foundations of mathematics. *HOPOS: The Journal of the International Society for the History of Philosophy of Science* 8: 62–86.

Irigoin, Jean. 2006. Les lecteurs royaux pour le grec (1530–1560). In *Histoire du Collège de France*, ed. A. Tuilier, vol. I, 233–256. Paris: Fayard.

James, Montague R. 1921. *Lists of manuscripts formerly owned by Dr. John Dee*. Oxford: The Bibliographical Society, Oxford University Press.

Jaouiche, Khalil. 1986. *La Théorie des parallèles en pays d'Islam. Contribution à la préhistoire des géométries non-euclidiennes*. Paris: Vrin.

Jesseph, Douglas M. 1999. *Squaring the circle: The war between Hobbes and Wallis*. Chicago: The University of Chicago press.

Johnson, A.F. 1928. Oronce Finé as an illustrator of books. *Gutenberg-Jahrbuch* 3: 107–109.

Johnston, Stephen. 2012. John Dee on geometry: Texts, teaching and the Euclidean tradition. *Studies in History and Philosophy of Science* 43: 470–479.

Josten, Conrad H. 1964. A Translation of John Dee's 'Monas Hieroglyphica' (Antwerp, 1564), with an Introduction and Annotations. *Ambix* 12 (2–3): 84–221.

Jugé, Clément. 1907. *Jacques Peletier du Mans, 1517–1582. Essai sur sa vie, son oeuvre, son influence*. Paris/Le Mans: A. Lemerre, A. Bienaimé-Leguicheux.

Jullien, Vincent. 1996. Descartes, la "Géométrie" de 1637. Paris: Presses Universitaires de France.

Karrow, Robert W. 1993. *Mapmakers of the sixteenth century and their maps: Bio-bibliographies of the cartographers of Abraham Ortelius, 1570: Based on Leo Bagrow's A. Ortelii Catalogus cartographorum*. Chicago: The Newberry Library by Speculum Orbis Press.

Kepler, Johannes. 1940 [1619]. Harmonices mundi. In *Johannes Kepler, Gesammelte Werke*, ed. M. Caspar. Munich: Beck.

———. 1993 [1597]. Mysterium cosmographicum. In *Johannes Kepler, Gesammelte Werke*, ed. M. Caspar. Munich: Beck.

Kessler, Eckhard. 1995. Clavius entre Proclus et Descartes. In *Les Jésuites à la Renaissance: Système éducatif et production du savoir*, ed. L. Giard, 285–308. Paris: Presses Universitaires de France.

Kessler-Mesguich, Sophie. 2006. L'enseignement de l'hébreu et de l'araméen par les premiers lecteurs royaux (1530–1560). In *Histoire du Collège de France*, ed. A. Tuilier, vol. I, 257–282. Paris: Fayard.

Killing, Wilhelm. 1898. *Einführung in die Grundlagen der Geometrie*. Paderborn: F. Schöningh.

Klein, Jacob. 1968. *Greek mathematical thought and the origin of algebra*, transl. E. Brann. Cambridge: The M.I.T. Press.

Knobloch, Eberhard. 1988. Sur la vie et l'oeuvre de Christopher Clavius (1583–1612). *Revue d'histoire des sciences* 41: 331–356.

———. 1990. Christoph Clavius. Ein Namen und Schriftenverzeichnis zu seinen Opera Mathematica. *Bollettino di storia delle scienze matematiche* 10: 13–189.

———. 1995a. Sur le rôle de Clavius dans l'histoire des Mathématiques. In *Christoph Clavius e l'attività scientifica dei gesuiti nell'età di Galileo*, ed. U. Baldini, 35–56. Bulzoni: Roma.

———. 1995b. L'oeuvre de Clavius et ses sources scientifiques. In *Les Jésuites à la Renaissance. Système éducatif et production du savoir*, ed. L. Giard, 263–283. Paris: Presses Universitaires de France.

———. 1997. Sur le développement de la géométrie pratique avant Descartes. In *Descartes et le Moyen Age*, ed. J. Biard and R. Rashed, 57–72. Paris: Vrin.

———. 2005. Géométrie pratique, Géométrie savante. *Albertiana* 8: 27–56.

Knoespel, Kenneth J. 1987. The Narrative matter of mathematics: John Dee's preface to the *Elements* of Euclid of Megara (1570). *Philological Quarterly* 66 (1): 26–46.

Knorr, Wilbur R. 1983. Construction as existence proof in ancient geometry. *Ancient Philosophy* 3: 125–148.

———. 1989. *Textual studies in ancient and medieval geometry*. Boston: Birkhauser.

Lachtermann, David R. 1989. *The ethics of geometry. A genealogy of modernity*. New York: Routledge.

Laird, Walter Roy. 1987. Robert Grosseteste on the subalternate sciences. *Traditio* 43: 147–169.

Lattis, James. 2010. *Between Copernicus and Galileo: Christoph Clavius and the Collapse of Ptolemaic Cosmology*. Chicago: University of Chicago Press.

Le Fèvre de la Boderie, Guy. 1578a. *La Galliade ou de la révolution des arts et sciences*. Paris: G. Chaudière.

———. 1578b. *L'Harmonie de monde divisee en trois cantiques. OEuvre singulier, et plein d'admirable erudition: Premiere composé en Latin par François Georges Venitien, & depuis traduict & illustré par Guy Le Fevre de la Boderine Secretaire de Monseigneur Frere unique du Roy, & son Interprete aux langues estrangeres. Plus l'Heptaple de Jean Picus Compte de la Mirande translaté par Nicolas Le Fevre de la Boderie*. Paris: J. Macé.

Lee, Sidney L. 1886. Billingsley, Sir Henry. In *Dictionary of national biography*, ed. S. Leslie, 33–34. New York/London: Macmillan, Smith, Elder, & co.

Lee, Eunsoo. 2018. Let the diagram speak: Compass arcs and visual auxiliaries in printed diagrams of Euclid's *Elements*. *Endeavour* 42 (2–3): 78–98.

Lefèvre, Jacques. 1494. *Pimander, seu De potestate et sapientia Dei, per Marsilium Ficinum traductus. Argumenta per Jacobum Fabrum*. Paris: W. Hopyl.

———. 1495. *Textus de sphaera Johannis de Sacrobosco, cum additione (quantum necessarium est) adiecta, novo commentario nuper edito ad utilitatem studentium philosophice parisiensis academie illustratus*. Paris: W. Hopyl.

———. 1505. *Pimander: Mercurii Trismegisti liber de sapienta & potestate Dei. Asclepius: ejusdem Mercurii liber de voluntate divina. Item Crater Hermetis, a Lazarelo Septempedano*. Paris: H. Estienne.

———. 1516. *Euclidis Megarensis Geometricorum elementorum libri XV. Campani Galli Transalpini in eosdem commentariorum libri XV. Theonis Alexandrini Bartholomæo Zamberto Veneto interprete, in tredecim priores, commentariorum libri XIII. Hypsiclis Alexandrini in duos posteriores, eodem Bartholomæo Zamberto Veneto interprete, commentariorum libri II*. Paris: H. Estienne.

Leibniz, G.W. 1995. *La Caractéristique géométrique*, ed. J. Echeverría and transl. M. Parmentier. Paris: Vrin.

Lewis, Neil. 2005. Robert Grosseteste and the continuum. In *Albertus magnus and the beginnings of the medieval reception of Aristotle in the Latin West*, ed. L. Honnefelder, R. Wood, M. Dreyer, and M.-A. Aris, 159–187. Münster: Aschendorff.

———. 2013. Robert Grosseteste's *On light*: An English translation. In *Robert Grosseteste and his intellectual milieu*, ed. J. Flood, J.R. Ginther, and J.W. Goering, 239–247. Toronto: Pontifical Institute of Mediaeval Studies.

Limbrick, Elaine. 1981. Hermétisme religieux au XVIe siècle: le "Pimandre" de François de Foix de Candale. *Renaissance and Reformation/Renaissance et Réforme, New Series/Nouvelle Série* 5 (1): 1–14.

Lindberg, David C. 1967. Alhazen's theory of vision and its reception in the West. *Isis* 58 (3): 321–341.

———. 1976. *Theories of vision from Al-Kindi to Kepler*. Chicago: The University of Chicago Press.

Livesey, Stephen J. 1989. *Theology and science in the fourteenth century. Three questions on the unity and subalternation of the sciences from John of Reading's commentary on the Sentences*. Leiden: Brill.

Lloyd, A.C. 1995. The later Neoplatonists. In *The Cambridge history of later Greek and early medieval philosophy*, ed. A.H. Armstrong, 269–326. Cambridge: Cambridge University Press.

Loget, François. 2000. *La Querelle de l'angle de contact (1554–1685). Constitution et autonomie de la communauté mathématique entre Renaissance et Âge baroque*. Ph.D. Thesis, EHESS, Paris.

———. 2002. Wallis entre Hobbes et Newton. La question de l'angle de contact chez les anglais. *Revue d'Histoire des Mathématiques* 8: 207–262.

———. 2004. Héritage et réforme du quadrivium au XVIe siècle. In *La Pensée numérique (Actes du colloque organisé à Peyresq, septembre 1999)*, ed. J.-L. Gardies, J.-C. Pont, F. Doridot, and J. Dhombres, 211–230. Turnhout: Brepols.

Lohr, Charles H. 1995. Les jésuites et l'aristotélisme du XVIe siècle. In *Les Jésuites à la Renaissance. Système éducatif et production du savoir*, ed. Luce Giard, 79–93. Paris: Presses Universitaires de France.

Lucretius, Carus Titus. 1924. *On the nature of things*, transl. W. H. D. Rouse and rev. M. E. Smith. Loeb Classical Library 181. Cambridge, MA: Harvard University Press.

Lukacs, Ladislaus. 1965–1992. *Monumenta Paedagogica Societatis Jesu*. Rome: Institutum historica Societatis Iesu.

Mahoney, Michael S. 1990. Barrow's mathematics: between ancients and modern. In *Before Newton: The life and times of Isaac Barrow*, ed. M. Feingold, 179–249. Cambridge: Cambridge University Press.

Maier, Anneliese. 1949. Kontinuum, Minimum und aktuell Unendliches. In *Die Vorläufer Galileis im 14. Jahrhundert. Studien zur Naturphilosophie der Spätscholastik I*, 155–215. Rome: Edizioni di Storia e Letteratura.

Maierù, Luigi. 1991. John Wallis: lettura della polemica fra Peletier e Clavio circa l'angolo di contatto. In *Atti del Convegno "Giornate di storia della matematica"*, ed. M. Galluzzi, 115–137. Editel: Rende.

———. 1999. La diffusione di Proclo, commentatore di Euclide, nel Cinquecento. *11o Annuario del Liceo Scientifico "B. G. Scorza"*, Cosenza, 49–68. Soveria Mannelli: Calabria Letteraria Editrice.

Malet, Antoni. 2006. Renaissance notions of number and magnitude. *Historia Mathematica* 33: 63–81.

———. 2012. Euclid's swan song: Euclid's *Elements* in early modern Europe. In *Greek science in the long run: Essays on the Greek scientific tradition (4th c. BCE–17th c. CE)*, ed. P. Olmos, 205–234. Newcastle upon Tyne: Cambridge Scholars Publishing.

Mancosu, Paolo. 1996. *Philosophy of mathematics and mathematical practice in the seventeenth century*. New York/Oxford: Oxford University Press.

———. 2008. Descartes and mathematics. In *A companion to Descartes*, ed. J. Broughton and J. Carriero, 103–123. Malden: Blackwell Publishing.

Mandosio, Jean-Marc. 1994. Entre mathématiques et physique: Note sur les sciences intermédiaires à la Renaissance. In *Comprendre et maîtriser la nature au Moyen âge. Mélanges d'histoire des sciences offerts à G. Beaujouan*, 115–138. Geneva: Droz.

———. 2003. Des 'mathématiques vulgaires' à la 'monade hiéroglyphique': les *Éléments* d'Euclide vus par John Dee. *Revue d'histoire des sciences* 56 (2): 475–491.

Marre, Aristide. 1881. *Le Triparty en la science des nombres par maistre Nicolas Chuquet parisien, publié d'après le manuscrit fonds français no. 1346 de la Bibliothèque nationale de Paris et précédé d'une notice par M. Aristide Marre*. Rome: Imprimerie des sciences mathématiques et physiques.

Masià, Ramon. 2016. A new reading of Archytas' doubling of the cube and its implications. *Archive for the History of Exact Sciences* 70: 175–204.

McConnell, Anita. 2008 [2004]. Billingsley, Sir Henry (d. 1606). In *Oxford dictionary of national biography*. Oxford: Oxford University Press.

McEvoy, James. 1978. The metaphysics of light in the middle ages. *Philosophical Studies* 26: 126–145.

———. 2000. *Robert Grosseteste*. Oxford: Oxford University Press.

Medina, José. 1985. Les mathématiques chez Spinoza et Hobbes. *Revue philosophique de la France et de l'Étranger* 175 (2): 177–188.

Mehl, Edouard. 2003. Euclide et la fin de la Renaissance: Sur le scholie de la proposition XIII.18. *Revue d'Histoire des Sciences* 56 (2): 439–455.

Menn, Stephen. 2015. How Archytas doubled the cube. In *The frontiers of ancient science. Essays in Honor of Heinrich von Staden*, ed. B. Holmes and K.-D. Fischer, 407–435. Berlin, Boston: De Gruyter.

Merkel, Ingrid, and Allen G. Debus, eds. 1988. *Hermeticism and the Renaissance: Intellectual history and the occult in early modern Europe*. Cranbury/London/Mississauga: Associated University Presses.

Molland, A. George. 1976. Shifting the foundations: Descartes's transformation of ancient geometry. *Historia Mathematica* 3: 21–49.

———. 1994. The philosophical context of medieval and Renaissance mathematics. In *Companion Encyclopedia of the history and philosophy of the mathematical sciences*, ed. I. Grattan-Guiness, vol. 1, 281–285. London: Routledge.

Molther, Johannes. 1619. *Problema Deliacum de cubi duplicatione hoc est de quorumlibet solidorum interventu Mesolabii secundi, quo duae capiantur mediae proportionales sub data ratione similium fabrica – nunc tandem post infinitos praestantissimorum mathematicorum conatus expedite et geometrice solutum*. Frankfurt: C. Schleich and D. & D. Aubry.

Montdoré, Pierre. 1551. *Euclidis elementorum liber decimus, Petro Montaureo interprete*. Paris: Michel de Vascosan.

Moreschini, Claudio. 2009. Il commento al *Corpus Hermeticum* di François Foix-Candale: Annotazioni storiche e filologiche. *Aries* 9 (1): 37–58.

Moretto, Antonio. 1984. *Hegel e la "Matematica dell' infinito"*. Trento: Verifiche.

Moscheo, Rosario. 1998. *I Gesuiti e le matematiche nel secolo XVI*. Messina: Società Messinese di Storia Patria.

Moyon, Marc. 2011. Le *De Superficierum Divisionibus Liber* d'al-Baghdādī et ses prolongements en Europe. In *Actes du IXème Colloque Maghrébin sur l'histoire des mathématiques arabes (Tipaza, 12–14 mai 2007)*, ed. A. Bouzari and Y. Guergour. Alger: Imprimerie Fasciné.

Mueller, Ian. 2006. *Philosophy of mathematics and deductive structure in Euclid's Elements*. Mineola: Dover Publications.

Mugler, Charles. 1958–1959. *Dictionnaire historique de la terminologie géométrique des grecs*. Paris: C. Klincksieck.

Murdoch, John. 1981. Henry of Harclay and the infinite. In *Studi sul XIV secolo in memoria di Anneliese Maier*, 219–261. Rome: Edizioni di storia e letteratura.

———. 1982. Infinity and continuity. In *The Cambridge history of later medieval philosophy*, ed. N. Kretzmann, A. Kenny, and J. Pinborg, 564–591. Cambridge, MA: Cambridge University Press.

———. 2009. Beyond Aristotle: Indivisibles and infinite divisibility in the later middle ages. In *Atomism in late medieval philosophy and theology*, ed. C. Grellard and A. Robert, 15–38. Leiden: Brill.

Nenci, Elio. 2020. Francesco Capuano di Manfredonia. In *De sphaera of Johannes de Sacrobosco in the early modern period. The authors of the commentaries*, ed. M. Valleriani, 91–110. Cham: Springer.

Netz, Reviel. 1999. *The shaping of deduction in Greek mathematics. A study in cognitive history*. Cambridge: Cambridge University Press.

Newton, Isaac. 1687. *Philosophiae Naturalis Principia Mathematica*. London: J. Streater.

———. 1964. *The mathematical works of Isaac Newton*, ed. D.T. Whiteside, vol. I. New York/ London: Johnson Reprint Corporation.

———. 1999. *The Principia: Mathematical principles of natural philosophy*. Introd. I. B. Cohen and transl. I. B. Cohen and A. Whitman. Berkeley: University of California Press.

Nicomachus of Gerasa. 1866. *Nicomachi Geraseni Pythagorei Introductionis arithmeticae libri II*, ed. R. Hoche. Leipzig: B.G. Teubner.

———. 1960. *Introduction to arithmetic*, ed. and transl. M. L. D'Ooge. Annapolis: St. John's college Press.

Nikulin, Dmitri. 2002. *Matter, imagination, and geometry. Ontology, natural philosophy, and mathematics in Plotinus, Proclus, and Descartes*. Aldershot: Ashgate.

———. 2008. Imagination and mathematics in Proclus. *Ancient Philosophy* 28: 153–172.

Norrgrén, Hilde. 2005. Interpretation and the Hieroglyphic Monad: John Dee's reading of Pantheus's Voarchadumi. *Ambix* 52 (3): 217–245.

O'Meara, Dominic. 2010. Plotinus. In *The Cambridge history of philosophy in late antiquity*, ed. L.P. Gerson, 301–324. Cambridge: Cambridge University Press.

Oosterhoff, Richard J. 2014. Idiotae, mathematics, and artisans: The untutored mind and the discovery. *Intellectual History Review* 24: 301–319.

———. 2016. Lovers in paratexts: Oronce Fine's Republic of mathematics. *Nuncius* 31: 549–583.

———. 2017. 'Secrets of Industry' for 'Common Men': Charles de Bovelles and early French readerships of technical print. In *Translating early modern science*, ed. S. Fransen, N. Hodson, and K.A.E. Enenkel, 207–229. Leiden/Boston: Brill.

———. 2018. *Making mathematical culture. University and print in the circle of Lefèvre d'Étaples*. Oxford: Oxford University Press.

———. 2020. A Lathe and the Material *Sphaera*: Astronomical Technique at the Origins of the Cosmographical Handbook. In *De sphaera of Johannes de Sacrobosco in the Early Modern Period*, ed. M. Valleriani, 25–52. Cham: Springer.

Pacioli, Luca. 1509a. *Euclidis megarensis philosophi acutissimi mathematicorumque omnium sine controversia principis opera a Campano interprete fidissimo translata [...]*. Venice: P. dei Paganini.

———. 1509b. *Divina proportione opera a tutti glingegni perspicaci e curiosi necessaria ove ciascun studioso si Philosophia, Prospectiva Pictura Sculptura, Architectura, Musica, e altre Mathematice, suavissima, sottile, e admirabile doctrina consequira, e delectarassi, con varie questione de secretissima scientia*. Venice: P. dei Paganini.

Palmieri, Paolo. 2009. Superposition: on Cavalieri's practice of mathematics. *Archive for the History of Exact Sciences* 63: 471–495.

Pantin, Isabelle. 1984. Microcosme et 'Amour volant' dans *L'Amour des Amours* de Jacques Peletier du Mans. *Nouvelle Revue du XVIe Siècle* 2: 43–54.

———. 1995. *La Poésie du ciel en France dans la seconde moitié du XVIe siècle*. Genève: Droz.

———. 2006. Teaching mathematics and astronomy in France: The Collège Royal (1550–1650). *Science and Education* 15: 189–207.

———. 2009a. Oronce Fine's role as Royal Lecturer. In *The worlds of Oronce Fine. Mathematics, instruments and print in Renaissance France*, ed. A. Marr, 13–30. Donington: Shaun Tyas.

———. 2009b. *Altior incubuit animus sub imagine mundi*. L'inspiration du cosmographe d'après un gravure d'Oronce Finé. In *Les Méditations cosmographiques à la Renaissance*, ed. F. Lestringant, 69–90. Paris: Presses de l'Université Paris-Sorbonne.

———. 2010. The astronomical diagrams in Oronce Finé's *Protomathesis* (1532): Founding a French tradition? *Journal for the History of Astronomy* 41: 287–310.

———. 2012. The first phases of the *Theoricae Planetarum* printed tradition (1474–1535): The evolution of a genre observed through its images. *Journal for the History of Astronomy* 43: 3–26.

———. 2013. Oronce Finé mathématicien et homme du livre: la pratique éditoriale comme moteur d'évolution. In *Mise en forme des savoirs à la Renaissance: à la croisée des idées, des techniques et des publics*, ed. I. Pantin and G. Péoux, 19–40. Paris: Armand Colin.

Panza, Marco. 2011. Rethinking geometrical exactness. *Historia Mathematica* 38: 42–95.

Pappus of Alexandria. 1876–1878. *Pappi Alexandrini collectionis quae supersunt*, transl. and comm. F. Hultsch. Berlin: Weidmann.

———. 2009. *Book 4 of the collection*, ed. and transl. H. Sefrin-Weis. London: Springer.

Peletier, Jacques. 1547. *Les Oeuvres poétiques de Jacques Peletier du Mans*. Paris: M. de Vascosan.

———. 1550. *Dialogue de l'Ortografe e Prononciation francoese, departi an deus livres par Jacques Peletier du Mans*. Poitiers: J. and E. de Marnef.

———. 1554. *L'Algèbre de Jaques Peletier du Mans, departie en deus livres*. Lyon: J. de Tournes.

———. 1557. *In Euclidis Elementa Geometrica Demonstrationum Libri sex*. Lyon: J. de Tournes and G. Gazeau.

————. 1563. *Commentarii tres. I. De dimensione circuli. II. De contactu linearum: et de duabus lineis in eodem plano neque parallelis, neque concurrentibus. III. De constitutione horoscopi.* Basel: J. Oporinus.

————. 1572. *De usu geometriae.* Paris: G. Gourbin.

————. 1573. *De l'Usage de geometrie.* Paris: G. Gourbin.

————. 1579a. *Christophorum Clavium, De contactu linearum, Apologia.* Paris: J. de Marnef.

————. 1579b. *Oratio Pictavii habita in praelectiones Mathematicas.* Poitiers: Bouchet. [repr. Sens: M. P. Laumonier, 1904].

————. 1581. *Euvres poétiques intituléz Louanges.* Paris: R. Coulombel.

————. 1610. *Euclidis Elementa Geometrica Demonstrationum Libri sex.* Geneva: J. de Tournes.

————. 1611. *Les Six premiers livres des Elements geometriques d'Euclide, avec les demonstrations de Jaques Peletier du Mans.* Geneva: J. de Tournes.

————. 1628. *Les six premiers livres des Elements geometriques d'Euclide, avec les demonstrations de Jaques Peletier du Mans.* Geneva: J. de Tournes.

————. 1904. *Oeuvres poétiques de Jacques Peletier du Mans, publiées d'après l'édition originale de 1547,* ed. L. Séché and comm. P. Laumonier. Paris: Revue de la Renaissance.

Perl, Eric. 2010. Pseudo-Dionysius the Areopagite. In *The Cambridge history of philosophy in late antiquity.* Cambridge: Cambridge University Press.

Philo of Alexandria. 1994. *De Opificio mundi. Philo,* vol. I, transl. F. H. Colson and G. H. Whitaker. Cambridge, MA: Harvard University Press.

Philoponus, John. 1897. *Ioannis Philoponi in Aristotelis De anima libros commentaria,* ed. M. Hayduck. Berlin: G. Reimer.

————. 2006. *Philoponus: On Aristotle's on the Soul 1.1–2,* transl. P.J. van der Eijk. Ithaca: Cornell University Press.

Pierre d'Ailly. 1531. Reverendissimi domini Petri de aliaco Cardinalis & Episcopi Cameracensis doctorisque celebratissimi 14 Quaestiones (ca. 1390). In *Tractatus de Sacrobosco.* Venice: L. Giunta.

Plato. 1929. *Plato, Timaeus. Critias. Cleitophon. Menexenus. Epistles.* Loeb Classical Library 234, ed. and transl. R.G. Bury. Cambridge, MA: Harvard University Press.

————. 2013. *Republic: Books 6–10.* Loeb Classical Library 276, ed. and transl. C. Emlyn-Jones and W. Preddy. Cambridge, MA: Harvard University Press.

Poulle, Emmanuel. 1978. Oronce Fine. In *Dictionary of scientific biography,* ed. C.C. Gillispie, vol. 15, 153–157. New York: Charles Scribner's Sons.

Proclus of Lycia. 1533. *Εὐκλείδου στοιχείων βιβλ. Ιε. ἐκ τῶν θεώνος συνουσιῶν. Εἰς τοῦ αὐτοῦ τὸ πρῶτον, ἐξηγημάτων Πρόκλου βιβλ. δ. Adjecta praefatiuncula in qua de disciplinis mathematicis nonnihil,* ed. S. Grynaeus. Basel: J. Hervagius.

————. 1873. *Procli Diadochi in primum Euclidis Elementorum librum commentarij,* ed. G. Friedlein. Leipzig: B.G. Teubner.

————. 1933. *The elements of theology,* transl. E.R. Dodds. Oxford: The Clarendon Press.

————. 1992. *A commentary on the first book of Euclid's Elements,* ed. and transl. G. Morrow. Princeton University Press: Princeton.

Quispel, Gilles. 2000. Reincarnation and magic in the Asclepius. In *From Poimandres to Jacob Böhme: Gnosis, Hermetism and the Christian tradition,* ed. R. van den Broek and C. van Heertum, 167–231. Amsterdam: In de Pelikaan.

Rabouin, David. 2010. What Descartes knew of mathematics in 1628. *Historia mathematica* 37 (3): 428–459.

————. 2018. Les mathématiques de Descartes avant la *Géométrie.* In *Cheminer avec Descartes. Concevoir, raisonner, comprendre, admirer et sentir,* 293–311. Paris: Classiques Garnier.

Rambaldi, Enrico. 1989. John Dee and Federico Commandino: An English and an Italian interpretation of Euclid during the Renaissance. *Rivista di Storia della Filosofia* 44: 211–247.

Rampling, Jennifer. 2011. The Elizabethan mathematics of everything: John Dee's 'Mathematicall praeface' to Euclid's *Elements*. *Journal of the British Society for the History of Mathematics* 26: 135–146.

Ramus, Petrus. 1545. *Elementa mathematica propositiones et definitiones librorum I XV edidit P. Ramus*. Paris: L. Grandin.

Rashed, Roshdi. 2005. La modernité mathématique: Descartes et Fermat. In *Philosophie des mathématiques et théorie de la connaissance. L'oeuvre de Jules Vuillemin*, ed. R. Rashed and P. Pellegrin, 239–252. Paris: Blanchard.

———. 2013. Le mouvement en géométrie classique. *Al-Mukhatabat* 7: 58–68.

Regiomontanus, Johannes (de). 1533. *Doctissimi viri et mathematicarum disciplinarum eximii professoris Joannis de Regio Monte De triangulis omnimodis libri quinque: quibus explicantur res necessariae cognitu, volentibus ad scientarum astronomicarum perfectionem devenire: quae cum nusquam alibi hoc tempore expositae habeantur, frustra sine harum instructione ad illam quisquam aspirarit. Accesserunt huc in calce pleraque D. Nicolai Cusani De quadratura circuli, deque recti ac curvi commensuratione: itemque Jo. de monte Regio eadem De re ἐλεγκτικά hactenus à nemine publicata*. Nürnberg: J. Petreius.

———. 2008. Oratio Johannis de Monteregio habita Patavij in praelectione Alfragani. In *Regiomontano e il rinnovamento del sapere matematico e astronomico nel quattrocento*, ed. Michela Malpangotto. Bari: Cacucci Editore.

Renn, Jürgen, and Peter Damerow. 2010. *Guidobaldo del Monte's Mechanicorum Liber*. Berlin: Edition Open Access.

Ribeiro do Rinascimento, Carlos A. 1974. Le statut épistémologique des 'sciences intermédiaires' selon S. Thomas d'Aquin. *Cahiers d'Études médiévales* 2: 33–95.

Riccardi, Pietro. 1974. *Saggio di una bibliografia Euclidea*. Hildesheim: Olms (1887–1892).

Robert, Aurélien. 2010. Atomisme et géométrie à Oxford au XIVe siècle. In *Mathématiques et connaissance du monde réel avant Galilée*, ed. S. Rommevaux, 17–86. Montreuil: Omniscience.

———. 2017. Atomisme pythagoricien et espace géométrique au Moyen Âge. In *Lieu, espace, mouvement: Physique, Métaphysique et Cosmologie (XIIe–XVIe siècles). Actes du colloque international Université de Fribourg (Suisse), 12–14 mars 2015*, ed. T. Suarez-Nani, O. Ribordy, and A. Petagine, 181–206. Turnhout: Brepols.

Roberts, R. Julian. 2006 [2004]. Dee, John (1527–1609). In *Oxford Dictionary of National Biography*. Oxford: Oxford University Press.

Roberval, Gilles Personne (de). 1730 [1693]. Observations sur la composition des mouvements et sur le moyen de trouver les touchantes des lignes courbes. In *Mémoires de l'académie royale des sciences depuis 1666 à 1699*, vol. 6. Paris: Compagnie des libraires.

Rochas, Adolphe. 1856–1860. *Biographie du Dauphiné*. Paris: Charavay.

Romano, Antonella. 1999. *La Contre-réforme mathématique. Constitution et diffusion d'une culture mathématique jésuite à la Renaissance*. Rome: École française de Rome.

Romano, Francesco. 2010. Hypothèse et définition dans l'*In Euclidem* de Proclus. In *Études sur le Commentaire de Proclus au premier livre des Éléments d'Euclide*, ed. A. Lernould, 181–196. Lille: Septentrion.

Rommevaux, Sabine. 2004. Les prologues aux éditions des *Éléments* d'Euclide au Moyen Age et à la Renaissance. In *Méthodes et statut des sciences à la fin du Moyen Age*, ed. C. Grellard, 143–158. Villeneuve-d'Ascq: Presses universitaires du Septentrion.

———. 2005. *Clavius: une clé pour Euclide au XVIe siècle*. Paris: Vrin.

———. 2006. Un débat dans les mathématiques de la Renaissance: le statut de l'angle de contingence. *Journal de la Renaissance* 4: 291–302.

———. 2012. Christoph Clavius, un promoteur des mathématiques à la Renaissance. *Seizième siècle* 8: 127–140.

Rose, Paul L. 1972. Commandino, John Dee, and the *De superficierum Divisionibus* of Machometus Bagdedinus. *Isis* 63 (1): 88–93.

———. 1973. Letters illustrating the career of Federico Commandino. *Physis* 15: 401–410.

———. 1975. *The Italian Renaissance of mathematics: Studies on humanists and mathematicians from Petrarch to Galileo*. Geneva: Droz.

Rosen, Edward. 1970. John Dee and Commandino. *Scripta mathematica* 28: 321–326.

———. 2008 [1971]. Commandino, Federico. In *Dictionary of scientific biography*, ed. C.C. Gillispie, vol. 3, 363–365. New York: Charles Scribner's Sons.

Ross, Richard P. 1971. *Studies on Oronce Fine, 1494–1555*. PhD dissertation, Columbia University.

Sacksteder, William. 1980. Hobbes: The art of the Geometricians. *Journal of the History of Philosophy* 18 (2): 131–146.

———. 1981. Hobbes: Geometrical objects. *Philosophy of Science* 48 (4): 573–590.

Sacrobosco, Johannes (de). 1478 [2003]. *Sphaera mundi. Venice: A. de Rottweil, 1478*, ed. R. de Andrade Martins. Campinas: Universidade estadual de Campinas.

———, et al. 1508. *Textus Sphaerae Ioannis De Sacro Busto*. Venice: G. and B. Rossi for G. Giunta.

———. 1531. *Sphaerae tractatus*. Venice: L. Giunta.

Sasaki, Chikara. 2003. *Descartes's mathematical thought*. Dordrecht: Springer.

Schmitt, Charles B. 1970. *Prisca Theologia* e *Philosophia Perennis*: due temi del Rinascimento italiano e la loro fortuna. In *Il Pensiero italiano del Rinascimento e il tempo nostro*, 211–236. Florence: Leo Olschki.

Serene, Eileen. 1982. Demonstrative science. In *The Cambridge history of later medieval philosophy: From the rediscovery of Aristotle to the disintegration of scholasticism, 1100–1600*, ed. N. Kretzmann, A. Kenny, J. Pinborg, and E. Stump, 496–517. Cambridge: Cambridge University Press.

Serfati, Michel. 1993. Les compas cartésiens. *Archives de philosophie* 56 (2): 197–230.

Sextus Empiricus. 1971. *Against the professors,* transl. R.G. Bury. Cambridge, MA: Harvard University Press.

Shea, William R. 1991. *The magic of numbers and motion. The scientific career of René Descartes*. Canton, MA: Watson Publishing.

Sherman, William H. 1995. *John Dee: The politics of reading and writing in the English Renaissance*. Amherst: University of Massachussetts Press.

Simpkins, Diana M. 1966. Early editions of Euclid in England. *Annals of Science* 22 (4): 225–249.

Smith, A. Mark. 1981. Getting the big picture in perspectivist optics. *Isis* 72 (4): 568–589.

———. 1987. Descartes's theory of light and refraction: A discourse on method. *Transactions of the American Philosophical Society* 77 (3): 1–92.

Smith, Helen. 2017. 'A unique instance of art': the proliferating surfaces of early modern paper. *Journal of the Northern Renaissance* 8: 1–39.

Smolarski, Dennis C. 2002. The Jesuit *Ratio Studiorum*, Christopher Clavius, and the study of mathematical sciences in universities. *Science in Context* 15 (3): 447–457.

Sozzi, Lionello. 1998. *Nexus caritatis*: l'Ermetismo in Francia nel Cinquecento. In *L'Ermetismo nell'Antichità e nel Rinascimento*, ed. L. Rotondi Secchi Tarugi, 113–126. Milan: Nuovi Orizzonti.

Staub, Hans. 1967. *Le Curieux désir: Scève et Peletier du Mans, poètes de la connaissance*. Geneva: Droz.

Steck, Max. 1981. *Bibliographia Euclideana*. Hildesheim: Gerstenberg.

Steuco, Agostino. 1540. *Augustini Steuchi Eugubini, Episcopi Kisami, Apostolicae Sedis Bibliothecarij, De perenni philosophia libri X. Idem De Eugubij, urbis suæ, nomine.* Lyon: S. Gryphius.

Stifel, Michael. 1544. *Arithmetica integra.* Nürnberg: J. Petreius.

Sylla, Edith Dudley. 2005. Swester Katrei and Gregory of Rimini: Angels, God, and mathematics in the fourteenth century. In *Mathematics and the divine: A historical study,* ed. T. Koetsier and L. Bergmans, 249–271. Amsterdam: Elsevier.

Tartaglia, Niccolò. 1543. *Euclide megarense philosopho: solo introduttore delle scientie mathematice.* Venice: V. Roffinelli.

———. (1565 [1543]). *Euclide megarense philosopho: solo introduttore delle scientie mathematice. Diligentemente rassettato, et alla integrità ridotto, per il degno professore di tal scientie Nicolo Tartalea Brisciano. Secondo le due tradottioni. Con una ampla espositione dello istesso tradottore di nuovo aggiunta. Talmente chiara, che ogni mediocre ingeno, senza la notitia, over suffragio di alcun'altra scientia con facilità serà capace poterlo intendere.* Venice: C. Troiano.

Taylor, Katie. 2011. Vernacular geometry: between the senses and reason. *BSHM Bulletin: Journal of the British Society for the History of Mathematics* 26 (3): 147–159.

Theodosius of Bithynia. 1852. *Sphaericorum libros tres,* ed. E. Nizze. Berlin: G. Reimer.

Theon of Smyrna. 1892. *Théon de Smyrne Philosophe Platonicien. Exposition des connaissances mathématiques utiles pour la lecture de Platon,* ed. and transl. J Dupuis. Paris: Hachette.

———. 1979. *Mathematics useful for understanding Plato,* transl. R. and D. Lawlor from the 1892 Greek/French edition of J. Dupuis (transl). San Diego: Wizards Bookshelf.

Thorndike, Lynn. 1949. *The sphere of Sacrobosco and its commentators.* Chicago: University of Chicago Press.

Tuilier, André, ed. 2006. *Histoire du Collège de France.* Paris: Fayard.

———, ed. 2006a. L'entrée en fonction des premiers lecteurs royaux. In Histoire du Collège de France, ed. A. Tuilier, vol. I, 145–163. Paris: Fayard.

———, ed. 2006b. Ce que signifie l'enseignement des mathématiques en 1530. In *Histoire du Collège de France,* ed. A. Tuilier, vol. I, 369–376. Paris: Fayard.

Turnèbe, Adrien. 1554. *Hermou tou Trismegistou Poimandrēs. Asklēpiou Horoi pros Ammona basilea. Mercurii Trismegisti Pœmander, seu De potestate ac sapientia divina. Aesculapii Definitiones ad Ammonem regem. Typis Regiis.* Paris: A. Turnèbe.

Turner, Anthony. 2009. Dropped out of sight: Oronce fine and the water-clock in the sixteenth and seventeenth centuries. In *The worlds of Oronce Fine. Mathematics, instruments and print in Renaissance France,* ed. A. Marr, 191–212. Donington: S. Tyas.

Tyard, Pontus (de). 1578. *Deux discours de la nature du monde et de ses parties, à sçavoir le premier Curieux traittant des choses materielles, et le second Curieux, des intellectuelles.* Paris: M. Patisson.

Valla, Giorgio. 1501. *Georgii Vallae De expetendis et fugiendis rebus opus: in quo haec continentur: de arithmetic, de musica, de geometria, de astrologia, de physiologia, de medicina, de grammatica, de dialectica...* Venice: A. Manutius.

Van den Broek, Roelof, and Cis van Heertum, eds. 2000. *From Poimandres to Jacob Böhme: Gnosis, Hermetism and the Christian tradition.* Amsterdam: In de Pelikaan.

Vermigli, Pietro Martire. 1568. *Most learned and fruitfull commentaries of D. Peter Martir Vermilius Florentine, professor of divinitie in the schole of Tigure, upon the Epistle of S. Paul to the Romanes: wherin are diligently and most profitably entreated all such matters and chiefe common places of religion touched in the same Epistle. With a table of all the common places and expositions upon divers places of the scriptures, and also an index to finde all the principall matters conteyned in the same. Lately tranlated out of Latine into Englishe, by Henry Billingsley.* London: J. Daye.

Vinel, Nicolas. 2010. La Rhusis mathématique. De l'ancien Pythagorisme à Proclus. In *Études sur le Commentaire de Proclus au premier livre des Éléments d'Euclide*, ed. A. Lernould, 111–124. Villeneuve d'Ascq: Presse Universitaire du Septentrion.

Vitrac, Bernard. 1990. *Euclide, Les Éléments*. Paris: Presse Universitaire de France.

———. 2005a. Quelques remarques sur l'usage du mouvement en géométrie dans la tradition euclidienne: de Platon et Aristote à Omar Khayyām. *Fahrang. Quarterly Journal of Humanities & Cultural Studies* 18: 1–56.

———. 2005b. Les classifications des sciences mathématiques en Grèce ancienne. *Archives de Philosophie* 68 (2): 269–301.

———. 2009. Mécanique et mathématiques à Alexandrie: le cas de Héron. *Oriens Occidens* 7: 155–199.

———. 2021. *La Traduction latine des Éléments d'Euclide par Federico Commandino: Sources, motivations.* Note from the research project "L'identification des sources des *Éléments* d'Euclide assistée par ordinateur: Sources manuscrites et imprimées" (with Alain Herreman), published online on September 19, 2021. https://hal.archives-ouvertes.fr/hal-03328386.

Vuillemin, Jules. 1960. *Mathématique et métaphysique chez Descartes*. Paris: Presses Universitaires de France.

Walker, D.P. 1954. The Prisca Theologia in France. *Journal of the Warburg and Courtauld Institutes* 17 (3–4): 204–259.

Wallis, John. 1655. *Elenchus geometriae hobbianae, sive Geometricorum, quae in ipsius Elementis Philosophiae a Thoma Hobbes Malmesburiensi proferuntur, refutatio.* Oxford: J. Crook.

———. 1656. Due correction for Mr Hobbes or Schoole discipline, for not saying his lessons right. In *In answer to his six lessons, directed to the professors of mathematicks*. Oxford: L. Lichfield.

Westman, Robert S. 1972. Kepler's theory of hypothesis and the 'realist dilemma'. *Studies in History and Philosophy of Science Part A* 3 (3): 233–264.

Yates, Frances A. 1964. *Giordano Bruno and the Hermetic Tradition*. London: Routledge and Kegan Paul.

———. 1967. The hermetic tradition in renaissance science. In *Art, science and history in the Renaissance*, ed. C.S. Singleton, 255–274. Baltimore: The John Hopkins Press.

Zamberti, Bartolomeo. 1505. *Euclidis megarensis philosophi platonici mathematicarum disciplinarum Ianitoris: Habent in hoc volumine quicunque ad mathematicam substantium aspirant: elementorum libros XIII cum expositione Theonis insignis mathematici. quibus multa quae deerant ex lectione graeca sumpta addita sub nec non plurima subversa & praepostere: voluta in Campani intepretatione : ordinata digesta & castigata sunt. Quibus etiam nonnulla ab illo venerando. Socratico philosopho mirando iudicio structa habent adiuncta. Deputatum scilicet Euclidi volumen XIIII cum expositione Hypsiclis Alexandrini. Itidemque & Phaenomena, Specularia & Perspectiva cum expositione Theonis. ac mirandus ille liber Datorum cum expositione Pappi Mechanici una cum Marini dialectici protheoria. Bartholomeo Zamberto Veneto Interprete.* Venice: I. Tacuini.

———. 1558. *Euclidis Megarensis mathematici clarissimi Elementorum geometricorum libri XV. Cum expositione Theonis in priores XIII à Bartholomaeo Zamberto Veneto latinitate donata, Campani in omnes, & Hypsiclis Alexandrini in duos postremos. His adjecta sunt Phaenomena, Catoptrica & Optica, deinde Protheoria Marini, & Data. Postremum verò, Opusculum de Levi & Ponderoso, hactenus non visum, ejusdem autoris.* Basel: J. Hervagium and B. Brand.

Zerlenga, Ornella. 2016. Federico Commandino (1509–1575). In *Distinguished figures in descriptive geometry and its applications for mechanism science from the middle ages to the 17th century*, M. Cigola, 99–128. Springer.

Zeuthen, H.G. 1896. Die geometrische Konstruktion als 'Existenzbeweis' in der antiken Geometrie. *Mathematische Annalen* 47: 222–228.

Index

© The Author(s), under exclusive license to Springer Nature Switzerland AG 2021
A. Axworthy, *Motion and Genetic Definitions in the Sixteenth-Century Euclidean Tradition*, Frontiers in the History of Science,
https://doi.org/10.1007/978-3-030-95817-6

Printed in the United States
by Baker & Taylor Publisher Services